MICROFLUIDICS AND NANOFLUIDICS

Edited by **Mohsen Sheikholeslami Kandelousi**

Microfluidics and Nanofluidics
http://dx.doi.org/10.5772/intechopen.71136
Edited by Mohsen Sheikholeslami Kandelousi

Contributors

Kai Seng Koh, Voon Loong Wong, Yong Ren, TranThe Long, Tran Minh Duc, Jian-Chiun Liou, Aminuddin Bin Ahmad Kayani, Mohd Anuar Md Ali, Burhanuddin Yeop Majlis, Jose Jaime Taha-Tijerina, Nadeem Ahmad Sheikh, Farhad Ali, Madeha Gohar, Ilyas Khan, Muhammad Saqib, Syed Aftab Alam Jan, Sudarmadji Sudarmadji, Bambang Sap, Santoso Santoso, Ishita Biswas, Mohtada Sadrzadeh, Aloke Kumar, Vassili Karanassios, Hafiz Muhammad Ali, Tayyab Raza Shah, Hamza Babar, Zargham Ahmad Khan, Zhibin Yan, Xiaoyang Huang, Chun Yang, Mingliang Jin, Lingling Shui, C H Li

Notice

Statements and opinions expressed in the chapters are these of the individual contributors and not necessarily those of the editors or publisher. No responsibility is accepted for the accuracy of information contained in the published chapters. The publisher assumes no responsibility for any damage or injury to persons or property arising out of the use of any materials, instructions, methods or ideas contained in the book.

First published in London, United Kingdom, 2018 by IntechOpen
IntechOpen is the global imprint of INTECHOPEN LIMITED, registered in England and Wales, registration number: 11086078, The Shard, 25th floor, 32 London Bridge Street
London, SE19SG – United Kingdom
Printed in Croatia

British Library Cataloguing-in-Publication Data
A catalogue record for this book is available from the British Library

Additional hard copies can be obtained from orders@intechopen.com

Microfluidics and Nanofluidics, Edited by Mohsen Sheikholeslami Kandelousi
p. cm.
Print ISBN 978-1-78923-540-1
Online ISBN 978-1-78923-541-8

For EU product safety concerns:
IN TECH d.o.o., Prolaz Marije Krucifikse Kozulić 3, 51000 Rijeka, Croatia,
info@intechopen.com or visit our website at intechopen.com.

Meet the editor

Dr. Mohsen Sheikholeslami works at the Babol Noshirvani University of Technology's Department of Mechanical Engineering in Iran. His research interests are CFD, nanofluid, mesoscopic modeling of fluid, nonlinear science, magnetohydrodynamics, ferrohydrodynamics, and electrohydrodynamics. He has written several papers and books in various fields of mechanical engineering. According to the reports of Thomson Reuters and Clarivate Analytics, he has been selected as a Web of Science *Highly Cited Researcher* (Top 0.01%) in 2016 and 2017. He is the editor of various journals and selected as the distinguished reviewer in various high-impact journals.

Contents

Preface

In this book, various aspects of microfluidics and nanofluidics are presented. Microfluidics and nanofluidics span a broad array of disciplines including mechanical, materials, and electrical engineering, surface science, chemistry, physics, and biology. In Chapter 1, the science and phenomena that become important when fluid flow is confined in microfluidic channels are discussed. In the second chapter, applications of nanofluid for thermal management of photovoltaic modules are reported. In the third chapter, nanofluid minimum quantity lubrication cooling (NMQLC) technique is summarized first; then a review on the mechanism of grinding thermodynamics under NMQLC condition is presented based on published literatures. In Chapter 4, each mechanism is presented in brief yet concise manner, for broad range of readers, which serves as a strong foundation for amateurs as well as a brainstorming source for experts, by description of: from the fundamental mechanism that underlies the phenomenon, covering the theoretical and schematic description; how the response is being tuned; and utmost practical, the understanding by specific implementation into bioparticle manipulation covering from micron-sized material down to molecular-level particles. Chapter 5 deals with transport and interactions of colloidal particles, and biomolecules in microchannels are of great importance to many microfluidic applications, such as drug delivery in life science, microchannel heat exchangers in electronic cooling and food processing industry. Chapter 6 aims to review and discuss the fluid flow behavior of the multiphase system, mathematical models, as well as the fundamental phenomena of driving force of microdroplet encapsulation and fission multiphase system. Chapter 7 presents the recent progress and applications of nanofluids in machining processes as well as some initial researches about microfluids. Nanofluids provide an excellent media in cutting zone for enhancing the thermal conductivity and tribological characteristics. Precisely addressed (DNA gene) spray microfluidic chip is investigated in Chapter 8. Chapter 9 aims to focus on a detailed description of the thermal transport behavior, challenges and implications that involve the development and use of HTFs under the influence of atomistic-scale structures and industrial applications. Heat and mass transfer analysis in magnetite molybdenum disulfide nanofluid of grade two is studied in Chapter 10. Chapter 11 deals with heat transfer enhancement as well as the pressure drop augmentation to determine whether nanofluid is feasible for use in practical applications. Chapter 12 presents an overview of the biofouling in membrane processes and different fabrication techniques of microfluidic membrane systems.

Mohsen Sheikholeslami Kandelousi (M. Sheikholeslami)
Department of Mechanical Engineering
Babol Noshirvani University of Technology
Babol, Islamic Republic of Iran

Chapter 1

Microfluidics and Nanofluidics: Science, Fabrication Technology (From Cleanrooms to 3D Printing) and Their Application to Chemical Analysis by Battery-Operated Microplasmas-On-Chips

Vassili Karanassios

Additional information is available at the end of the chapter

http://dx.doi.org/10.5772/intechopen.74426

Abstract

The science and phenomena that become important when fluid-flow is confined in micro-fluidic channels are initially discussed. Then, technologies for channel fabrication (ranging from photolithography and chemical etching, to imprinting, and to 3D-printing) are reviewed. The reference list is extensive and (within each topic) it is arranged chronologically. Examples (with emphasis on those from the authors' laboratory) are highlighted. Among them, they involve plasma miniaturization via microplasma formation inside micro-fluidic (and in some cases millifluidic) channels fabricated on 2D and 3D-chips. Questions addressed include: How small plasmas can be made? What defines their fundamental size-limit? How small analytical plasmas should be made? And what is their ignition voltage? The discussion then continues with the science, technology and applications of nanofluidics. The conclusions include predictions on potential future development of portable instruments employing either micro or nanofluidic channels. Such portable (or mobile) instruments are expected to be controlled by a smartphone; to have (some) energy autonomy; to employ Artificial Intelligence and Deep Learning, and to have wireless connectivity for their inclusion in the Internet-of-Things (IoT). In essence, those that can be used for chemical analysis in the field for *"bringing part of the lab to the sample"* types of applications.

Keywords: microfluidics, nanofluidics, wet chemical etching, embossing, polymeric substrates, 3D printing, rapid prototyping, microplasmas, portability, postage stamp sized 2D-chips, 3D-chips, Lab-on-a-chip, MEMS, NEMS

1. Science, technology and advantages of microfluidics

As is the case in many fields of scientific research, the field of microfluidics has three main components: a science, a technology and an applications component.

For microfluidics, a common thread between all of these components is that they are micro-sized, so size will be briefly discussed first. The dimensions shown in **Figure 1** are approximate because size of naturally-occurring objects (and of some manufactured-things) varies, for example the diameter of a human hair is between 50 and 100 μm; the diameter of the tip of a rollerball pen is between fine, medium and bold (e.g., between 0.5 and 0.7 mm); and of a 1 cent coin with its diameter varying slightly depending on the jurisdiction the penny was minted (typically around 20 mm or somewhat more).

1.1. Microfluidics as a science

Microfluidics has been defined [1–17] as the study of the behavior of fluids (or whatever is in them, e.g., colloids, discrete nanoparticles or individual cells), **in micro or in sub-millimeter channels** or around microstructures. Although microchannels can be relatively long (e.g., several 10's of mm), they are still called microchannels as long as one **critical dimension** (e.g., channel-width or channel-depth or tube radius) is in the micro scale. Microfluidic channels can be used for example to confine or to guide or to mix or to manipulate fluids.

- **The science of scaling as applied to microfluidics:** a number of physical properties of fluids change as size gets smaller [1–47], to quote *"smaller brings new capability"* [31]. These changes are often non-linear and have been discussed in books [1–17] and in journal papers [18–29]. A non-exhaustive list of size-dependent phenomena and effects is outlined below.

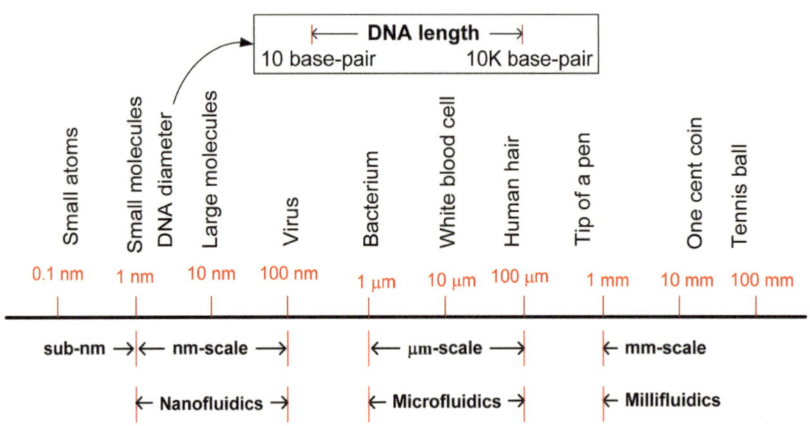

Figure 1. Examples of an approximate scale of things. The boundaries between micro and nanofluidics and between micro and millifluidics are fuzzy. In many cases, the strict definition adopted by the National Science Foundation (NSF) of the US for nano as anything with one critical dimension ≤100 nm is not strictly adhered to, thus there is a gap between 100 nm and 1 μm. Similar arguments apply to the NSF definition for micro (defined as one with a critical dimension between 1 and 100 μm). In many cases, the micro-scale is arbitrarily widened to ~1 mm and sometimes slightly more. The term millifluidics has recently been used for channels (or structures) with one critical dimension of a few mm.

- **The length-cube relationship:** the geometrical scale of length varies linearly but volume varies as length-to-the-power-of-three. As a consequence, volume changes rapidly as length decreases. Typical volumes of fluids in microfluidic channels range between nano-liter (nL) and femtoliter (fL). At the μm-scale, some properties of fluids change (as compared to a **mesoscale**, arbitrarily defined as the intermediate scale between the micro-scale and the macro-scale). Example properties that dominate at a micro-scale and that are different than those observed at the meso and macroscales include dominance of laminar-flow, diffusion-dominated mixing and capillary action. To highlight one such effect, a counter-intuitive example (from an every-day scale point of view) involves two parallel-flowing fluid-streams that come into contact in a microchannel. Since there are no eddy currents or turbulence (due to laminar flow), the only mixing that occurs is a result of slow-occurring diffusion at the interface between the two fluid-flows. Since there is no bulk mixing, mixture-separations in microchannels are faster and have shorter separation times.
- **The square-cube law:** states that volume increases faster than surface area. In microfluidics, fluid-flows in microchannels are influenced or controlled or are a function of surface area (e.g., surface tension) while others (e.g., weight) are a function of volume. Typically, in microfluidics there are no gravity effects but dominance of surface tension and of interface effects.
- **Examples of other phenomena influenced by size and expressed by dimensionless numbers:** these include laminar flow expressed by the Reynolds number; surface tension expressed by the Bond number; transient thermal effects expressed by the Fourier number; viscous heating by the Brinkman number; and fluid compressibility by the Mach number.
- As a result of channel-size, microfluidics enables one to probe individually whatever it is in a fluid constrained in a microchannel (e.g., a single cell), thus providing additional avenues for scientific inquiry and discovery (important especially in the bio-analytical sciences).

Overall, the relevant literature [1–47] describes efforts at exploring and understanding the Physics of flow-related phenomena. Developments enabled by microfluidics will be highlighted in this chapter, with emphasis on ionized gases (e.g., Paschen's law for electrical gas breakdown; plasma sheaths and the Debye length) as applied to microplasmas formed inside fluidic channels.

1.2. Microfluidics as a technology

Microfluidics refers to a variety of approaches that enable exploitation of the phenomena mentioned above by fabricating microfluidic channels on a variety of substrates. For instance, on crystalline Silicon (of c-Si) wafers, on amorphous glass or on polymeric substrates. Due to the advantages of confining flow in microfluidic channels, several fabrication technologies have been developed and tested and will be briefly reviewed. These technologies are often collectively called micro Total Analysis Systems (μTAS) or Lab-on-a-Chip (LoC) or Micro Electro Mechanical Systems (MEMS). Microfluidics or whatever acronym is used to describe it, has attracted significant attention in books [1–17] and in journals [18–29]. While in the topic of publications, older references have been purposely included in this chapter followed by some recent publications. Where possible, the citations in the reference list have been grouped either according to fabrication technology or according to the type of substrate used (e.g., c-Si, amorphous, polymeric) or according to application. Within each technology, the reference list has been sorted out chronologically to help interested readers follow the origin and evolution of ideas and technologies. Despite of the relatively large number of references included, this is not a comprehensive review. The reference list simply offers starting points. Getting back to the main theme, the question still remains: why does microfluidics continue to receive increased attention? What are the advantages of using microfluidics, especially for chemical analysis applications?

1.3. Advantages and selected applications of microfluidics

The science and technology mentioned above are widely exploited and applied to give microfluidics a host of advantages. A brief list includes use of small volumes of sample and reagents (thus reducing cost per analysis and minimizing waste disposal); rapid sample processing; potential for automation (thus reducing cost); reduced risk of contamination; short analysis time (e.g., by increasing speed of separations); small footprint and light-weight thus enabling development of future portable microfluidic-based, portable micro-instruments that can be employed *on-site* or for personal use or for personal dosimetry; potential for massive parallelism (for high sample throughput); and overall, lower ownership and operating costs (vis-à-vis conventional, lab-sized systems). Application areas (to name but a few), include analytical chemistry, synthetic chemistry (including nanomaterials synthesis), microbiology, biotechnology, point-of-care diagnostics, drug delivery, immunoassays and medicine, health-monitoring and health-diagnostics, agriculture, food safety and environmental monitoring [30–47].

2. Technology for fabrication of microfluidic channels

2.1. Fabrication using either crystalline Si (c-Si) or other substrates

Microchannel fabrication technology has been borrowed from the semiconductor industry. Initially, *bulk micromachining* [1–17, 48–51] was employed on crystalline Si (c-Si) substrates and on amorphous glass. To use it, a photolithographically patterned wafer was dipped into a chemical etching solution to etch-away (or subtract) material from the substrate, thus forming microchannels of desired geometry. This method is often referred to as *wet chemical etching* [48–51]. Inadequate control of channel depth (resulting unevenly etched channels) due to spatial etch-rate variations and to pyramid formation when crystalline-Si (c-Si) substrates and deep microchannels were etched are two key disadvantages. In contrast, *surface micromachining* [52–54] involves repetitive patterning, thin layer deposition and selective etching of sacrificial layers. The challenge here stems from the many photolithography steps involved and from the precautions required so that previously deposited layers are not damaged.

We used (as far back as the 1990's) cleanroom-based photolithography, bulk micromachining and **wet chemical etching** [48–51] to fabricate shallow-depth microchannels (with relatively low width-to-depth **aspect ratio**). This approach is often referred to as 2D sculpting of Manhattan-like structures and it offers a planar, 2D- rather than a 3D-perspective. Some examples will be briefly discussed later.

For completeness, other methods of microchannel fabrication on inorganic substrates (either crystalline or amorphous) have been described. A short list includes **laser machining** [55–58]; lithographie galvanoformung adformung (LIGA) or lithography electroplating molding [59–61] which is well suited for fabrication of high aspect ratio channels; deep reactive ion etching (DRIE) [62–65] often used for fabrication of microchannels with a high aspect ratio; and, **SU-8** (an epoxy-based negative photoresist) and its variants such as SU-8 series 2000 and SU-8 Series 3000) [66–68].

Technologies involving polymeric substrates [69–83] include replication via **imprinting** [69–73] or **embossing** [74–76]. Polymeric substrates are selected due to their bio-compatibility or to reduce cost of ownership. Examples will be shown later. The terms disposable or recyclable microfluidic devices is often used for microfluidic channels on polymeric substrates. **Soft lithography** [77–83] (defined as a collection of fabrication techniques for replication of microchannels) is a technology that does not require access to a clean room. It is called soft because it uses soft and flexible (primarily) elastomeric materials such as poly di methyl siloxane (PDMS) and often cyclic olefin copolymer (COC).

There are other techniques that are rather difficult to classify either according to fabrication technology or according substrate. Despite of being brief, the list includes **droplet microfluidics** [84–89], in which discrete droplets or small volumes of immiscible liquids are guided through microchannels. In the early literature, this approach was often called digital microfluidics. As it is known now, **digital microfluidics** [90–95] is an outgrowth of electrowetting [90, 92] and it involves use of discrete droplets on arrays of electrodes, with individual droplets manipulated by electrical means. The list also includes **centrifugal microfluidics** [96–101], a technique that enables micro-flow manipulation by using rotational forces (e.g., Coriolis) obtained by spinning a CD on top of which there are microfluidic channels. This technique is often called *"lab on a CD"*. It also includes **paper microfluidics** [102–108], a technique that uses paper for development of microfluidic approaches intended for use in resource limited situations (e.g., remote geographical areas or resource-limited locations).

Rapid prototyping via 3D-printing [109–122] involves both a technology (e.g., a 3D printer) and a materials platform (e.g., a polymer) for formation (primarily) of mill-sized fluidic (and recently) micro-sized channels [115, 117, 120]. An example of 3D printing will be discussed later in this chapter.

2.2. Fabrication technology examples

To highlight substrate-dependence of fabrication, the fabrication steps required for microchannels on c-Si and on amorphous glass or quartz substrates are compared and contrasted in **Figure 2**. It should be noted that depending on crystallographic orientation of the substrate and of the chemical cocktail used in the etching solution, isotropic or anisotropic etching may be obtained [48–51].

Example 1: ***Planar 2D-chips*** **and wet chemical etching for fabrication of microchannels on crystalline and amorphous substrates (Figure 2).**

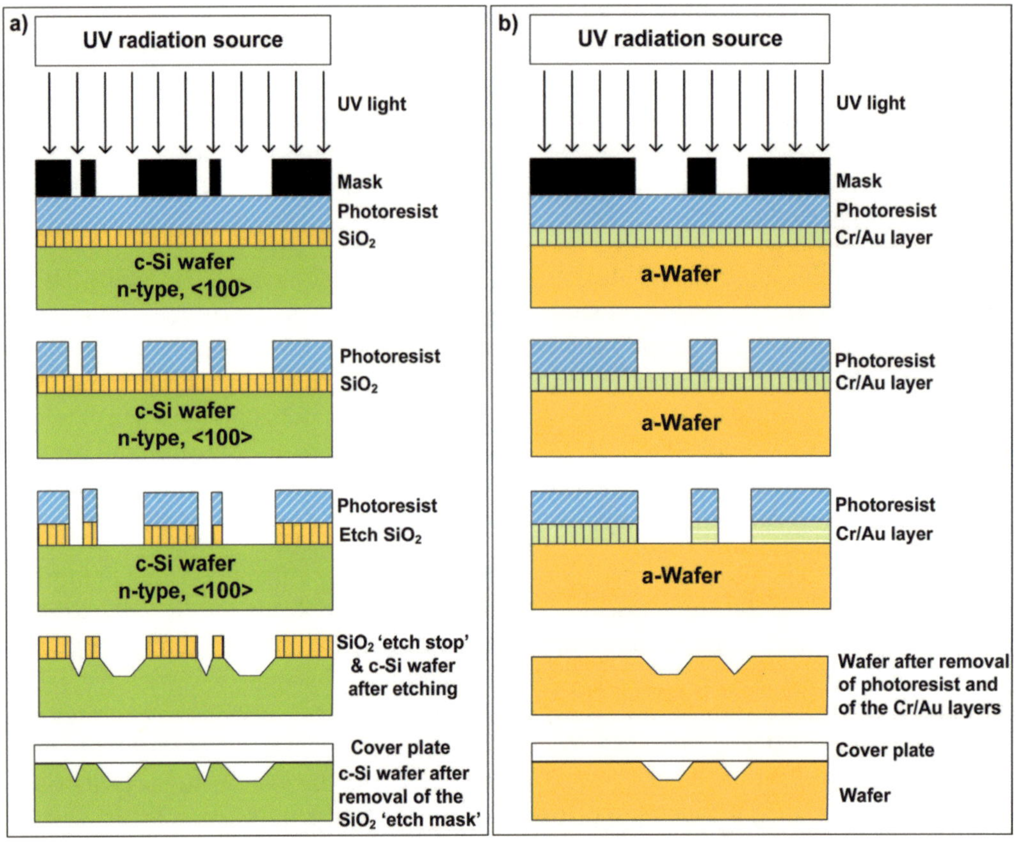

Figure 2. Simplified steps used for fabrication of microchannels on a) a c-Si wafer as a substrate and on b), a wafer made from an amorphous material (abbreviated as a-wafer-above, such as glass).

For completeness, an example of wet chemically etched microchannels on glass is shown in **Figure 3**.

The quality of the etched microchannels depended on the composition of the etching solution and on the geometric-primitives that were used to define the channels. To enclose the microchannel of **Figure 3**, a cover plate was used (but is not shown in **Figure 3**). Depending on the required optical transparency, a UV-transparent quartz cover plate was employed for most of the work described here. Furthermore, depending on the substrate (e.g., c-Si or glass), a variety of bonding methods can be employed [2–17].

Despite of the ability to fabricate low aspect ratio microchannels, wet chemical etching has shortcomings arising from costs, from limited access by many to photolithography and to cleanrooms, and from time-delays between mask-design (**Figure 3b**) and receipt of finished prototype (e.g., **Figure 3a**). At present, access to cleanrooms is not required because microfluidic chips can now be ordered from specialized foundries. In spite of this, there are still costs

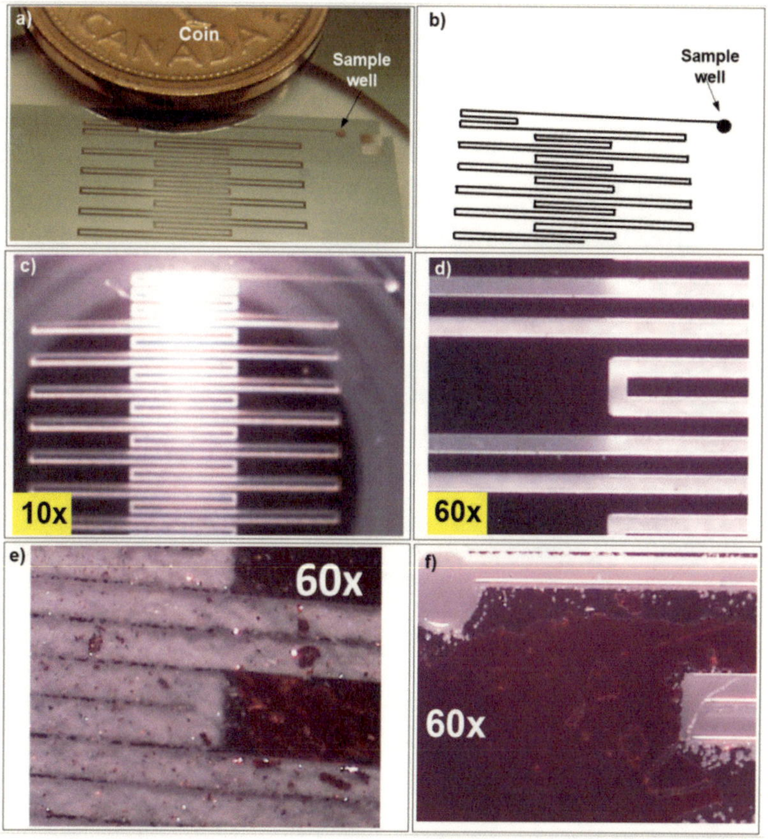

Figure 3. (a) Part of a 14.5 mm by 25.6 mm chip of an etched microfluidic channel on corning 7059 glass with the photoresist removed and (for clarity) without a cover plate. Also omitted are pipette-tips used as sample reservoirs that are attached to the sample well. A coin was included for size. (b) Part of a Mylar mask used for photo-lithography. (c) Part of a washed meandering microchannel shown under 10x magnification and (d) shown under 60-fold magnification. (e) an unwashed microchannel immediately after etching showing etching by-products inside the microchannels, thus requiring their removal. (f) a sample-well and a washed microchannel showing the quality of etching, in particular for the round sample-well. For (d), (e) and (f) the photoresist was not removed to provide contrast for the photographs.

and time-delays involved. There is another limitation if microchannels are to be used with biological samples, because many biosamples adhere to substrates. Thus, functionalized surfaces or microfluidic channels on polymeric substrates are preferred.

Example 2. Imprinting microchannels on planar polymeric 2D-chips. 2D-microchannel fabrication on polymeric substrates is one way of overcoming some of the limitations mentioned above. But polymers may contain additives, fillers or plasticizers that may contaminate the samples, and they may display auto-fluorescence. As for fabrication (**Figure 4**), it may be achieved by using Si-stamp imprinting (**Figure 4**) or by imprinting (by pressing) a wire on a substrate [69] (**Figure 5**). In the example shown in **Figure 4**, a c-Si stamp (or master or hard mold) was developed and was employed for replication by imprinting.

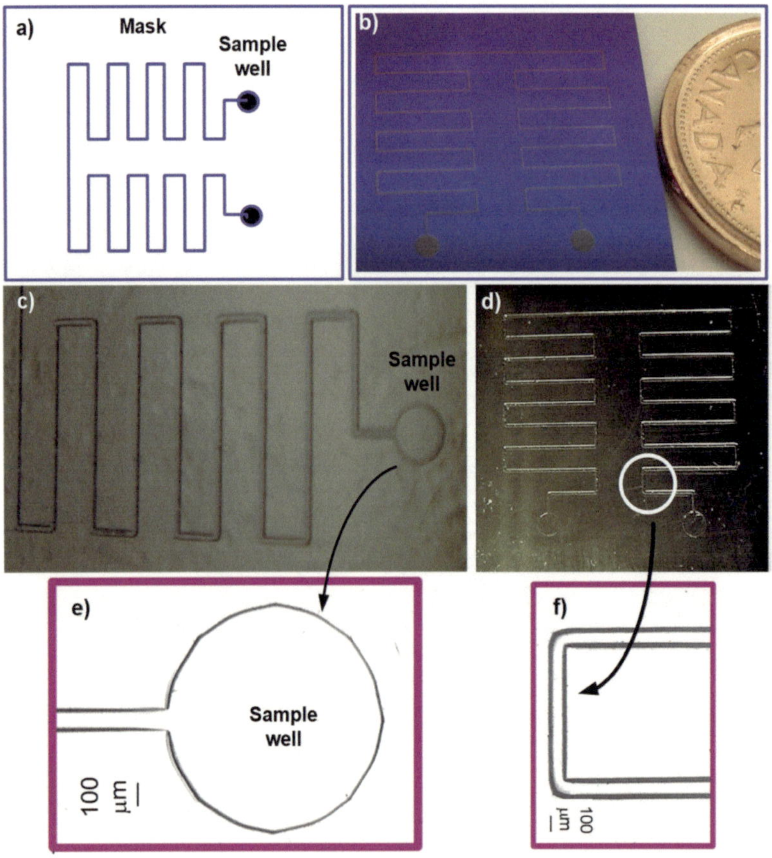

Figure 4. (a) Mask; (b) mask on c-Si chip, coin has been added for size; (c) chemically etched c-Si chip (serving as a stamp), the meandering pattern is protruding from the surface of the chip; (d) imprint generated by pressing the stamp and the polymeric chip together by placing them in a hydraulic press and by applying pressure at room temperature; (e) imprinted sample-well on a polymer chip shown under magnification; and (f), part of an imprinted meandering channel shown under magnification. For (d) and (e) and (f) different polymeric materials were used.

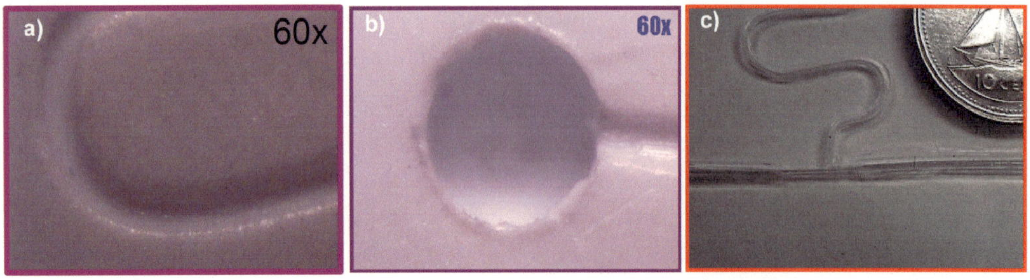

Figure 5. (a) Imprinted channel on a polymeric chip (60x magnification), (b) sample-well (60x magnification) and (c) Venturi micropump with no moving parts and no electrical power requirements fabricated by imprinting (coin included for size) [73].

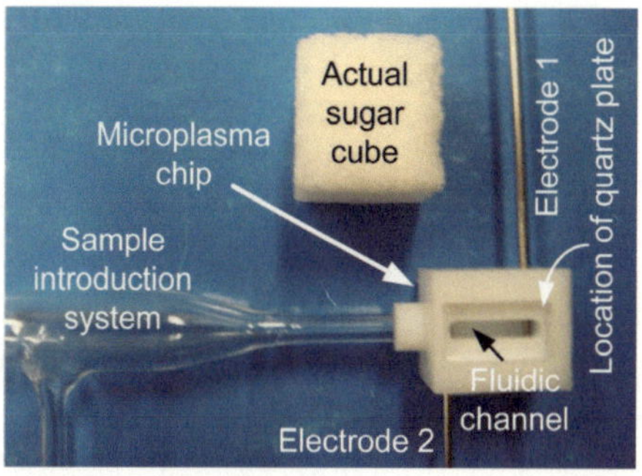

Figure 6. Sugar cube-sized, 3D-printed hybrid-chip with a millifluidic channel to be fitted with a quartz cover plate (selected for UV transparency). A sample introduction system is also shown and it has been included to provide an overall size for this "critical" component of a potential future micro-instrument. An actual sugar-cube (~1 cm by ~1 cm) has been included for scale comparisons.

Example 3: 3D-printed, milli-sized fluidic channels on polymeric materials for hybrid 3D chips. 3D printing technology [109–122] using polymeric materials is receiving attention for rapid prototyping [109] including fabrication of mm channels (often called **millifluidics**) and more recently for sub-mm channels (using specialized printers) [120, 121]. We used 3D-printing due to reduced fabrication and ownership costs and due to quick turn-around times (often from concept to prototype in hours). A simple, hybrid, 3D-printed 3D-chip containing a millifluidic channel is shown in **Figure 6**. The word *hybrid* was used because the two needle electrodes and the quartz cover plate were not 3D-printed.

In my laboratory, some of the fabrication technologies discussed thus far have been used to constrain plasmas in microfluidic or in millifluidic channels. But why plasmas and why microplasmas?

3. Why plasmas?

There are four states of matter: gases, liquids, solids and plasmas [123–131]. To generalize, atmospheric pressure plasmas are ionized gases that are either hot or cold (about room temperature or somewhat above it). Plasmas occur in nature, for example those found in inter-stellar space, in the ionosphere, in auroras and in lightening. There are also artificially-generated plasmas that are being used in many every-day-life applications. Neon signs and fluorescent lights in which low-pressure plasmas are formed either in Neon (Ne) gas or in Argon (Ar) gas) are two such examples. Other examples include plasmas employed for device fabrication by the semiconductor industry or for materials synthesis in nanoscience and nanotechnology [129–131]. **It has been estimated that over 50%** of whatever goes inside any electronic device (e.g., a tablet, a smartphone, TV) is fabricated using a low-pressure plasma.

Conventional-scale (or **lab-scale**) atmospheric pressure plasmas are widely used in **chemical analysis**, primarily in the form of atmospheric-pressure, **6000–10,000 K hot** Inductively Coupled Plasmas or ICPs [132]. Due to their size and weight (e.g., in the few 100's of pounds), their gas consumption (e.g., ~20 L/min), their power usage (e.g., 1–2 kW) and their need for cooling, ICPs are primarily used in a lab.

3.1. Some fundamental aspects of plasma science

A plasma is an ionized gas [123–131]. The term *plasma* was coined by Langmuir in the 1920's and it is derived from the ancient Greek word *πλάσμα* (plasma), freely translated to something "*moldable*". A plasma consists of ions (with ion number density n_i) and electrons (with an electron number density n_e), and on the average it is quasi-neutral, and for singly ionized gases $n_e \approx n_i$. Thus, a prerequisite for plasma formation is ionization. Singly-charge ionization (in the form of ion-electron pair formation) is done by detaching an electron from a neutral gaseous atom or molecule. Although there are other ways of detaching an electron (e.g., thermally), one way doing it is by placing a gas between two electrodes and by applying an electric field with a sufficiently field-strength to ionize the gas (**Figure 7**), thus forming an *electrical gas discharge*. Because neutral gaseous atoms or molecules (ordinarily insulators) become ion-electron pairs, they also become (partial) conductors. Partial because to an approximation, conductivity depends on the degree of ionization (this is important for weakly ionized plasmas).

To obtain electrical gas breakdown, the dielectric strength of the gas must be exceeded. The dielectric strength is the maximum electric field-strength (in V/m) an insulating gas can endure without breaking down into ions and electrons. If there is a sufficiently large field-strength, breakdown of the dielectric strength will cause formation of (typically) a low-current **spark** (i.e., a momentary electrical discharge, an example is electrostatic discharge from static electricity), or formation of a continuous electric-**arc** requiring continuous application of an electric field from an external power supply (**Figure 7**) capable of providing high-current (often in the Amp range). Arcs find applicability in welding of metals.

Conditions for sustaining continuous plasma operation: Following gas breakdown, there must be continuous application of external power to sustain a plasma. Other criteria include an electrode distance d that must be >> λ_D and that $n_e\lambda^3_D$ must be >> 1 (this is easy to satisfy

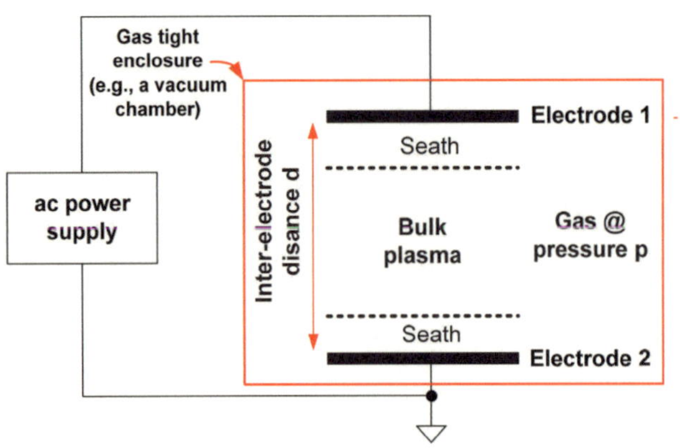

Figure 7. Ideal plasma formed in a gas-tight and pressure-controlled enclosure. The plasma is formed between two conducting plates or electrodes positioned at a distance (or gap) d from each other. For dc operation, pertinent literature should be consulted [124].

for the plasmas of interest to this work), where λ_D is the Debye length [133–137]. These will be briefly discussed later in this section.

For microplasmas formed inside fluidic microchannels, in addition to gas breakdown and to continuous application of power, a microplasma must be formed in a constrained microchannel.

3.2. Scaling of lab-size, ambient-pressure plasmas to microplasmas

Arbitrarily defined, microplasmas are those with **one critical dimension** in the micro-meter (μm) or in the sub-milli-meter regime [138, 139]. The words *"critical dimension"* (i.e., one dimension such as channel depth or width or radius) are important here: an **atmospheric** pressure microplasma in a microfluidic channel can range in length from μm to a 10's of mm, as long as its critical dimension fits the definition above. But as the critical dimension is reduced to sub-mm and depending on operating conditions, atmospheric pressure plasmas transition from **thermal and 10,000°C hot** (e.g., lab-scale ICP [132]) to **non-thermal and cold** [133–139] (e.g., microplasmas). They also transition from equilibrium to non-equilibrium (to an approximation, to those with gas temperature $T_g \ll T_e$ (electron T). There are scientific implications due to these transitions (e.g., for nanomaterials synthesis) and for excitation mechanisms (e.g., for chemical analysis). In terms of technology-implications, cold plasmas enable use of inexpensive polymeric substrates that do not melt because microplasmas are cold and they do not require cooling; and they allow use of inexpensive 3D printing technology for fabrication.

Why miniaturize atmospheric-pressure plasmas? Operation at (or near) atmospheric-pressure is preferred because it obviates the need for heavy-weight and power-consuming vacuum pumps. By reducing weight and power consumption, atmospheric-pressure operation enables microplasma portability for chemical analysis *on-site* (i.e., in the field). By bringing a microplasma-based instrument to the field, microplasmas are expected to cause a paradigm shift in classical chemical analysis in which samples are collected in the field and are brought to a lab for analysis [140–148].

Due to plasma miniaturization, a number of questions arise. For example, how small **can** microplasmas be made? And, how small **analytical** microplasmas **should** be made? From a technology perspective, what is the minimum voltage required to ignite and sustain a microplasma? Would substrates tolerate the required high voltage? And, what is the preferred fabrication technology?

3.3. How small atmospheric-pressure microplasmas can be made?

A plasma (**Figure 7**, regardless of its size) consists of two plasma sheaths (located in the vicinity of two electrodes bathed in a gas-of-interest in a gas-tight container) and a bulk plasma [133–137]. Shielding (or damping or screening) of the electric field arises from the presence of charged species in the plasma and from the unequal mobility of ions and electrons in the vicinity of the electrodes. Inside the plasma sheath, macroscopic electrical neutrality is likely not maintained. But outside of it (labeled bulk plasma in **Figure 7**), macroscopic neutrality is maintained and (time-averaged) electron and ion fluxes are roughly equal. Thus (on a time-average and per unit-volume), $n_e \approx n_i$ (for singly charged species). The distance (or thickness) a sheath screens electric fields is called the *Debye length* (λ_D), given by Eq. 1.

$$\lambda_D = \left[\frac{\epsilon_0 kT}{n_e e^2}\right]^{1/2} \qquad (1)$$

where k is the Boltzmann constant, T is the electron temperature, ϵ_0 is the permeability in vacuum, n_e is the electron number density and e is the charge of an electron.

To generalize, a key assumption is that sheath thickness is about the same magnitude as the Debye length. A few, what-if type thought-experiments will be used to obtain an indication on how λ_D changes as T and n_e vary. For example, for an atmospheric pressure plasma when T = 10,000 K (with 1 eV = 11,600 K) and $n_e = 10^{16}$ m^{-3}, then λ_D = 110 μm. But when T = 5000 K and assuming that there is no thermal ionization (thus the degree of ionization is constant and the same as in the example above) with $n_e = 10^{16}$ m^{-3}, then λ_D = 80 μm. For less than atmospheric pressure operation and assuming that $n_e = 5 \times 10^{14}$ m^{-3} and (for simplicity, assuming that the degree of ionization is unchanged) and that T = 5000 K, then λ_D = 350 μm. Because plasmas cannot be made smaller than their boundary layers (per conditions outlined in Section 3.1), plasma sheaths (and Debye length) set a **fundamental limit** as to how small the inter-electrode distance d (**Figure 7**) can become, in other words, **how small a microplasma can be made.**

Since inter-electrode distance d must be >> λ_D, and for the example with λ_D = 110 μm and for a two-electrode operation, then the microplasma must be larger (or much larger) than 2 times λ_D or (for this example) it must be >> 220 μm). As d becomes ~2 times the length of the sheath, the sheath-bulk plasma structure must disappear and thus the plasma must become devoid of a bulk plasma (**Figure 7**), that is to become a sheath-only plasma. But in a strict interpretation of the definition of a plasma, can such an ionized gas still be called a *"plasma"* [134]? There are published reports of microplasmas formed in constrained cavities that are smaller than 10 μm by 10 μm [133–137]. This has been explained by considering that sheath-thickness scales as inter-electrode distance decreases. Several open-ended questions in this research area still remain unanswered for instance, would microplasmas the size of 10's of μm be useful for chemical analysis? To obtain insights, perhaps this question must be re-phrased to read *"how small analytical, atmospheric pressure microplasmas* **should** *be made"?*

3.4. How small analytical, ambient-pressure plasmas should be made?

There are two answers to this question. One is that microchannels can be 10's of mm long ([48–51] and cited literature). Since there does not seem to be a fundamental reason why microplasmas should be constrained in μm-size cavities, microplasmas can occupy part of mm-long microchannels (**Figure 8**). Therefore, such microplasmas are not limited by Debye length or by plasma sheaths.

The other answer involves residence time of an analyte in a microplasma (**analyte = the chemical species of interest in a sample to be used for chemical analysis**). Residence time (important in elemental chemical analysis) is defined as the time an analyte resides in, or is in contact with or it interacts with a microplasma. In general, as microplasma length (dictated by the inter-electrode distance or gap) decreases, so does residence time. But as residence time decreases, so does signal intensity from an analyte introduced into a microplasma. This

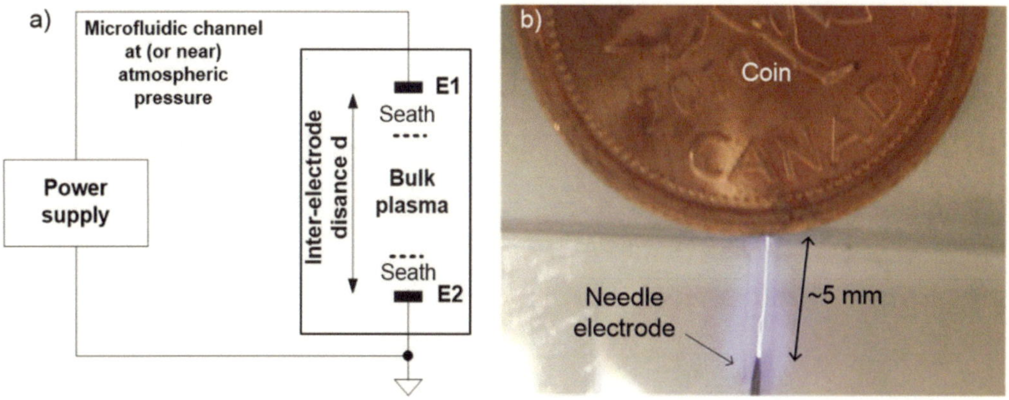

Figure 8. (a) Simplified diagram of a microplasma and (b) microplasma formed at the end of a needle electrode (OD: 470 μm, ID: 130 μm) inside a microfluidic channel on a microfluidic chip. A Canadian 1 cent coin (about the same diameter as that of a US one-cent coin, or UK's one-pence, or a one-cent euro) has been included for size.

is mainly due to a reduced interaction-time between an analyte and a microplasma. Thus, from an elemental analysis viewpoint, decreasing the length of a microplasma (e.g., by fabricating microplasmas in μm cavities) may not necessarily be beneficial in terms of signal intensity. This is significant because as signal intensity worsens, signal-to-noise ratio (SNR) degrades, thus degrading the detection limit (defined as the minimum amount or concentration that can be detected with a stated statistical confidence). The detection limit is a key figure of merit in chemical analysis. From the foregoing it can be concluded that mm-long microplasmas formed inside microfluidic channels (e.g., **Figure 8**) will likely be beneficial for elemental chemical analysis.

3.5. Igniting and sustaining a microplasma at atmospheric pressure

Conceptually, there are two steps involved in forming and sustaining a continuously-operated atmospheric-pressure microplasma. For instance, a microplasma must be first initiated (or "ignited") and then it must be sustained. The minimum "ignition" (or sparking) voltage (V_b) for which the entire discharge gap is fully formed (often called "bridged") when a uniform electric field is applied between two flat electrodes at a distance or gap (d) immersed in a gas of interest under pressure (p) can be determined using Paschen's law (Eq. 2).

$$V_b = \frac{Bpd}{\ln\left[\frac{Apd}{\ln(1/\gamma)}\right]} \tag{2}$$

A and B are constants that depend on the properties of the gas in which the electrodes are immersed in (not accounting for any ionization due to background radiation). The values of A and B are either determined experimentally or they are calculated from literature values [139]. The coefficient γ (also known as Townsend's coefficient) incorporates properties of the

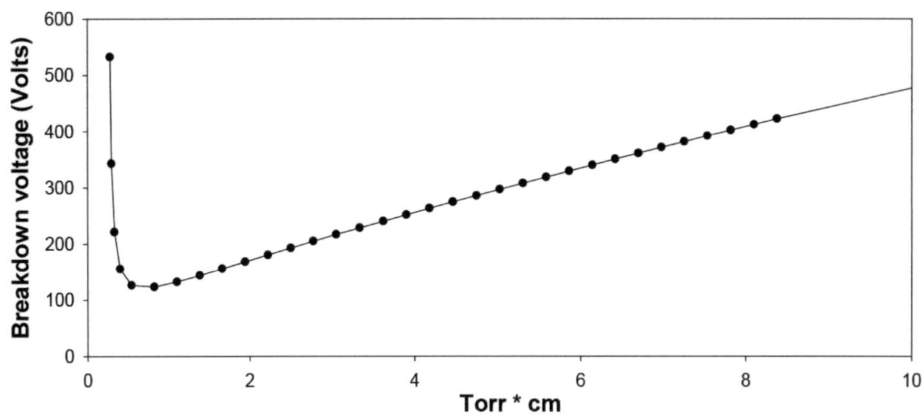

Figure 9. Paschen curve for argon gas and for a 2.8 mm inter-electrode gap (d) as a function of pd.

electrode material (e.g., work function) and it assumes that gas breakdown is predominantly a function of electron emission from the electrodes. In short, the two key variables in this equation are pressure (p) and inter-electrode distance (d). The product of p times d is often called "**pd scaling.**" An example of a Paschen curve is shown in **Figure 9**.

Paschen's law applies to electrical discharges formed at low-pressures. In high-vacuum or at high pressures (e.g., atmospheric), Paschen's law fails ([139] and references herein). There are also deviations from the behavior predicted by Eq. 2 when kHz or MHz ac voltages are used or when μm inter-electrode distances (or gaps d) are employed [139]. Undeniably, there are limits to applicability of Paschen's law. Despite of these limitations, Paschen's law (**presumably, the only choice**) can be used to obtain rough estimates of the magnitude of the voltage required to ignite (or initiate) an atmospheric pressure plasma. Thus it can be used as an aid in the design of appropriate power supplies. For instance, when the electrodes are made from Iron (Fe) and the inter-electrode distance d is 2.8 mm, and the discharge gas is Argon (Ar) at (or near) atmospheric pressure, the minimum voltage (V_b) required for gas breakdown (or for microplasma ignition) is about 6000 V. As the inter-electrode distance d decreases from 2.8 to 1 mm (and by keeping all else constant), V_b drops to about 2400 V, and when d further decreases to 0.5 mm, V_b drops to about 1400 V. It should be emphasized that gas breakdown at the minimum voltage V_b is not always necessary and that (once ignited), to sustain a microplasma lower voltages are typically required. An example is the ballast used in fluorescent lights.

4. Microplasma formation inside fluidic channels

The key idea behind microplasma miniaturization [138] is to obtain analytical performance about equal to that of lab-scale ICP-optical emission spectrometry (ICP-OES) systems [132] but by using self-igniting, low-power, low-cost, small-size, light-weight, continuous-flow and low gas-consumption (e.g., 250 mL/min) atmospheric-pressure microplasmas. The expectation is that such microplasmas can be used for *"taking part of the lab to the sample"* types of applications [140–142].

Based on these ideas, we fabricated and tested a variety of battery-operated, atmospheric pressure, self-igniting, mm-length microplasmas in fluidic channels [143–156]. Due to their mm-length, plasma sheath and Debye length are not of a concern. In addition to being "cold", their high surface area-to-volume ratio further facilitates heat dissipation, thus facilitating use of polymeric substrates and 3D-printing fabrication methods. Example microplasmas fabricated in a variety of substrates will be discussed next.

4.1. Microplasmas in fluidic channels on amorphous substrates

For microplasmas formed inside a **microfluidic channel on a chip**, a dual substrate approach was used (**Figure 10**). Briefly, cleanroom-technologies (**Figure 2**) were employed to define and to sputter-deposit Au electrodes E1 and E2 (**Figure 10a**). Holes were drilled for the inlet and the outlet. On the bottom wafer, a chemically etched microchannel was formed. The top and bottom wafers (**Figure 10a** and **b**) were aligned so that the central part of the etched channel matched the protruding part of electrodes E1 and E2. Then the wafers were bonded together (**Figure 10c**) [143] and glass-tubes were affixed to the inlet and outlet holes (**Figure 10d**). The inlet was connected to a gas-supply (Ar-3%H_2) that was used as the microplasma gas and as the sample-introduction carrier-gas. Upon application of electrical power, the microplasma self-ignited, it was formed between electrodes E1 and E2 and was sustained by continuous application of electrical power (~10 W). To avoid electrode breakage, a high-voltage ac [143] was used.

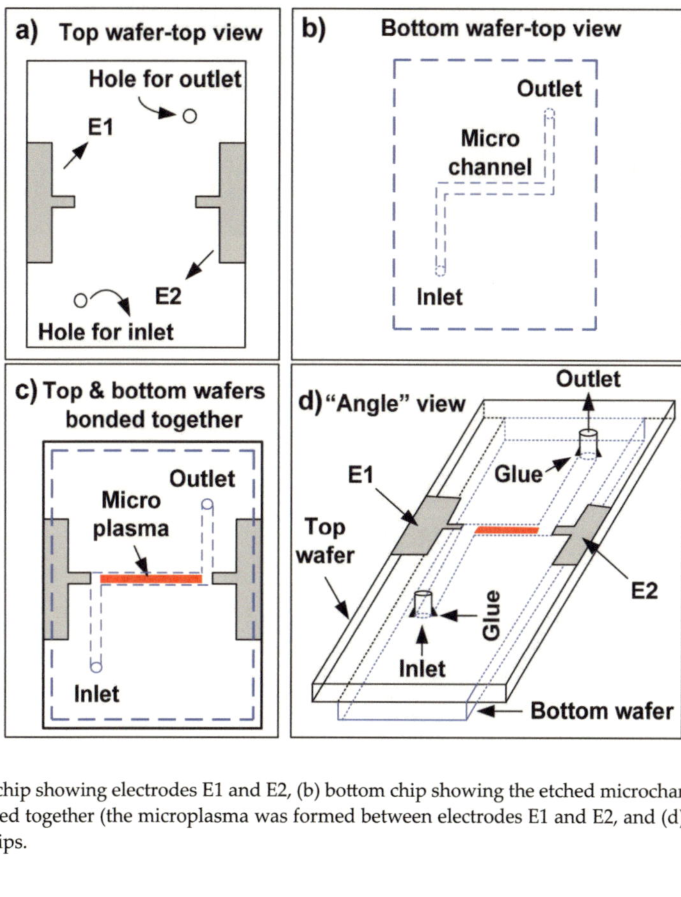

Figure 10. (a) Top chip showing electrodes E1 and E2, (b) bottom chip showing the etched microchannel, (c) the top and bottom chips bonded together (the microplasma was formed between electrodes E1 and E2, and (d) an *"angle"* view of the two bonded chips.

4.2. Postage stamp-sized microplasmas on polymeric substrates

To reduce ownership, operation and fabrication costs, we developed and evaluated a variety of microplasmas on polymeric substrates (e.g., **Figures 11** and **12**) [144–146]. Although a critical microplasma dimension was in μm-meter regime (**Figure 11**), these microplasmas were formed inside **millifluidic channels** (e.g., ~2 mm wide). This was done for rapid prototyping [109] and to avoid accidental contact of the microplasma with the channel-walls

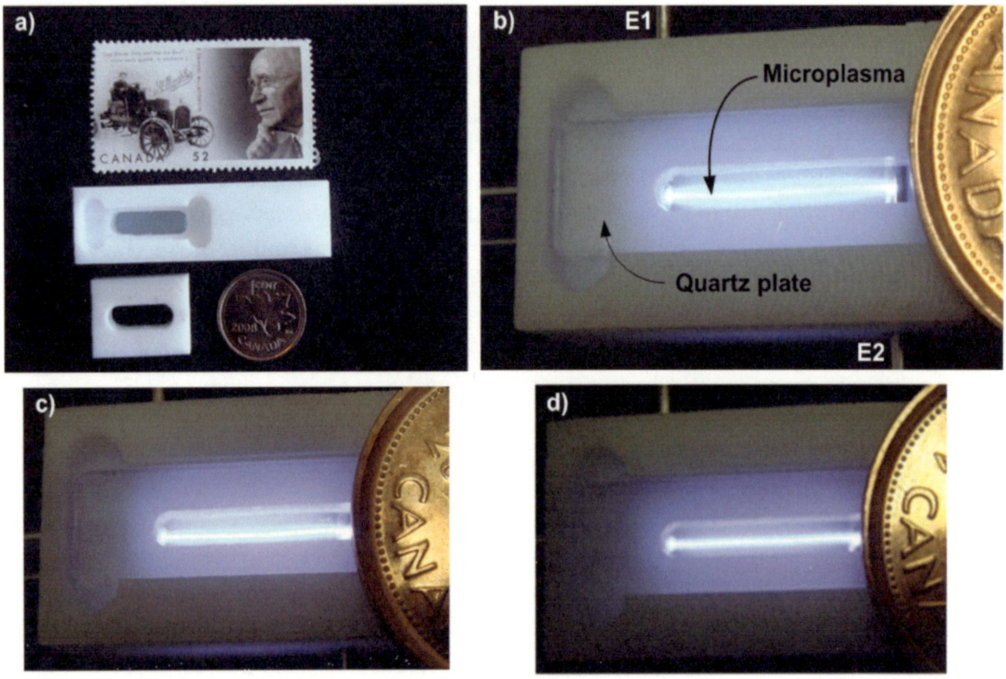

Figure 11. (a) Postage stamp-sized polymeric 3D-chips and (b) microplasma formed between electrodes E1 and E2. Depending on operating conditions, microplasmas with diameters of (b) ~750 μm, (c) ~400 μm and (d) ~200 μm were formed. A 1 cent coin was included for size, the microplasma fit inside the letter a of the coin.

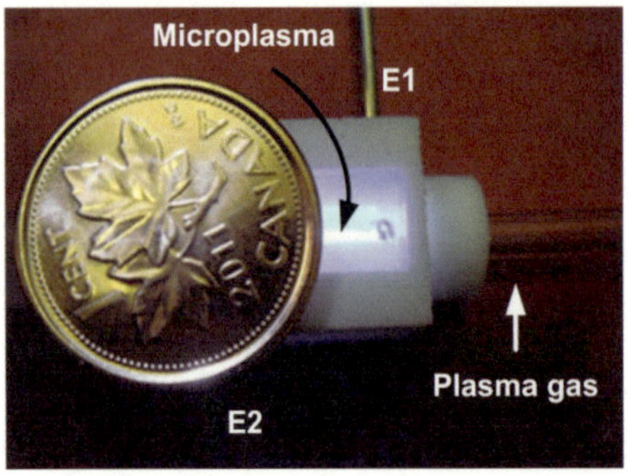

Figure 12. 3D printed microplasma on a hybrid 3D-chip formed between electrodes E1 and E2 (coin has been included for size, the microplasma fit inside the letter a of the 1 cent coin).

(important during testing). Once prototypes were produced, channel width was never revisited. Although polymeric substrates have high dielectric strength, to address poor transmission of polymers in the UV, the channels were fitted with a quartz plate (**Figure 11b**).

4.3. Millifluidic channels in 3D-printed chips for microplasmas

3D-printing [109–122] was accomplished using a 3D-printer (~$1000) to rapidly prototype 3D-chips in a few hours (or less), thus obviating the need for cleanrooms and lithography. We used 3D-printing to fabricate hybrid chips (fitted with a quartz plate and needle electrodes) for microplasma formation in millifluidic channels [146, 147, 149]. An example is shown in **Figure 12**.

5. Nanofluidics

The nanoscale [157–194] is a natural extension of the microscale (**Figure 1**) and it is defined as the science, technology and application of transport phenomena and of fluid-flow in channels ≤100 nm or around nano-size objects [158, 170]. This is not universally accepted, many consider nano-size as anything with one critical dimension ≤1 μm. The range between 100 nm and 1 μm is sometimes referred to as *"extended nanofluidics"* [181]. Nanofluidics is not new, although the name is [159, 160].

5.1. The science of nanofluidics

In nanofluidics, size (or scale) is important, likely more so than in microfluidics. For instance, at the nano-scale many dimensions of molecules are of similar size as the nano-fluidic channels that constrain them (**Figure 1**). A few scientific questions that being addressed include: How do properties of individual atoms, ions or molecules, manifest themselves as they are confined in spaces (roughly) of their own size? Would quantum effects become important [173]? Since pressure is not used to force fluids through nanochannels, should electrokinetic flow be preferred? And, as surface-to-volume ratio increases significantly (over microchannels), what is the effect of surface-charge on ions or molecules confined in nanochannels? What is the effect of surface roughness on fluid-flow? And, how do surfaces interact with ions or molecules so close to them? What are the best surface modification approaches? What is the effect of van der Waals forces and of the electric double layer (EDL) at the nm-scale? Some questions arising from technology include: how would one introduce very small volumes of analytical samples into nanofluidic channels? To facilitate discussion, assume a cylindrical nanochannel with 100 nm diameter and 1 μm length. In this case, the volume is 100 **a**tto **L**iter (**aL**). How would one introduce an aL volume sample into a nanochannel without evaporation of some of the analyte or of the solvent? Due to the infinitesimal volumes used, would single atom, ion or molecule measurement techniques be essential? In support of this, it has been estimated that in a liquid the volume of a cube with dimensions 100 nm by 100 nm by 100 nm, there are only ~6 analytes when the concentration of the analyte is 1 μm [160]. Would sample separation, pre-concentration and use of highly-sensitive detection techniques (e.g., laser induced fluorescence or LIF) become essential?

5.2. Technology for fabrication of nanofluidic channels

According to the National Science Foundation (NSF) in the US and its National Nanotechnology Initiative (NNI), nanotechnology involves *"the application of scientific knowledge to manipulate and control matter in the nanoscale"* [158, 170], more or less arbitrarily defined at ≤100 nm [158, 170]. Nanofluidics often falls under nano electro mechanical systems (NEMS) [164, 177–179, 184] typically fabricated using complementary metal oxide semiconductor (CMOS) technology [177, 178].

For nanofabrication, many technologies have been described [159–162, 165, 166, 169, 171, 172, 180]. Some of them are nano-specific [159, 169] for example, scanning probe lithography (SPL) [161], etching using a focused ion beam (FIB) [171] and nanoimprinting [159]. In many cases use of a cross-sectional area of a nanochannel is preferred (e.g., 10 nm by 10 nm) rather than aspect ratio. Nanofluidic channels can be nanofabricated using either top-down or bottom-up approaches.

Top down methods of fabrication of nanochannels: By analogy to micromachining, these fabrication methods include bulk nanomachining; surface nanomachining; and, imprinting (as is typical of soft-lithography) [159, 161–165, 169]. A top plate is typically used to cover nanochannels but due the nanosize of the channels and unless precautions are taken, channels may plug-up during bonding.

Bottom up methods of nanostructure formation: in some cases molecules can be "convinced" to self-assemble into nanostructures by controlling chemical conditions [160, 162].

Associated nanofabrication technologies include scanning probe lithography (**SPL**) [161], electron beam lithography (EBL) [159] and dip-pen nanolithography [185]. Such approaches are typically used to bypass the diffraction-limit of photolithography or to provide new capabilities.

5.3. Applications of nanofluidics

In addition to enabling fundamental studies of fluid-flow and of transport phenomena (with many studies aimed at the study of naturally occurring processes in biological nanochannels), many applications are aimed at bio-sciences, bio-nano-technology and bio-analytical chemistry where applications exist in abundance [166–169]. Applications outside of classical nano-fluidics include nano-pores (e.g., for bio-applications and DNA sequencing) [186–192] and even for the study of fluid-flow in nano-porous media [193, 194]. For chemical analysis, NEMS have been developed for single protein mass spectrometry [174] and for airborne nanoparticle detection [176]. From this short list it can be concluded that nanofluidics has the potential to become a disruptive technology worthy of further investigation.

6. Conclusions

Microfluidics continues to receive attention in science and technology due to its many applications. And as shown, it has the potential to find applicability in constraining atmospheric-pressure microplasmas in 2D-microfluidic channels (**Figures 8** and **10**) or in 3D-millifluidic chips (**Figures 11** and **12**). Future developments include coupling of standard CMOS fabrication technology [179, 183, 184, 195–197] with microfluidics or millifluidics, thus allowing integration of fluidics with electronics. Microinstruments are those with at least one critical (or essential) component operating in the micro-regime. For nanofluidics as may be applied to chemical analysis, it appears that it will be best if nanofluidic channels was packaged alongside microfluidic channels.

It is envisioned that future fluidics (**Figure 1**) will be embedded within portable micro- or nano-instruments for measurements *on-site* (i.e., in the field). Such instruments will have (some) energy autonomy [198–200], will incorporate some *"smarts"* [201] (e.g., based on Artificial Intelligence and Deep Learning) and will have wireless capability [202] so that they can become a part of the **I**nternet **o**f **T**hings (**IoT**) [200–203]. Clearly, fluidics (e.g., milli-, micro- or nano-) have the potential to become critical components of mobile (or even wearable) instruments that are *"cheap, smart and under wireless control"* [139].

Acknowledgements

Financial assistance from NSERC (Natural Sciences and Engineering Research Council) of Canada is gratefully acknowledged. A special thank you to Professor (now Emeritus, ETH Zurich, Switzerland) Dr. Henry Baltes for the many enlightening discussions we had on MEMS and on miniaturization.

Author details

Vassili Karanassios

Address all correspondence to: vkaranassios@uwaterloo.ca

Department of Chemistry and Waterloo Institute for Nanotechnology,
University of Waterloo, Canada

References

[1] Bruus H. Theoretical Microfluidics. Oxford, UK: Oxford University Press; 2008

[2] Kumar CS. Microfluidics in Nanotechnology. New York: Wiley; 2010

[3] Tabeling P. Introduction to Microfluidics. Oxford, UK: Oxford University Press; 2011

[4] Mitra SK, Chakraborty S, editors. Microfluidics and Nanofluidics. Florida: CRC Press; 2011

[5] Madou MJ. Fundamentals of Microfabrication. Florida: CRC Press; 2011

[6] Kirby BJ. Micro- and Nano-Scale Fluid Mechanics. Cambridge, UK: Cambridge University Press; 2013

[7] Li X(J), Zhou Y. Microfluidics for Biomedical Applications. Cambridge, UK: Woodhead; 2013

[8] Conlisk AT. Essentials of Micro- and Nano-Fluidics. Cambridge, UK: Cambridge University Press; 2013

[9] Lagally E editor. Microfluidics and Nanotechnology: Biosensing to the Single Molecule Limit (Devices, Circuits, and Systems). Florida: CRC Press; 2014

[10] Li D. Encyclopedia of Microfluidics and Nanofluidics. Vol. five volume series. New York: Springer; 2015. DOI: 10.1007/981-1-4614-5491-5

[11] Berthier J, Brakke KA, Berther E. Open Microfluidics. Massachusetts: Scrivener publishing; 2016

[12] Dixit CK, Kaushik A. Microfluidics for Biologists. Germany: Springer; 2016

[13] Panigrahi PK. Transport Phenomena in Microfluidic Systems. New York: Wiley; 2016

[14] Giri B. Laboratory Methods in Microfluidics. Netherlands: Elsevier; 2017

[15] Lei K-M, Mak P-I, Law M-K, Martins RP. Handheld Total Chemical and Biological Analysis Systems: Bridging NMR, Digital Microfluidics, and Semiconductors. Germany: Springer; 2017

[16] Chu L-Y, Wang W. Microfluidics for Advanced Functional Polymeric Materials. Germany: Wiley-VCH; 2017

[17] Piraino F, Selimović Š, editors. Diagnostic Devices with Microfluidics. Florida: CRC Press; 2017

[18] Petersen KE. Silicon as a mechanical material. Proceedings of the IEEE. 1982;**70**(5):420-457. DOI: 10.1109/PROC.1982.12331

[19] Angell JB, Terry SC, Barth PW. Silicon micromechanical devices. Scientific American. 1983;**248**(4):44-55. DOI: 10.1038/scientificamerican0483-44

[20] Manz A, Harrison DJ, Verpoorte EMJ, Fettinger JC, Lüdi APH, Widmer HM. Planar chips for miniaturization and integration of separation techniques into monitoring systems: CE on a chip. Journal of Chromatography A. 1992;**593**(1-2):253-258. DOI: 10.1016/0021-9673(92)80293-4

[21] Harrison DJ, Fluri K, Seiler K, Fan Z, Effenhauser CS, Manz A. Micromachining a miniaturized capillary electrophoresis. Science. 1993;**1261**(5123):895-897. DOI: 10.1126/science.261.5123.895

[22] Khandurina J, McKnight TE, Jacobson SC, Waters LC, Foote RS, Ramsey JM. Integrated system for rapid PCR-based DNA analysis in microfluidic devices. Analytical Chemistry. 2000;**72**(13):2995-3000. DOI: 10.1021/ac991471a

[23] Oleschuk RD, Shultz-Lockyear LL, Ning Y, Jed Harrison D. Trapping of bead-based reagents within microfluidic systems. Analytical Chemistry. 2000;**72**(3):585-590. DOI: 10.1021/ac990751n

[24] Verpoorte E, De Rooij NF. Microfluidics meets MEMS. Proceedings of the IEEE. 2003;**91**(6): 930-953. DOI: 10.1109/JPROC.2003.813570

[25] Squires TM, Quake SR. Microfluidics: Fluid physics at the nanoliter scale. Reviews of Modern Physics. 2005;**77**(3):977-1026. DOI: 0.1103/RevModPhys.77.977

[26] Whitesides GA. The origins and the future of microfluidics. Nature. July 27, 2006;**442**:368-373. DOI: 10.1038/nature05058

[27] Janesik D, Franzke J, Manz A. Scaling and the design of miniaturized chemical-analysis systems. Nature. 27 July, 2006;**442**:374-380. DOI: 10.1038/nature05059

[28] Yuan X, Oleschuk RD. Advances in microchip liquid chromatography. Analytical Chemistry. 2018;**90**(1):283-301. DOI: 0.1021/acs.analchem.7b04329

[29] Qin N, Wen JZ, Ren CL. Highly pressurized partially miscible liquid-liquid flow in a micro-T-junction. II. Physical Review E. 2017;**95**(4):043111. DOI: 10.1103/PhysRevE.95.043111

[30] Ramsey JM. The burgeoning power of the shrinking laboratory. Nature Biotechnology. 1999;**17**:1061-1062. DOI: 10.1038/15044

[31] Whitesides GM. The 'right' size in nanobiotechnology. Nature Biotechnology. 2003;**21**: 1161-1165. DOI: 10.1038/nbt872

[32] McClain MA, Culbertson CT, Jacobson SC, Allbritton NL, Sims CE, Ramsey JM. Microfluidic devices for the high-throughput cell analysis. Analytical Chemistry. 2003;**75**(21):5646-5655. DOI: 10.1021/ac0346510

[33] Neethirajan S, Kobayashi I, Nakajima M, Wu D, Nandagopal S, Lin F. Microfluidics for food, agriculture and biosystems industries. Lab on a Chip. 2011;**11**:1574-1586. DOI: 10.1039/C0LC00230E

[34] Morin SA, Shepherd RF, Kwok SW, Stokes AA, Nemiroski A, Whitesides GM. Camouflage and display for soft machines. Science. 2012;**337**(6096):828-832. DOI: 10.1126/science.1222149

[35] Yin H, Marshall D. Microfluidics for single cell analysis. Current Opinion in Biotechnology. 2012;**23**(1):110-119. DOI: 10.1016/j.copbio.2011.11.002

[36] Jokerst JC, Emory JM, Henry CS. Advances in microfluidics for environmental analysis. Analyst. 2012;**137**:24-34. DOI: 10.1039/C1AN15368D

[37] Sun Y, Jaffray DA, Yeow JTW. The design and fabrication of carbon nanotube based X-ray cathode. IEEE Transactions on Electronic Devices. 2013;**60**(1):464-470. DOI: 10.1109/TED.2012.2226036

[38] Zalesskiy SS, Danieli E, Blümich B, Ananikov VP. Miniaturization of NMR systems. Chemical Reviews. 2014;**114**(11):5641-5694. DOI: 10.1021/cr400063g

[39] Sackmann EK, Fulton AI, Beebe DJ. The present and future role of microfluidics in biomedical research. Nature. 2014;**507**:181-189. DOI: 10.1038/nature13188

[40] Sun J-Y, Keplinger C, Whitesides GM, Suo Z. Ionic skin. Advanced Materials. 2014;**26**(45): 7608-7614. DOI: 10.1002/adma.201403441

[41] Syms RRA, Wright S. MEMS mass spectrometers: The next wave of miniaturization. Journal of Micromechanics and Microengineering. 2016;**26**:023001. DOI: 10.1088/0960-1317/26/2/023001

[42] Zhang J, Yan S, Yuan D, Alici G, Nguyen N-T, Warkiani ME, Li W. Fundamentals and applications of inertial microfluidics: Review. Lab on a Chip. 2016;**16**:10-34. DOI: 10.1039/C5LC01159K

[43] Sohi AN, Nieva PM. Frequency response of curved bilayer microcantilevers with applications to surface stress measurement. Journal of Applied Physics. 2016;**119**(4):044503. DOI: 10.1063/1.4940951

[44] Chiu DT, deMello AJ, Di Carlo D, Doyle PS, Hansen C, Maceiczyk RM, Wootton RCR. Small but perfectly formed? Journal of Chemistry. 2017;**2**(2):201-223. DOI: 10.1016/j.chempr.2017.01.009

[45] Rane AS, Rutkauskaite J, deMello A, Stavrakis S. High-throughput multi-parametric imaging flow cytometry. Chem. 2017;**3**:588-602. DOI: 10.1016/j.chempr.2017.08.005

[46] Miled A, Greener J. Recent advancements towards full-system microfluidics. Sensors. 2017;**17**:1-7. DOI: 10.3390/s17081707

[47] Weiss M, Frohnmayer JP, Benk LT, Haller B, Janiesch J-W, Heitkamp T, Börsch M, Lira RB, Dimova R, Lipowsky R, Bodenschatz E, Baret J-C, Vidakovic-Koch T, Sundmacher K, Platzman I, Spatz JP. Sequential bottom-up assembly of mechanically stabilized synthetic cells by microfluidics. Nature Materials. October 16, 2017 (online). DOI: 10.1038/nmat5005

[48] Karanassios V, Sharples JT. Microchannels and microcells for gaseous microsamples. Sensors & Materials. 1997;**9**(6):363-378. http://myukk.org/SM2017/sm_pdf/SM298.pdf

[49] Karanassios V, Sharples JT, Nathan A. In situ ultrasound assisted etching of <100> Si wafers by KOH. Sensors & Materials. 1997;**9**(7):427-436. http://myukk.org/SM2017/sm_pdf/SM303.pdf

[50] Karanassios V, Mew G. Anisotropic wet chemical etching of Si for chemical analysis applications. Sensors & Material. 1997;**9**(7):395-416. http://myukk.org/SM2017/sm_pdf/SM301.pdf

[51] Swart NR, Stevens M, Nathan A, Karanassios V. A flow-insensitive thermal conductivity microsensor. Sensors & Materials. 1997;**9**(6):387-394. http://myukk.org/SM2017/sm_pdf/SM300.pdf

[52] French PJ, Sarro PM. Surface versus bulk micromachining: The contest for suitable applications. Journal of Micromechanics and Microengineering. 1998;**8**:4552. DOI: 10.1088/0960-1317/8/2/002

[53] Bustillo JM, Howe RT, Muller RS. Surface micromachining for microelectromechanical systems. Proceedings of the IEEE. 1998;**86**(8):1552-1574. DOI: 10.1109/5.704260

[54] Kovacs GTA, Maluf NI, Petersen KE. Bulk micromachining of silicon. Proceedings of the IEEE. 1998;**86**(8):1536-1551. DOI: 10.1109/5.704259

[55] Pethig R, Burt JPH, Parton A, Rizvi N, Talary MS, Tame JA. Development of biofactory-on-a-chip. Journal of Micromechanics and Microengineer. 1998;**8**(2):57-63. DOI: 10.1088/0960-1317/8/2/004

[56] Klank H, Kuttera JP, Geschke O. CO_2-laser micromachining for rapid production of PMMA-based microfluidic systems. Lab on a Chip. 2002;**2**:242-246. DOI: 10.1039/B206409J

[57] Cheng J-Y, Wei C-W, Hsu K-H, Young T-H. Direct-write laser micromachining. Sensors Actuators B. 2004;**99**(10):186-196. DOI: doi.org/10.1016/j.snb.2003.10.022

[58] Osellame R, Cerullo G, Ramponi R. Femtosecond Laser Micromachining: Photonic and Microfluidic Devices in Transparent Materials. Germany: Springer; 2012

[59] Becker EW, Ehrfeld W, Hagmann P, Maner A, Münchmeyer D. Fabrication of microstructures by LIGA. Microelectronic Engineering. 1986;**4**(1):35-56. DOI: 10.1016/0167-9317(86)90004-3

[60] Lorenz H, Despont M, Fahrni N, Brugger J, Vettiger P, Renaud P. High-aspect-ratio, negative-tone near-UV photoresist and its applications for MEMS. Sensors Actuators A. 1998;**64**(1):33-39. DOI: 10.1016/S0924-4247(98)80055-1

[61] Malek CK, Saile V. Applications of LIGA to precision manufacturing of high-aspect-ratio micro-components: A review. Microelectronics Journal. 2004;**35**(2):131-143. DOI: 10.1016/j.mejo.2003.10.003

[62] Li X, Abe T, Esashi M. Deep reactive ion etching of pyrex glass using SF_6 plasma. Sensors Actuators A. 2001;**87**(3):139-145. DOI: 10.1016/S0924-4247(00)00482-9

[63] Laermer F, Urban A. Challenges, developments and applications of silicon deep reactive ion etching. Microelectronic Engineering. 2003;**67-68**:349-355. DOI: 10.1016/S0167-9317(03)00089-3

[64] Marty F, Rousseau L, Saadany B, Mercier B, Français O, Mita Y, Bourouina T. Advanced etching of silicon based on DRIE etching of Si high aspect ratio microstructures and 3D micro- and nanostructures. Microelectronics Journal. 2005;**36**(7):637-677. DOI: 10.1016/j.mejo.2005.04.039

[65] Laermer F, Franssila S, Sainiemi L, Kolari K. Chapter 21: Deep Reactive Ion Etching. In: Lindroos V, Motooka T, Franssila S, Paulasto-Krockel M, Tilli M, Airaksinen V-M, editors. Handbook of Silicon Based MEMS Materials and Technologies. 2nd ed. 2015. pp. 444-469. DOI: 10.1016/B978-0-323-29965-7.00021-X

[66] Lorenz H, Despont M, Fahrani N, LaBianca N, Renaud P, Vettiger P. SU-8: A low-cost negative resist for MEMS. Journal of Micromechanics and Microengineering. 1991;**7**:121-124. DOI: 10.1088/0960-1317/7/3/010

[67] Lee KY, LaBianca N, Rishton SA. Micromachining applications of a high resolution ultrathick photoresist. Journal of Vacuum Science and Technology B. 1995;**13**(6):3000-3006. DOI: 10.1116/1.588295

[68] Tay FHE, van Kan JA, Watt F, Choong WO. A novel micro-machining method for the fabrication of thick-film SU-8 micro-channels. Journal of Micromechanics and Microengineering. 2001;**11**:27-32. DOI: 10.1088/0960-1317/11/1/305

[69] Xu J, Locascio L, Gaitan M, Lee CS. Room-temperature imprinting method for plastic microchannel fabrication. Analytical Chemistry. 2000;**72**(8):1930-1933. DOI: 10.1021/ac991216q

[70] Becker H, Locascio LE. Polymeric microfluidic devices. Talanta. 2002;**86**(2):267-287. DOI: 10.1016/S0039-9140(01)00594-X

[71] Mills CA, Martinez E, Bessueille F, Villanueva G, Bausells J, Samitier J, Errachid A. Production of structures for microfluidics using polymer imprint techniques. Microelectronic Engineering. 2005;**78-79**:695-700. DOI: 10.1016/j.mee.2004.12.087

[72] Bridget XS, Weichun AP, Hector Y, Becerril A, Woolley AT. Rapid prototyping of PMMA microfluidic systems using solvent imprinting and bonding. Journal of Chromatography A. 2007;**1162**(2):162-166. DOI: 10.1016/j.chroma.2007.04.002

[73] Curtis C, Eshaque B, Badali K, Karanassios V. Rapid prototyping of a microfluidic Venturi pump imprinted on polymeric chips. Proceedings of SPIE. 2012;**8366**:83660. DOI: 10.1117/12.91956

[74] Becker H, Heim U. Hot embossing as a method for the fabrication of polymer high aspect ratio structures. Sensors Actuators A. 2000;**83**(1-3):130-135. DOI: 10.1016/S0924-4247(00)00296-X

[75] Lee G-B, Chen S-H, Huang G-R, Sung W-C, Lin Y-H. Microfabricated plastic chips by hot embossing methods. Sensors Actuators B. 2001;**75**(1-2):142-148. DOI: 10.1016/S0925-4005(00)00745-0

[76] Qi S, Liu X, Ford S, Barrows J, Thomas G, Kelly K, McCandless A, Lian K, Goettert J, Soper SA. Microfluidic devices fabricated in PMMA using hot-embossing with integrated sampling capillary. Lab on a Chip. 2002;**2**:88-95. DOI: 10.1039/B200370H

[77] Xia Y, Whitesides GM. Soft lithography. Annual Review of Material Science. 1998;**28**:153-184. DOI: 10.1146/annurev.matsci.28.1.153

[78] Unger MA, Chou H-P, Thorsen T, Scherer A, Quake SR. Monolithic microfabricated valves and pumps. Science. 2000;**288**(5463):113-116. DOI: 10.1126/science.288.5463.113

[79] Yang P, Wirnsberger G, Huang HC, Cordero SR, McGehee MD, Scot B. Mirrorless lasing from mesostructured waveguides in COC patterned by soft lithography. Science. 2000;**287**(5452):465-467. DOI: 10.1126/science.287.5452.465

[80] Whitesides GM, Ostuni E, Takayama S, Jiang X, Ingber DE. Soft lithography in biology and biochemistry. Annual Review of Biomedical Engineering. 2001;**3**:335-373. DOI: 10.1146/annurev.bioeng.3.1.335

[81] Odom TW, Love JC, Wolfe DB, Paul KE, Whitesides GM. Improved pattern transfer in soft lithography. Langmuir. 2002;**18**(13):5314-5320. DOI: 10.1021/la020169l

[82] Qin D, Xia Y, Whitesides GM. Soft-lithography for micro- and nanoscale patterning. Nature Protocols. 2003;**5**(3):491-502. DOI: 10.1038/nprot.2009.234

[83] Li C, Yang Y, Craighead HG, Lee KH. Isoelectric focusing in COC microfluidic channels. Electrophoresis. 2005;**26**(9):1800-1806. DOI: 10.1002/elps.200410309

[84] Teh S-Y, Lin R, Hung L-H, Lee AP. Droplet microfluidics. Lab on a Chip. 2008;**8**:198-220. DOI: 10.1039/B715524G

[85] Theberge AB, Courtois F, Schaerli Y, Fischlechner M, Abell C, Hollfelder F, Huck WTS. Microdroplets in microfluidics. Angewandte Chemie International. 2010;**49**:5846-5868. DOI: 10.1002/anie.200906653

[86] Seemann R, Brinkmann M, Pfohl T, Herminghaus S. Droplet based microfluidics. Reports on Progress in Physics. 2011;**75**:016601. DOI: 10.1088/0034-4885/75/1/016601

[87] Zhua Y, Fang Q. Analytical detection techniques for droplet microfluidics: A review. Analytica Chimica Acta. 2013;**787**:24-35. DOI: 10.1016/j.aca.20sssss13.04.064

[88] Courtney M, Chen X, Chan S, Mohamed T, Rao PN, Ren CL. Droplet microfluidic system for drug screening applications. Analytical Chemistry. 2017;**89**:910-915. DOI: 10.1021/acs.analchem.6b04039

[89] Ding Y, Choo J, deMello AJ. From single-molecule detection to next-generation sequencing. Microfluidics Nanofluidics. 2017;**21**(58). DOI: 10.1007/s10404-017-1889-4

[90] Pollack MG, Fair RB. Electrowetting-based actuation of liquid droplets for microfluidic applications. Applied Physics Letters. 2000;**77**:1725. DOI: 10.1063/1.1308534

[91] Fair RB. Digital microfluidics: Is a true lab-on-a-chip possible? Microfluidics and Nanofluidics. 2007;**3**:245-281. DOI: 10.1007/s10404-007-0161-8

[92] Wheeler AR. Putting electrowetting to work. Science. October 24, 2008;**322**:539-540. DOI: 10.1126/science.1165719

[93] Yang H, Luk VN, Abelgawad M, Barbulovic-Nad I, Wheeler AR. A world-to-chip interface for digital microfluidics. Analytical Chemistry. 2009;**81**(3):1061-1106. DOI: 10.1021/ac802154h

[94] Choi K, Ng AHC, Fobel R, Wheeler AR. Digital microfluidics. Annual Review of Analytical Chemistry. 2012;**5**:413-440. DOI: 10.1146/annurev-anchem-062011-143028

[95] Berthier J. Micro-drops and Digital Microfluidics. 2nd ed. Netherlands: Elsevier; 2013

[96] Madou M, Zoval J, Jia G, Kido H, Kim J, Kim N. Lab on a CD. Annual Review of Biomedical Engineering. 2006;**8**:601-628. DOI: 10.1146/annurev.bioeng.8.061505.095758

[97] Gorkin R, Park J, Siegrist J, Amasia M, Lee BS, Park J-M, Kim JKH, Madou M, Cho Y-K. Centrifugal microfluidics. Lab on a Chip. 2010;**10**:1758-1773. DOI: 10.1039/B924109D

[98] Gorkin R, Clime L, Madou M, Kido H. Pneumatic pumping in centrifugal microfluidic platforms. Microfluidics and Nanofluidics. 2010;**9**(2-3):541-549. DOI: 0.1007/s10404-010-0571-x

[99] Kong MCR, Salin ED. Pneumatically pumping fluids radially inward on centrifugal microfluidic platforms in motion. Analytical Chemistry. 2010;**82**(19):8039-8041. DOI: 10.1021/ac102071b

[100] Wang L, Li PCH. Microfluidic DNA microarray analysis: A review. Analytica Chimica Acta. 2011;**687**(1):12-27. DOI: doi.org/10.1016/j.aca.2010.11.056

[101] Schwemmer F, Zehnle S, Mark D, von Stetten F, Zengerle R, Paust N. A microfluidic timer for centrifugal microfluidics. Lab on a Chip. 2015;**15**:1545-1553. DOI: 10.1039/C4LC01269K

[102] Martinez AW, Phillips ST, Whitesides GM. Three-dimensional microfluidic devices fabricated in layered paper and tape. PNAS. 2008;**105**(50):19606-19611. DOI: 10.1073/pnas.0810903105

[103] Carrilho E, Martinez AW, Whitesides GM. Understanding wax printing. Analytical Chemistry. 2009;**81**(16):7091-7095. DOI: 0.1021/ac901071p

[104] Martinez AW, Phillips ST, Whitesides GM, Carrilho E. Diagnostics for the developing world: Microfluidic paper-based analytical devices. Analytical Chemistry. 2010;**82**(1):3-10. DOI: 10.1021/ac9013989

[105] Nie Z, Nijhuis CA, Gong J, Chen X, Kumachev A, Martinez AW, Narovlyansky M, Whitesides GM. Electrochemical sensing in paper-based microfluidic devices. Lab on a Chip. 2010;**10**:477-483. DOI: 10.1039/B917150A

[106] Dungchai W, Chailapakul O, Henry CS. A low-cost, simple, and rapid fabrication method for paper-based microfluidics. Analyst. 2011;**136**:77-82. DOI: 10.1039/C0AN00406E

[107] Mao X, Huang TJ. Microfluidic diagnostics for the developing world. Lab on a Chip. 2012;**12**:1412-1416. DOI: 10.1039/C2LC90022J

[108] Hamedi MM, Ainla A, Güder F, Christodouleas DC, Fernández-Abedul MT, Whitesides GM. Integrating electronics and microfluidics on paper. Advanced Materials. 2016;**28**(25):5054-5063. DOI: 10.1002/adma.201505823

[109] Weagant S, Li L, Karanassios V. Rapid Prototyping of Hybrid, Plastic-Quartz 3D-Chips for Battery-Operated Microplasmas. London, UK: InTech Publishing; 2011. Chapter 10. pp. 1-18. DOI: 10.5772/24994

[110] Kitson PJ, Rosnes MH, Cronin L. Configurable 3D-printed millifluidic and microfluidic 'lab on a chip' reactionware devices. Lab on a Chip. 2012;**12**:3267-3271. DOI: 10.1039/c2lc40761b

[111] Au AK, Lee W, Folch A. Mail-order microfluidics: Evaluation of stereolithography for the production of microfluidic devices. Lab on a Chip. 2014;**14**:1294-1301. DOI: 10.1039/C3LC51360B

[112] Bhargava KC, Thompson B, Malmstadt N. Discrete elements for 3D microfluidics. PNAS. 2014;**111**(42):15013-15018. DOI: 10173/pnas/14114764111

[113] Meng C, Ho B, Ng SH, Lia KHH, Yoon Y-J. 3D printed microfluidics for biological applications. Lab on a Chip. 2015;**15**:3627-3637. DOI: 10.1039/C5LC00685F

[114] Au AK, Huynh W, Horowitz LF, Folch A. 3D-printed microfluidics. Angewandte Chemie International. 2016;**55**(12):3862-3881. DOI: 10.1002/anie.201504382

[115] Ahmadian A, Adam Y, William P, Wong W, Nguen T, Pan Y, Xu J. 3D printing. Microfluidics Nanofluidics. 2016;**20**(50). DOI: 10.1007/s10404-016-1715-4

[116] Shatford R, Karanassios V. 3D printing in chemistry: Past, present and future. Proceedings of SPIE. 2016;**9855**:98550B-98560. DOI: 10.1117/12.2224404

[117] Singh M, Tong Y, Webster K, Cesewski E, Haring AP, Laheri S, Carswell B, O'Brien TJ, Aardema CH, Senger RS, Robertson JL, Johnson BN. 3D printed conformal microfluidics for isolation and profiling of biomarkers. Lab on a Chip. 2017;**17**:2561-2571. DOI: 10.1039/C7LC00468K

[118] Mattio E, Robert-Peillard F, Puzio CBK, Margaillan A, Brach-Papa C, Knoery J, Boudenne J-L, Coulom B. 3D-printed flow system for determination of lead in natural waters. Talanta. 2017;**168**:298-302. DOI: 10.1016/j.talanta.2017.03.059

[119] Macdonald NP, Cabot JM, Smejkal P, Guijt RM, Paull B, Breadmore MC. Comparing microfluidic performance of three-dimensional (3D) printing platforms. Analytical Chemistry. 2017;**89**(7):3858-3866. DOI: 10.1021/acs.analchem.7b00136

[120] Beauchamp MJ, Nordin GP, Woolley AT. Moving from millifluidic to truly microfluidic sub-100-μm cross-section 3D printed devices. Analytical and Bioanalytical Chemistry. 2017;**409**:4311-4319. DOI: 10.1007/s00216-017-0398-3

[121] Gong H, Bickham BP, Woolley AT, Nordin GP. Custom 3D printer and resin for 18 μm × 20 μm microfluidic flow channels. Lab on a Chip. 2017;**17**:2899-2909. DOI: 10.1039/C7LC00644F

[122] Salentijn GI-J, Oleschuk RD, Verpoorte E. 3D-printed paper spray ionization cartridge with integrated desolvation feature and ion optics. Analytical Chemistry. 2017; **89**(21):11419-11426. DOI: 10.1021/acs.analchem.7b02490

[123] Howston AM. Introduction to Gas Discharges. Oxford, UK: Pergamon Press; 1965

[124] Marcus RK. Glow Discharge Spectroscopies. New York: Plenum Press; 1993

[125] Goldston RJ, Rutherfrd PH. Plasma Physics. Pennsylvania: Institute of Physics Publishing; 1995

[126] Smirnov BM. Physics of Ionized Gases. New York: Wiley; 2001

[127] Bittecourt JA. Fundamentals of Plasma Physics. 3rd ed. New York: Springer-Verlag; 2004

[128] Lieberman MA, Lichtenberg AJ. Principles of Plasma Discharges and Materials Processing. New Jersey: Wiley-Interscience; 2005

[129] Ostrikov K. Plasma Nanoscience. New York: Wiley; 2008

[130] Ostrikov K, Cvelbar U, Murphy AB. Plasma nanoscience: Setting directions, tackling grand challenges. Journal of Physics D: Applied Physics. 2011;**44**(17):174001. DOI: 10.1088/0022-3727/44/17/174001

[131] Ostikov K, Neyts EC, Meyyappan M. Plasma nanoscience: From nano-solids in plasmas to nano-plasmas in solids. Advances in Physics. 2013;**02**(02):113-0224. DOI: 10.1080/00018732.2013.808047

[132] Montaser A, Golightly DW, editors. Inductively Coupled Plasmas in Analytical Atomic Spectrometry. New York: VCH; 1992

[133] Schulze A, Jeong JY, Babayan SE, Park J, Selwyn GS, Hicks RF. The atmospheric-pressure plasma jet. IEEE Transactions on Plasma Science. 1998;**16**(6):1685-1694. DOI: 10.1109/27.747887

[134] Eden JG, Park S-J. New opportunities for plasma science in non-equilibrium, low-temperature plasmas confined to microcavities: There's plenty of room at the bottom. Physics of Plasmas. 2006;**13**:057101. DOI: 10.1063/1.2179413

[135] Shin JJ, Kong MG. Evolution of discharge structure in capacitive radio-frequency atmospheric microplasmas. Physical Review Letters. 2006;**96**:105009. DOI: 10.1103/PhysRevLett.96.105009

[136] Wagner AJ, Mariotti D, Yurchenko KJ, Das KT. Experimental study of a planar atmospheric-pressure plasma operating in the microplasma regime. Physical Review E. 2009;**80**:065401R. DOI: 10.1103/PysRevE80.065401

[137] Lu X, Wu S, Gou J, Pan Y. An atmospheric-pressure, high-aspect ratio, cold micro-plasma. Scientific Reports. 2014;**4**(7488). DOI: 10.1038/srep07488

[138] Karanassios V. Microplasmas for chemical analysis: Analytical tools or research toys? Spectrochimica Acta Part B. 2004;**59**:909-928. DOI: 10.1016/j.sab.2004.04.005

[139] Weagant S, Smith AT, Karanassios V. Mobile micro- and nano-instruments: Fast, cheap and under wireless control. ECS Transactions. 2010;**28**(14):1-6. DOI: 10.1149/1.3490180

[140] Badiei HR, Lai B, Karanassios V. Micro- and nano-volume samples by electrothermal, near-torch vaporization (NTV) using removable, interchangeable and portable rhenium coiled-filament assemblies and ICP-AES. Spectrochimica Acta Part B. 2012;**77**:19-30. DOI: 10.1016/j.sab.2012.07.025

[141] Badiei HR, McEnaney J, Karanassios V. Bringing part of the lab to the field: On-site Cr speciation by electrodeposition of Cr(III)/Cr(VI) on portable coiled-filaments and measurement in a lab by NTV-ICP-AES. Spectrochimica Acta Part B. 2012;**78**:42-49. DOI: 10.1016/j.sab.2012.10.002

[142] Badiei HR, Liu C, Karanassios V. Taking part of the lab to the sample: On-site electrodeposition of Pb followed by measurement in a lab using electrothermal, near torch vaporization (NTV) sample introduction and ICP-AES. Microchemical Journal. 2013;**108**:131-136. DOI: 10.1016/j.microc.2012.10.013

[143] Karanassios V, Johnson K, Smith AT. Micromachined, planar-geometry, atmospheric-pressure, battery-operated microplasma devices (MPDs) on chips for liquids, gases or solids by optical emission spectrometry. Analytical and Bioanalytical Chemistry. 2007;**388**:1595-1604. DOI: 10.1007/s00216-007-1273-4

[144] Weagent S, Karanassios V. Helium-hydrogen microplasma device (MPD) on postage-stamp-size plastic-quartz chips. Analytical and Bioanalytical Chemistry. 2009;**395**:577-589. DOI: 10.1007/s00216-009-2942-2

[145] Weagant S, Chen V, Karanassios V. Battery-operated, Ar-H_2 microplasma on hybrid, postage stamp-size plastic-quartz chips for elemental analysis using a portable optical emission spectrometer. Analytical and Bioanalytical Chemistry. 2011;**401**:2865-2880. DOI: 10.1007/s00216-011-5372-x

[146] Weagant S, Karanassios V. Battery-operated, planar-geometry microplasma on a postage-stamp size chips: Some fundamentals. Proceedings of SPIE. 2011;**8024**:80240L. DOI: 10.1117/12.884329

[147] Zhang X, Karanassios V. Rapid prototyping of solar-powered, battery-operated, atmospheric-pressure, sugar-cube size microplasma on hybrid, 3D chips using a portable optical emission spectrometer. Proceedings of SPIE. 2012;**8366**:83660D. DOI: 10.1117/12.919550

[148] Abbaszadeh S, Karim KS, Karanassios V. Measurement of UV from a microplasma by a microfabricated amorphous selenium detector. IEEE Transactions on Electronic Devices. 2013;**60**(2):880-883. DOI: 10.1109/TED.2012.2231682

[149] Devathasan D, Trebych K, Karanassios V. 3D-printed, sugar cube-size microplasma on a hybrid chip used as a spectral lamp to characterize UV-vis transmission of polycarbonate chips for microfluidic applications. Proceedings of SPIE. 2013;**8718**:87180B. DOI: 10.1117/12.2016222

[150] Nguon O, Gauthier M, Karanassios V. Determination of the loading and stability of Pd in an arborescent copolymer in ethanol by microplasma-optical emission spectrometry. RSC Advances. 2014;**4**:8978-8984. DOI: 10.1039/C3RA46232C

[151] Shatford R, Karanassios V. Microplasma fabrication: From semiconductor technology for 2D-chips and microfluidic channels to rapid prototyping to 3D-printing of microplasma devices. Proceedings of SPIE. 2014;**9106**:9106H1-9106H7. DOI: 10.11171/12.2050538

[152] Nguon O, Huang S, Gauthier M, Karanassios V. Microplasmas: From applications to fundamentals. Proceedings of SPIE. 2014;**9105**:9101061-9101067. DOI: 10.1117/12.2050348

[153] Weagent S, Dulai G, Li L, Karanassios V. Characterization of rapidly-prototyped, battery-operated, Ar-H_2 microplasma on a chip for elemental analysis of microsamples by portable optical emission spectrometry. Spectrochimica Acta Part B. 2015;**106**:75-80. DOI: 10.1016/j.sab.2015.01.009

[154] Shatford R, Kim D, Karanassios V. Microfluidics for spectrochemical applications. Proceedings of SPIE. 2015;**9486**:94860N1-94960N6. DOI: 10.1117/12.2177513

[155] Yang R, Sazonov A, Karanassios V. Flexible, self-powered, visible-light detector characterized using a battery-operated, 3D-printed microplasma operated as a light source. 2016 IEEE Sensors. 2016:997-999. ID 978-1-4799-8287-5/16. DOI: 10.1109/ICSENS.2016.7808738

[156] So H, Cebula DA, Karanassios V. Towards chromium speciation in lake-waters by microplasma optical emission spectrometry. Proceedings of SPIE. 2017;**10215**:102150I. DOI: 10.1117/12.2262955

[157] Feynman RP. There is plenty of room at the bottom. Engineering and Science at Caltech. 1960;**13**(5):22-36. http://calteches.library.caltech.edu/47/3/ES.23.5.1960.0.pdf

[158] NSF "Partnership in Nanotechnology", Program announcement, societal implications of nanoscience and nanotechnology. Arlington, VA (1997), http://www.nsf.gov/nano (Accessed Dec. 2017)

[159] Gates BD, Xu Q, Stewart M, Ryan D, Willson CG, Whitesides GM. New approaches to nanofabrication. Chemical Reviews. 2005;**105**(4):1171-1196. DOI: 10.1021/cr030076o

[160] Eijkel JC, van der Berg A. Nanofluidics: What is it and what can we expect from it?. Microfluid Nanofluidics. 2005;**1**:249-267. DOI: 10.1007/s10404-004-0012-9

[161] Tsenga AA, Notargiacomo A. Nanofabrication by scanning probe microscope lithography: A review. Journal of Vacuum Science and Technology B. 2005;**23**:877. DOI: 10.1116/1.1926293

[162] Mijatovic D, Eijek JCT, Van der Tert VA. Technologies for nanofluidic systems: Top-down vs. bottom-up: A review. Lab on a Chip. 2005;**5**:402-500. DOI: 10.1039/b416951d

[163] Huh D, Mills KL, Zhu X, Burns MA, Thouless MD, Takayama S. Tuneable elastomeric nanochannels. Nature Materials. 2007;**6**:424-428. DOI: 10.1038/nmat1907

[164] Li M, Tang HX, Roukes ML. Ultra-sensitive NEMS-based cantilevers for sensing, scanned probe. Nature Nanotechnology. 2007;**2**:114-120. DOI: 10.1038/nnano.2006.208

[165] He Q, Chen S, Su Y, Fang Q, Chen H. Fabrication of 1D nanofluidic channels on glass substrate by wet etching. Analytica Chimica Acta. 2008;**628**:1-8. DOI: 10.1016/j.aca.2008.08.040

[166] Prakash S, Piruska A, Gatimu EN, Bohn PW, Sweedler JV, Shannon MA. Nanofluidics: Systems and applications. IEEE Sensors Journal. 2008;**8**(5):441-450. DOI: 10.1109/JSEN.2008.918758

[167] Abgrall P, Nguyen NT. Nanofluidic devices and their applications. Analytical Chemistry. 2008;**80**(7):2326-2341. DOI: 10.1021/ac702296u

[168] Sparreboom W. van den Berg A, Eijkel JCT. Principles and applications of nanofluidic transport. Nature Nanotechnology. 2009;**4**:713-720. DOI: 10.1038/nnano.2009.332

[169] Piruska A, Gong M, Sweedler JV, Boha PW. Nanofluidics in chemical analysis. Chemical Society Reviews. 2010;**39**:1060-1072. DOI: 10.1039/b900409m

[170] Roco MC. The long view of nanotechnology development: The National Nanotechnology Initiative at 10 years. Journal of Nanoparticle Research. 2011;**13**:427-445. DOI: 10.1007/s11051-010-0192-z

[171] Menard LD, Ramsey JM. Fabrication of sub-5 nm nanochannels in insulating substrates using focused ion beam milling. Nano Letters. 2011;**11**(2):512-517. DOI: 10.1021/nl103369g

[172] Chantiwas R, Park S, Soper SA, Kim BC, Takayama S, Sunkara V, Hwangc H, Cho Y-K. Flexible fabrication and applications of polymer nanochannels and nanoslits. Chemical Society Reviews. 2011;**40**:3677-3702. DOI: 10.1039/C0CS00138D

[173] Poot M, van der Zant HSJ. Mechanical systems in the quantum regime. Physics Reports. 2012;**511**(5):273-335. DOI: 10.1016/j.physrep.2011.12.004

[174] Hanay MS, Kelber S, Naik AK, Chi D, Hentz S, Bullard EC, Colinet E, Duraffourg L, Roukes ML. Single-protein nanomechanical mass spectrometer in real time. Nature Nanotechnology. 2012;**7**:602-608. DOI: 10.1038/nnano.2012.119

[175] Duan C, Wang W, Xin Q. Fabrication of nanofluidic devices. Biomicrofluidics. 2013;**7**. DOI: 026501, 10.1063/1.4794973

[176] Schmid S, Kurek M, Adolphsen JQ, Boisen A. Real-time single airborne nanoparticle detection with nanomechanical resonant filter-fiber. Scientific Reports. 2013;**3**(1288). DOI: 10.1038/srep01288

[177] Uranga A, Verd J, Marigó E, Giner J, Muñóz-Gamarra JL, Barniol N. Exploitation of non-linearities in CMOS-NEMS. Sensors Actuators A. 2013;**197**:88-95. DOI: 10.1016/j.sna.2013.03.032

[178] Muñoz-Gamarra JL, Alcaine P, Marigó E, Giner J, Uranga A, Esteve J, Barniol N. Integration of NEMS resonators in CMOS. Microelectronics Engineering. 2013;**110**:246-249. DOI: 10.1016/j.mee.2013.01.038

[179] Berman D, Krim J. Surface science, MEMS and NEMS. Progress in Surface Science. 2013;**88**(2):171-211. DOI: 10.1016/j.progsurf.2013.03.001

[180] Bocquet L, Tabeling P. Physics and technological aspects of nanofluidics. Lab on a Chip. 2014;**14**:3143-3158. DOI: 10.1039/c4lc00325j

[181] Mawatari K, Kazoe Y, Shimizu H, Pihosh Y, Kitamori T. Extended-nanofluidics. Analytical Chemistry. 2014;**86**(9):4068-4077. DOI: 10.1021/ac4026303

[182] Sage E, Brenac A, Alava T, Morel R, Dupré C, Hanay MS, Roukes ML, Duraffourg L, Masselon C, Hentz S. Neutral particle mass spectrometry with nanomechanical systems. Nature Communications. 2015;**6**(6482). DOI: 10.1038/ncomms7482

[183] Hui Y, Nan T, Sun NX, Rinaldi M. High resolution magnetometer based on a MEMS-CMOS oscillator. Journal of Microelectromechanical Systems. 2015;**24**(1):134-143. DOI: 10.1109/JMEMS.2014.2322012

[184] Schmid S, Villanueva LG, Roukes ML. Fundamentals of Nanomechanical Resonators. New York: Springer; 2016

[185] Kandemir AC, Erdem D, Ma H, Reiser A, Spolenak R. Polymer nanocomposite patterning by dip-pen nanolithography. Nanotechnology. 2016;**27**(13):135303. DOI: 10.1088/0957-4484/27/13/135303

[186] Hou X, Guoa W, Jiang L. Biomimetic smart nanopores and nanochannels. Chemical Society Reviews. 2011;**40**:2385-2401. DOI: 10.1039/C0CS00053A

[187] Schneider GF, Dekker C. DNA sequencing with nanopores. Nature Biotechnology. 2012;**30**:326-328. DOI: 10.1038/nbt.2181

[188] Ayub M, Hardwick SW, Luisi BF, Bayley H. Nanopore-based identification of individual nucleotides for direct RNA sequencing. Nano Letters. 2013;**13**(12):6144-6150. DOI: 10.1021/nl403469r

[189] Miles BN, Ivanov AP, Wilson KA, Doğan F, Japrung D, Edel JB. Single molecule sensing with solid-state nanopores. Chemical Society Reviews. 2013;**42**:15-28. DOI: 10.1039/C2CS35286A

[190] Taniguchi M. Selective multidetection using nanopores. Analytical Chemistry. 2015;**87**(1):188-199. DOI: 10.1021/ac504186m

[191] Deng T, Wang Y, Chen Q, Chen H, Liu Z. Massive fabrication of silicon nanopore arrays with tunable shapes. Applied Surface Science. 2016;**390**:681-688. DOI: doi.org/10.1016/j.apsusc.2016.07.171

[192] Taniguchi M. Single-molecule sequencing. In: Kiguchi M, editor. Single-Molecule Electronics. Singapore: Springer; 2016. DOI: 10.1007/978-981-10-0724-8_9

[193] Sheikholeslami M. Influence of coulomb forces on Fe_3O_4-H_2O nanofluid thermal improvement. International Journal of Hydrogen Energy. 2017;**42**:821-829. DOI: 10.1016/j.ijhydene.2016.09.185

[194] Sheikholeslami M, Seyednezhad M. Simulation of nanofluid flow in a porous media. International Journal of Heat and Mass Transfer. 2018;**120**:772-781. DOI: 10.1016/j.ijheatmasstransfer.2017.12.087

[195] Hagleitner C, Hierlemann A, Lange D, Kummer A, Kerness N, Brand O, Baltes H. Smart single-chip gas sensor microsystem. Nature. November 15, 2001;**414**:293-296. DOI: 10.1038/35104535

[196] Hierlemann A, Baltes H. CMOS-based chemical microsensors. Analyst. 2003;**128**:15-28. DOI: 10.1039/B208563C

[197] Graf M, Barrettino D, Baltes H, Hierlemann A. Microtechnology and MEMS series. In: CMOS Hotplate Chemical Microsensors. Germany: Springer; 2007. DOI: 10.1007/978-3-540-69562-2

[198] Lee D, Dulai G, Karanassios V. Survey of energy harvesting and energy scavenging for powering wireless networks. Proceedings of SPIE. 2013;**8028**:8720S1. DOI: 10.1117/12.2016238

[199] Yang R, Sazonov A, Karanassios V. Flexible, self-powered, visible-light detector characterized using a microplasma. Proceedings of the IEEE Sensors. 2016;**978**:997-999. DOI: 10.1109/ICSENS.2016.7808738

[200] Trizcinski P, Nathan A, Karanassios V. Approaches to energy harvesting and energy scavenging for energy autonomy. Proceedings of SPIE. 2017;**10194**:10194A1-10194A9. DOI: 10.1117/12.2262957

[201] Li Z, Karanassios V. Artificial Neural Networks (ANNs) for Spectral Interference Correction Using a Large-Size Spectrometer and ANN-Based Deep Learning for a Miniature one. London, UK: InTech Publishing; 2018. Chapter 12. pp. 227-249. DOI: 10.5772/intechopen.71039

[202] Trzcinski P, Weagent S, Karanassios V. Wireless data acquisition of transient signals for mobile spectrometry applications. Applied Spectroscopy. 2016;**70**:905-915. DOI: 10.1177/0003702816638304

[203] Trzcinski P, Karanassios V. How can wireless, mobile data acquisition be used for taking part of the lab to the sample, and how can it join the internet of things?. Proceedings of SPIE. 2016;**9855**:985503. DOI: 10.1117/12.2224400

Chapter 2

Application of Nanofluids for Thermal Management of Photovoltaic Modules: A Review

Hafiz Muhammad Ali, Tayyab Raza Shah,
Hamza Babar and Zargham Ahmad Khan

Additional information is available at the end of the chapter

http://dx.doi.org/10.5772/intechopen.74967

Abstract

Mounting temperature impedes the conversion efficiency of photovoltaic systems. Studies have shown drastic efficiency escalation of PV modules, if cooled by nanofluids. Ability of nanofluids to supplement the efficiency improvement of PV cells has sought attention of researchers. This chapter presents the magnitude of improved efficiency found by different researchers due to the cooling via nanofluids. The effect of factors (such as, nanoparticle size, nanofluid concentration, flowrate of nanofluid and geometry of channel containing nanofluid) influencing the efficiency of PV systems has been discussed. Collective results of different researchers indicate that the efficiency of the PV/T systems (using nanofluids as coolant) increases with increasing flowrate. Efficiency of these systems increases with increasing concentration of nanofluid up to a certain amount, but as the concentration gets above this certain value, the efficiency tends to decline due to agglomeration/clustering of nanoparticles. Pertaining to the most recent studies, stability of nanoparticles is still the major unresolved issue, hindering the commercial scale application of nanofluids for the cooling of PV panels. Eventually, the environmental and economic advantages of these systems are presented.

Keywords: PV systems, Nanofluids, efficiency, concentration, Flowrate, stability

1. Introduction

Exceeding energy demands and swiftly eliminating conventional energy resources have compelled the researchers to find the alternative means of power generation. To date, only 14% of the world's power demands are being met via renewable energy means. Sun is the most vital source of energy, almost 1.8x 10^{11} MW energy from the sun intercepts the earth's surface [1]. According to the estimate of International Energy Agency (IEA), quarter of world's power demands could be fulfilled by solar energy by 2050 [2]. Silicon-based photovoltaic cells are used to convert the solar radiations into electricity. But the issue with these PV solar cells is that almost 85% of the solar energy reaching the surface of the PV unit is either reflected or absorbed as heat energy [3]. Al-shamani et al. [4] reviewed that only 5–20% of the solar radiation reaching to the PV cell surface is converted into electrical energy. Whereas, rest of the radiations are either reflected back or absorbed by the cell in the form of heat. Absorbed heat can increase its temperature up to $70^{\circ}C$. Oruc et al. [5] found that the electrical efficiency of PV module drops by 0.5% with every unit degree increment in the temperature of the module above $25^{\circ}C$ due to the contraction of the band gap and increased number of carriers. Increased number of carriers cause the saturation current to increase whereas the open circuit voltage to decrease thus lowering the electrical power output. Cooling of PV units depicts electrical efficiency enhancement as per the experimental results obtained by the researchers. Underdeveloped countries like Pakistan, with hot and sunny days throughout the year, are well suitable for power production via solar energy. According to research, during summer, temperature of the module can elevate in a devastating way (about $20^{\circ}C$ higher), in turns destructing the conversion efficiency of PV modules [6, 7]. Bashir et al. [8] reported that cooling of PV modules via water minimized heat losses and module's temperature elevation, thus, improving the efficiency by 13% and 6.2% for monocrystalline and polycrystalline PV modules respectively. Ali et al. [9] experimentally showed that cooling of PV modules by using micro-channels increased the efficiency of PV modules by 3%.

There are several methods of PV cooling such as, air cooling (natural air circulation and forced air circulation), water cooling, heat pipe cooling, cooling with Phase Change Materials (PCMs) and cooling via nanofluids [10, 11]. A PV/T system consists of PV module coupled with a heat absorbing unit in which a liquid (water or nanofluid) is circulated to absorb the heat of PV unit to improve the efficiency. The researches show that a PV/T system performs way better than conventional PV systems [12, 13]. Lelea et al. [14] investigated the effect of cooling via Al_2O_3 on the performance of concentrated PV/T system. The results showed a decrement in the temperature of module, when cooled by nanofluid and water.

Mixture of solid particles (metallic oxides, metals or carbon nanotubes) of less than 100 nm size at least in one dimension (nanoparticles) disseminated in the liquid fluids like water and polyethylene glycol etcetera (base fluid), is known as nanofluid. Nanofluids can be employed as a coolant as well as optical filters within PV/T systems [15]. PV/T system using nanofluid as coolant can produce far better results than the water cooled system. Al-Waeli et al. [16] conducted an experimental study and they found that cooling of PV module via SiC increased the electrical efficiency by 24.1%, thermal efficiency by 100.19% and overall efficiency by 88.9% as compared to the water-cooled PV/T system. Xu and Kleinstreuer [17] suggested nanofluid based silicon PV/T systems as a useful option for domestic applications as its overall efficiency reached up to 70% (11% electrical efficiency and 59% thermal efficiency).

This chapter reviews the efficiency of PV systems being cooled by various nanofluids. The common ways of cooling PV system via nanofluids are stated in detail along with the parameters influencing the efficiency of the PV/T systems such as irradiance, concentration and flow rate of nanofluid, size of nanoparticles and geometry of micro-channels. Impact of other factors such as the type of nanoparticles and base fluid on the system efficiency are discussed. Eventually, economic and environmental advantages are described.

2. Methods of cooling of PV systems via nanofluids

There are several methods of extracting heat from the PV units via nanofluids. The most common ways are, employing heat collector at the rear end of the panel and using nanofluid as a liquid in spectral splitting filter joined on the front surface of PV module. Sometimes, both methods are used simultaneously in order to increase the efficiency.

2.1. Rear end cooling

In rear-end cooling a thermal collector is coupled at the back end of the PV module to extract the heat. Nanofluid is set to flow through the collector thus taking up the heat of the cells and increasing its own temperature. Nanofluid gets warmed and its heat is further employed for useful purposes. Nanofluid is able to extract major part of the heat energy because of its improved thermophysical properties. The most important thermophysical property is the thermal conductivity. A schematic display of such an arrangement is depicted in the **Figure 1**.

Energy balance of such PV/T systems is evaluated by the following equation by [19].

$$\dot{E}_{in} = \dot{E}_{el} + \dot{E}_{th} + \dot{E}_{loss} \tag{1}$$

Here, " $\dot{E}_{in}$ " is the incident irradiation, " $\dot{E}_{el}$ " is output electrical power, " $\dot{E}_{th}$ " is useful thermal energy obtained by the collector and " $\dot{E}_{loss}$ " presents the energy losses from the control volume. Overall efficiency of the system is found by the following formula.

$$\eta_{pvt} \cong \frac{\dot{E}_{el} + \dot{E}_{th}}{\dot{E}_{th}} \Rightarrow \eta_{pvt} = \eta_{pvt} = \eta_{th} + r \times \eta_{el} \tag{2}$$

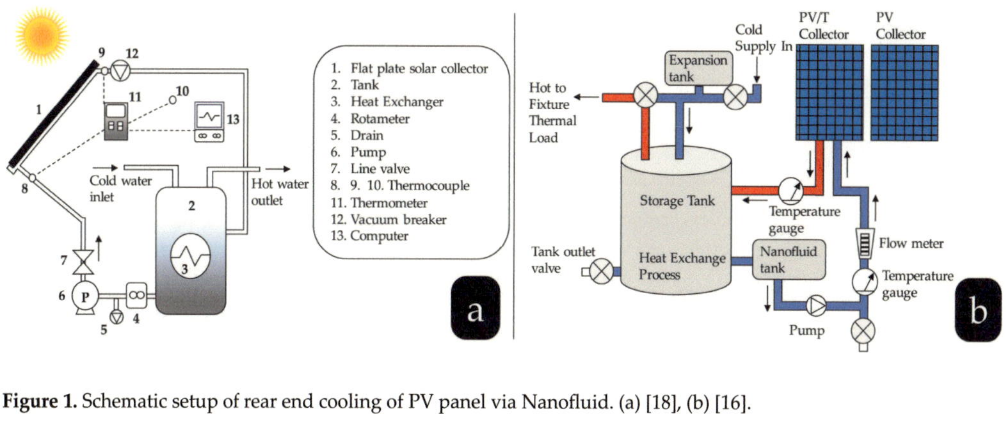

Figure 1. Schematic setup of rear end cooling of PV panel via Nanofluid. (a) [18], (b) [16].

Here,"r" is the packing factor.

$$r = \frac{A_{pv}}{A_c} \tag{3}$$

Here, "A_c" is the collector area and "A_{pv}" is area of PV cells.

Area of PV to produce a certain amount of electrical power is calculated by the following formula.

$$A_p = \frac{R\,E_{out,max}}{E_{out,1m^2}} \tag{4}$$

Here, "$E_{out,1m^2}$" is output electrical power per unit area and "$R\,E_{out,max}$" is the required output power.

Thermal output energy is found by the following equation.

$$\dot{E}_{th} = m_f \times C_{pf} \times (T_{fo} - T_{in}) \tag{5}$$

Here, "m_f" is the mass flowrate of the fluid through the collector, "C_{pf}" is the fluid's specific heat and "T_{in}" and "T_o" depicts the fluid's inlet and outlet temperature respectively. The formulas to determine "C_{pf}" are given in Ref [20].

Electrical efficiency is found by the following formula.

$$\eta_{el} = \frac{\dot{E}_{el}}{\dot{E}_{in}} = \frac{V_{oc} \times I_{sc} \times FF}{G_{eff}} \tag{6}$$

Here, "V_{oc}" is the open circuit voltage, "I_{sc}" is the short circuit current, and "G_{eff}" is the effective absorbed solar irradiation by the PV module. "FF" represents the fill factor and it is defined as the maximum power conversion efficiency.

$$FF = f \times \left(\frac{V_{oc}}{T}\right) \tag{7}$$

Using the aforementioned formulas, the efficiency of a PV/T system is determined.

Radwan et al. [20] examined the cooling effect of Al_2O_3, SiC nanoparticles and water on the performance of concentrated PV system. Pertaining to the results, SiC-water nanofluid produced better impact as compared to Al_2O_3 and water. It was observed that at higher concentration ratio (area of aperture/area of cell) and smaller Re, higher electrical efficiency was found. Using the pure water at CR = 40, the cell temperature reached a maximum of 68° C. Were as, for 4vol% SiC, the maximum temperature of the cell was found to be 60° C.

2.2. Optical filter cooling

Extensive work has been carried out on efficiency improvement by using nanofluid flowing through optical filters [21, 22]. Silicon-based Photovoltaic cells can generate electricity by absorbing the part of solar radiation with 400 nm to 1200 nm wavelength. Rest of the solar radiation's part is either reflected back or absorbed by the PV cells as heat. In optical filter cooling, an optical filter containing nanofluid is held above the front surface of cells to split the spectrum of radiation. Nanofluid-based optical filters separate the part of solar radiation for the PV cells from the radiation part that is more useful for heat generation. There are two kinds of proposed configurations of these systems.

1. Single Pipe System
2. Two Pipe System

In single pipe system, there are two sections of pipe; primary section and secondary section. Primary section is set underneath the rear surface of the Photovoltaic module having aluminum sheet in between. Primary section further elongates above the upper surface of PV module. Nanofluid enters from the inlet of primary section, thus, absorbing heat of the module. Heated nanofluid further passes over the PV's upper surface, in turns filtering the solar radiation. Part of radiation having wavelength equal to silicon bandgap is filtered and rest of the section is absorbed by the nanofluid flowing in the secondary channel which gets out of the secondary pipe at secondary outlet. Air exists between upper surface of PV module and secondary channel section. As the air gets hot, it flows in upward direction and the cool air still remains in contact with PV surface. It is assumed that no convection current is produced in the air. The results indicated, 83% and 80% overall and 76.5% and 74% thermal efficiency for Ag/water and Cu/water nanofluid respectively for above configuration [23]. Schematic diagram of such system is shown in the following **Figure 2**.

Wei An et al. [24] designed a spectral splitting Polypyrrole nanofluid based PV/T system in order to impede thermal losses and escalate the system's efficiency. Nanofluid used in spectral splitting filter is capable to absorb the part of solar radiations that cannot be utilized by PV cell and converts it into medium temperature thermal energy. The efficiency of PV/T system was found to be 25.2% for nanofluid based spectral splitting filter whereas, its value was 13.3% when there was no filter employed. Hjerrild et al. [25] worked on the cooling of PV system by the help of optical filters, they used Silver as nanoparticle (50 nm diameter) with coating of Silica. The results showed that, base fluid absorbed the ultraviolet part of solar radiation thus decreasing the heat losses whereas, nanoparticles absorbed visible portion of radiation, in turns increasing the overall efficiency of the system. Water showed highest electrical efficiency (85% higher than unfiltered PV) whereas highly diluted nanofluid ($Ag-SiO_2$) showed highest

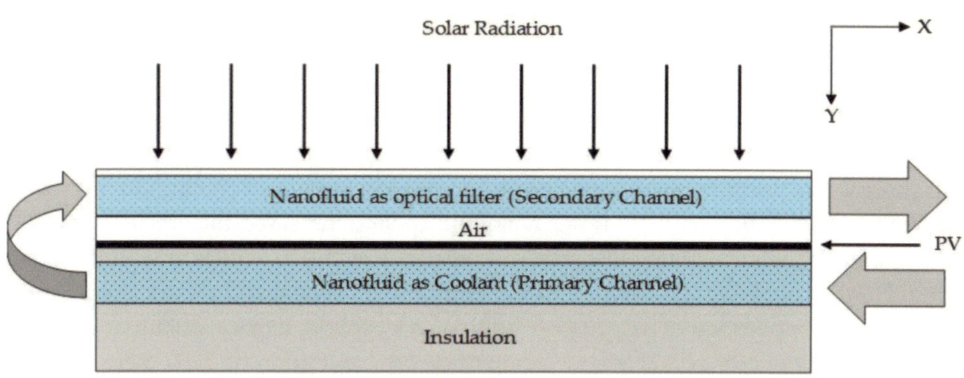

Figure 2. Schematic diagram of Nanofluid based spectral splitting filter PV/T system (reproduced) [23].

overall efficiency as well as greatest merit function. Hassani et al. [26] numerically investigated the effect of cooling on PV performance. The results revealed that PV system with optical filter (containing Ag-Water nanofluid) held above the PV surface along with thermal receiver (containing CNTs) at the rear end of PV, performed best in terms of high-grade energy as compared to conventional PV, PV being cooled by water only, PV being cooled by CNTs and PV being cooled by CNTs at rear end and optical filter containing water held at upper surface of the panel. Optical filter containing nanofluid was able to absorb both UV and IR spectrum and it only allowed the radiation in range of PV absorptivity spectrum (400-1200 nm). Whereas, optical filter containing water could only absorb IR spectrum. Saroha et al. [27] tested the effect of silver and gold based nanofluid working as optical filters in PV/T system. The results revealed that unwanted wavelengths were more absorbed by silver as compared to gold based nanofluid. Silver/water nanofluid based PV/T system approached 9.6% electrical, 67.8% thermal and 78.4% overall efficiency. Whereas, gold/water nanofluid based PV/T system achieved 9% electrical, 67.6% thermal and 76.6% overall efficiency. Jin et al. [28] investigated the effect of liquid optical filter based on magnetic electrolyte nanofluid for PV/T system. Electrolyte nanofluid is prepared by dispersing Fe_3O_4 nanoparticle in 50% water and 50% EG solutions containing methylene blue or copper sulfate, in this way they obtained two stable ENF filters. By adjusting the volume fraction of nanoparticles and molar fraction, more optimized ENF is produced. This ENF presents more better results compared to the simple liquid filters. Merit function of this newly developed ENF is found to be much more than the conventional liquid optical filter.

An arrangement in which nanofluids flows in separated channels outperforms the single channel through which the nanofluid is set to flow. In this arrangement a channel is placed underneath the rear surface of PV panel whereas, a separate channel is held above the front surface of the module. Upper channel nanofluid is made to achieve high liquid filter performance whereas the nanofluid flowing beneath the surface achieves higher thermal performance (working as a coolant). This technique achieved 8.5% higher electrical efficiency as compared to the double pass channel in which fluid flows in a single channel [29].

3. Efficiency improvement using nanofluid

Integrating the heat receivers with the conventional PV system is found to elevate both electrical and thermal efficiencies. Several fluids such as water or nanofluids can be used in these receivers to remove heat so as to improve the efficiency of the system. Studies have proved that nanofluid based PV/T system outperforms conventional PV system and water-based PV/T system. Soltani et al. [30] used five different methods for PV cooling (natural cooling, forced air cooling, water cooling, SiO_2-water nanofluid cooling and Fe_3O_4-water nanofluid cooling) to improve the performance. They found that SiO_2-water nanofluid cooling increased the efficiency by 3.35% and Fe_3O_4-water nanofluid cooling increased the efficiency by 3.13% as compared to the natural cooling. Hussien et al. [31, 32] found enhancement in the thermal and electrical efficiency of PV/T system by application of $Al_2O_3/water$ nanofluid as a coolant. Experimentation was carried out at constant flow rate of 0.2L/s and nanoparticles concentration of 0.3%. Results showed the increase in thermal and electrical efficiency when temperature was decreased from 79.1 to 42°C. Thermal and electrical efficiency of system enhanced up to 34.4% and 12.1% respectively using nanofluid. Ebaid et al. [33] used TiO_2 water-polyethylene glycol nanofluid and Al_2O_3 water cetyltrimethylammonium bromide nanofluid (with 0.01, 0.05, and 0.1 wt% concentration at a flowrate of 500–5000 ml/min) to test the efficiency enhancement of PV module via the cooling process. Pertaining to the results, Al_2O_3 nanofluid decreased the cell temperature by 13.83% and TiO_2 reduced the temperature by 11.2% at 5000 ml/min relative to water cooling. The best performance was witnessed in case of TiO_2 nanofluid cooling, it produced 50% more average efficiency compared to the water cooling (0.82% for TiO_2 and 0.48% for water cooling compared with no cooling). Karami and Rahimi [34] performed experiments to investigate the enhancement in the efficiency of PV module being cooled by the Boehmite (AlOOOH-xH_2O) based nanofluid flowing inside microchannel at the rear end of the PV module. The results showed that the maximum increase in the electrical efficiency due to cooling as compared to the without cooling power output was found to be 27.12% at a concentration ratio of 0.01 wt.% and 300 ml/min flowrate. Similarly, Sardarabadi et al. [64] observed as much as 9.75% electrical efficiency increment for silica/water nanofluid based PV/T system as compared to uncooled system. **Figures 3** and **4** depict the maximum efficiencies of PV/T systems obtained by different researchers.

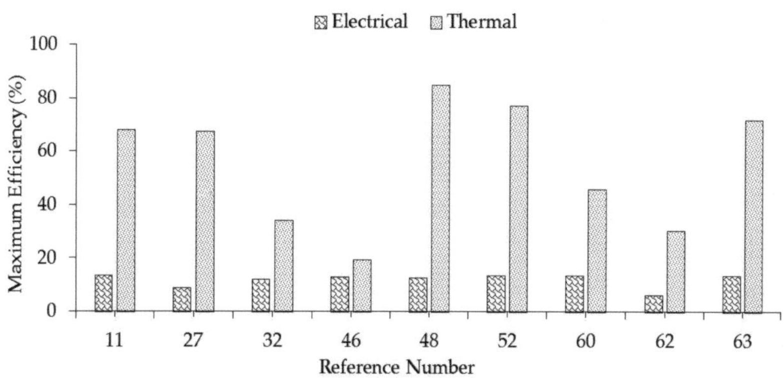

Figure 3. Maximum efficiency for obtained by researchers with Nanofluid cooling.

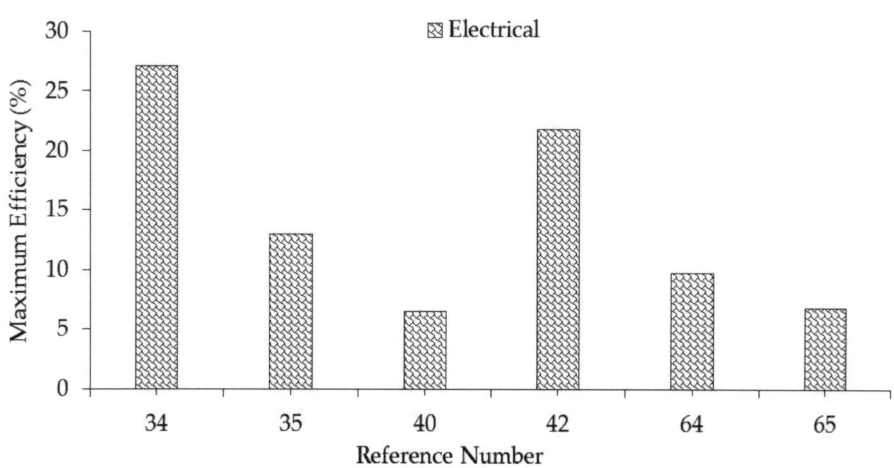

Figure 4. Maximum efficiency improvement by Nanofluid cooling compared to conventional PV.

Authors	Nanoparticle	Base Fluid	Concentration	Flowrate	Module Type, Irradiation (W/m²)	Ambien Temp	Module Temp	Electrical Efficiency	Thermal Efficiency	Overall Efficiency	
										Energy	Exergy
Al-Waeli et al. [16]	SiC	Deionized Water	3 wt%	—	—	—	—	100.19% Increase compared to Water Cooled PV System	24.1% increase compared to Water Cooled PV System	88.9% increase compared to conventional PV	—
Sardarabadi et al. [19]	No Cooling	—	—	—	845.42, Monocrystalline		—	—	—	—	10.90%
	—	Deionized Water	—	—			—	—	—	—	12.23%
	ZnO	Deionized Water	—	—			10°C Reduction compared to conventional PV	—	—	—	12.29%
	—	PCM + Deionized water	—	—			6°C Reduction compared to conventional PV	—	—	—	13.17%
	PCM + ZnO	Deionized Water	—	—			—	—	—	—	13.42%

Authors	Nanoparticle	Base Fluid	Concentration	Flowrate	Module Type, Irradiation (W/m²)	Ambien Temp	Module Temp	Electrical Efficiency	Thermal Efficiency	Overall Efficiency	
										Energy	Exergy
Soltani et al. [30]	—	Water	—	—	Silicon Crystalline PV Module	—	—	—	—	3.051% increase compared to natural cooling	—
	Fe_3O_4	Water	0.5 wt.%	—			—	—	—	3.13% increase compared to natural cooling	—
	SiO_2	Water	0.5 wt.%	—			—	—	—	3.35% increase compared to natural cooling	—
Hussien et al. [32]	—	—	—	—	1000, Monocry-stalline	—	79.1 °C	8%	—	—	—
	Al_2O_3	Water	0.30wt.%	0.2 L/s			42.2 °C	12.10%	34.40%	—	—
Ebaid et al. [33]	—	Water	—	5000 ml/min	750, Monocry stalline	—	16.58% Decrease compare to conventional PV	.61% Increase compared to conventional PV	—	—	—
	Ti_2O_3	Water-polyethylene glycol	0.1 wt%	5000 ml/min			22.9%% Decrease compare to conventional PV	0.82% Increase compared to conventional PV	—	—	—

Authors	Nanoparticle	Base Fluid	Concentration	Flowrate	Module Type, Irradiation (W/m²)	Ambien Temp	Module Temp	Electrical Efficiency	Thermal Efficiency	Overall Efficiency	
										Energy	Exergy
Karami and Rahimi [34]	ALOOH-XH_2O	Water	0.01 wt.%	300 ml/min	1000 Monocrystalline	25°C	Decrease from 62°C to 32.5°C	27.12% Increase compared to conventional PV	—	—	—
Sardarabadi et al. [37]	No Cooling	—	—	—	855, Monocry stalline	33°C	—	—	—	11%,	11.53%
	—	Deionized Water	—	30L/h			—	8.2% Increase compared to conventional PV	35.60%	47.20%	13.54%
	SiO_2	Water	1 wt%	—			—	9.01% Increase compared to conventional PV	—	49.80%	13.85%
	SiO_2	Water	3 wt.%	—			—	9.75% Increase compared to conventional PV	—	52.40%	14.02%
Sardarabadi and Passandideh. [40]	TiO_2	Deionized Water	0.2 wt.%	30 kg/h	917 Monocry stalline	33.4 °C	11.48°C Reduction as compared to Conventional PV	6.54% Increase compared to conventional PV	—	—	—
	ZnO		0.2 wt.%	30 kg/h			11.85°C Reduction as compared to Conventional PV	6.46% Increase compared to conventional PV	—	—	—
	Al_2O_3		0.2 wt.%	30 kg/h			11.03°C Reduction as compared to Conventional PV	6.36% Increase compared to conventional PV	—	—	—

Authors	Nanoparticle	Base Fluid	Concentration	Flowrate	Module Type, Irradiation (W/m^2)	Ambien Temp	Module Temp	Electrical Efficiency	Thermal Efficiency	Overall Efficiency	
										Energy	Exergy
Abd-Allah et al. [42]	Boehmite (ALOOH-xH_2O)	Water	0.1 wt.%	200 ml/min	—	—	21.6°C Reduction compared to without cooling	21.87% Increase compared to without cooling	—	—	—
Sathieshkumar et al. [46]	No cooling	—	—	—	Monocry stalline	–	—	11.31%	—	—	—
	—	Water	—	0.02 kg/s		–	—	12.42%	18.43%	—	—
	$CuTiO_2$	Water	0.2 wt.%	0.02 kg/s		–	—	12.87%	19.50%	—	—
Hasan et al. [48]	—	Water	—	0.167 kg/s	1000 Polycry stalline	30 °C	Decreased from 87–57°C	11.40%	—	—	—
	SiC	Water	1 wt.%	0.167 kg/s			Decreased from 87–41°C	12.75%	85%	97.75%	—
	TiO_2	Water	1 wt.%	0.167 kg/s			Decreased from 87–45°C	12.30%	—	—	—
	SiO_2	Water	1 wt.%	0.167 kg/s			Decreased from 87–50 °C	11.80%	—	—	—
Maadi et al. [54]	Al_2O_3	Water	10 wt%	30 kg/h	Monocry stalline	—	—	6.23% Increase compared to pure water	—	—	—
	TiO_2	Water	10 wt%	30 kg/h		—	—	6.02% Increase compared to pure water	—	—	—
	ZnO	Water	10 wt.%	30 kg/h		—	—	6.88% Increase compared to pure water	—	—	—
	SiO_2	Water	10 wt.%	30 kg/h		—	—	5.77% Increase compared to pure water	—	—	—

Authors	Nanoparticle	Base Fluid	Concentration	Flowrate	Module Type, Irradiation (W/m^2)	Ambien Temp	Module Temp	Electrical Efficiency	Thermal Efficiency	Overall Efficiency	
										Energy	Exergy
Sahini at el. [58]	—	Deionized Water	—	0.026 kg/s	Polycry stalline PV Module	—	—	8.5% Increase compared with conventional PV system	—	—	—
	Silver with 1 vol.% potassium oleate surfactant	Deionized Water	0.5 vol.%	0.026 kg/s		—	—	0.9% Increase compared to water cooled system	—	—	—
Sardarabadi et al. [61]	No Cooling	—	—	—	917 Monocry stalline	34.42°C	—	12.73%	—	12.73%	10.29%
	—	Water	0.2 ey%	30 kg/h			11% decrease compared to conventional PV	13.41%	34.12%	47.53%	11.56%
	ZnO	Water	0.2 wt%	30 kg/h			11.85% decrease compared to conventional PV	13.59%	46.05%	59.64%	12.17%
	TiO2	Water	0.2 wt%	30 kg/h			11.48% Decrease compared to conventional PV	13.63%	44.34%	57.97%	11.93%
	Al2O3	Water	0.2 wt%	30 kg/h			11.03% Decrease compared to conventional PV	13.44%	36.66%	50.10%	11.88%

Authors	Nanoparticle	Base Fluid	Concentration	Flowrate	Module Type, Irradiation (W/m²)	Ambien Temp	Module Temp	Electrical Efficiency	Thermal Efficiency	Overall Efficiency	
										Energy	Exergy
J.J. Michael and S. Inyan. [62]	No Cooling	—	—	—	—	—	—	8.98%	—	—	—
	—	Water	—	0.01 kg/s	Without Glazing	—	—	8.77%	19.36%	—	—
	—		—	0.01 kg/s	With Glazing	—	—	6.40%	21%		—
	CuO	Water	0.05%	0.01 kg/s	Without Glazing	—	—	7.62%	28.22%	—	—
		Water	0.05%	0.01 kg/s	With Glazing	—	—	6.18%	30.43%	—	—
Al-Waeli et al. [63]	—	—	—	—	—	25°C	68.3°C	7.11%	—	—	—
	—	Water	—	0.175 kg/s	—		45.22°C	9.92%	35.40%	—	—
	—	PCM + Water	—	0.175 kg/s	—		42.22°C	12.32%	50.50%	—	—
	PCM + SiC	Water	—	0.175 kg/s	—		39.52°C	13.70%	72%	—	—
Hamdan and Kardasi [65]	No Cooling	—	—	—	—	—	46.9	10.04%	—	—	—
	—	Water	—	—	—		25.47	11.20%	—	—	—
	Al_2O_3		0.4 wt.%	—	—		22.67	12.06%	—	—	—
	No Cooling		—	—	—		48.49	12.57%	—	—	—
	—		—	—	—		24.93	11.39%	—	—	—
	CuO		0.6 wt.%	—	—		22.13	10.23%	—	—	—

Table 1. Effect of Nanofluids on PV/T System's performance.

Gangadevi et al. [35] experimentally examined that the electrical, thermal and overall efficiency of PV module being cooled by Al_2O_3/water nanofluid got increased by 13%, 45% and 58% respectively as compared to water based and 1 wt% Al_2O_3 nanofluid based cooling. Mustafa et al. [36] numerically tested the effect of mass flowrate and concentration of nanofluid (TiO_2/water) on the efficiency of PV/T system. As per the results, electrical and thermal efficiency of this system is linearly proportional to mass flowrate. Best results are obtained at low concentration of nanofluid.

Electrical, thermal and overall efficiencies of the various PV/T systems working with different nanofluids is expressed in the **Table 1**.

Efficiency enhancement of PVT systems being cooled by the nanofluids is due to the enhanced thermal conductivity of the nanofluids. Increase in thermal conductivity is dependent on concentration, size and type of the nanoparticle [4].

4. Factors affecting efficiency of nanofluid-based PV/T systems

Various factors such as the concentration of nanofluid, flowrate of nanofluid, size of the nanoparticle, geometry of microchannel, type of base fluid and irradiance influence the efficiency of nanofluid-based PV/T system. Effects of these factors are discussed in the subsequent sections.

4.1. Irradiance

Increase in irradiance cause the module temperature to escalate as more heat reaches the surface. Khanjari et al. [2] investigated environmental parameters that affect the efficiency of a PV/T system cooled by nanofluids ($Al_2O_3/water$) via CFD simulation. As the absorbed solar radiation increased from 200 W/m^2k to 800 W/m^2k the electrical efficiency of system decreased from 11.41% to 10.12% for pure water and 11.4% to 10.23% for alumina nanofluid whereas, thermal efficiency increased from 65–79% for pure water and 76–91% for alumina nanofluid. As the absorber plate temperature increased from 291 K to 324 K the electrical efficiency decreased from 11.1% to 9.4% for water and 11.2% to 9.5% for alumina nanofluid whereas, the thermal efficiency did not change with increasing inlet temperature of fluid after reaching a primary value. Similarly, the system efficiency was found to escalate with decreasing irradiation i.e. the maximum overall efficiency of the system was found to increase from 78.60% to 80.58% and 73.58% to 75.93% for 1 wt% and 3 wt% respectively, when the irradiation value decreased from 1100 W/m^2 to 600 W/m^2 [37]. Effect of irradiance found by Al-Waeli et al. [38] has been presented in **Table 2**.

4.2. Concentration

Researchers have found contradictory results when it comes to concentration enhancement of nanofluids. Manikandan and Rajan [39] harnessed sand for the cooling of PV/T system in order to enhance the efficiency. They tested 0.5, 1 and 2 vol% concentration and the collection efficiency ratio for these concentrations was found to be 3.6%, 11.2% and 26.9% whereas the solar collection efficiency increased by 9% and 16.5% for 0.5% and 2% respectively. Sardarabadi and Fard [40] also examined that increasing the mass fraction of nanoparticles from 0.05 to 10 wt%, the thermal performance of the system increased by four times. Wei An. [24] examined the effect of nanofluid concentration in spectral splitting filter based PV/T system. They observed that increasing the concentration of the nanofluid increased the nanofluid temperature and system's electrical efficiency, but the thermal efficiency gets decreased in this way.

The maximum overall efficiency of the system was found to be 75.93% and 80.58% when the ferrofluid concentration was increased from 1 wt% to 3 wt% respectively [37]. Khanjari et al. [41] observed that increasing volumetric concentration of the nanoparticle (from 1–5%) increased the heat transfer coefficient and thus the overall efficiency (from 1.33% to 11.54% for silver and 0.72% to 4.26% for alumina). Radwan et al. [20] observed efficiency escalation with increasing concentration. But some researchers witnessed contradictory results. Karami and Rahimi [34] examined that increasing concentration of nanoparticles reduces the efficiency because of agglomeration or clustering of the suspended particles. Abd-Allah, [42] found best results at 0.1 wt% amongst (0.01, 0.1, 0.5 wt%).

Cieslinski et al. [43] found no impact of nanoparticle concentration on the performance of the PV/T system. They observed that 1 wt% of $Al_2O_3/water$ rather decreased the thermal efficiency compared to the distilled water and 3 wt% and 3 wt% did not change the thermal efficiency as compared to the distilled water thermal efficiency. Whereas, the overall efficiency of the system reached up to 80%.

In order to obtain best results, there is always a need to determine the optimum concentration of nanoparticles in base fluid instead of using high volume fraction of nanofluid [43, 44]. However, instead of increasing the concentration of the same kind of nanoparticle, blending a different kind of nanoparticles can help improve the efficiency of PV module in a more efficient way [45].

4.3. Flowrate

Sathieshkumar et al. [46] concluded that both electrical and thermal efficiency of the PV/T system increases with increasing flow rate but after a certain flowrate magnitude the efficiencies of the system start to decline. Overall energy efficiency is found to be higher in turbulent regime whereas overall exergy efficiency is higher in laminar regime [47]. Mustafa et al. [36] numerically tested the effect of mass flowrate and concentration of nanofluid (TiO_2/water) on the efficiency of PV/T system. As per the results, the electrical and thermal efficiency of this system was found to be linearly proportional to mass flowrate.

Hasan et al. [48] observed that increasing the mass flowrate increased the cell efficiency linearly. As the mass flowrate increased from 0 to 1.666 kg/s the electrical efficiency of the cell increased from 8% to 16.5% at $500\ W/m^2$ solar irradiance in case of SiC-water nanofluid. Mean photovoltaic temperature decreased from $87^\circ C$ to $41^\circ C$ as the mass flowrate changed from 0 to 1.666 kg/s at 1000 W/m^2 solar irradiance in case of SiC. Karami and Rahimi. [34] observed that temperature of the module decreased from $62^\circ C$ to $32.5^\circ C$ when the flow rate increased from zero to 300 ml/min. Khanjari et al. [41] observed that increase in inlet fluid velocity (from 0.05 m/s to 0.23 m/s) increase the first law (energy) efficiency but decreases the second law (exergy) efficiency (from 15.40% to 12.50% for silver). Lelea et al. [14] observed lower maximum module temperature for nanofluid based cooling as compared to water cooling at lower Re number. Whereas, at higher Re ($Re > 1000$) the maximum module temperature overlaps for nanofluid based cooling and water-based cooling of PV module.

PV/T system in laminar regime outperforms turbulent regime. More PV efficiency can be achieved in turbulent regime but it requires higher pumping power thus making the overall system efficiency lesser [15]. Although heat transfer in case of higher Reynolds numbers is seemed to increase because of greater Brownian motion of particles but too high a Reynolds number requires higher pumping power which eventually reduces the overall performance of the microchannels containing nanofluids [49]. Xu and Kleinstreur [50] concluded that increased concentration elevates the system efficiency when cooled by Al_2O_3/water nanofluid. Higher inlet Reynolds number yields higher cell efficiency but too high a Reynolds number is not favorable. Low inlet temperature of nanofluid is capable to produce pronounced cooling effect. Height of channel containing nanofluid is also of much consideration, slight variation in channel height varies the required pumping power and significant change in entropy generation rate.

4.4. Nanoparticle size

Due to the smaller size, nanoparticles have large surface area which is attributed to higher heat transfer rates. Nanoparticles have high thermal conductivity, but heat capacity is low. Nanoparticles are stable in the base fluid at high temperatures and they do not agglomerate in the water as well [51]. Energy and exergy efficiency of the system can be increased by increasing the size of the nanoparticle in the turbulent regime but in laminar regime the case is opposite. Yazdanifard et al. [15] interestingly found no effect of particle size on the efficiency. They used Titanium dioxide nanofluid and Aluminum oxide nanofluid for the cooling purpose but no significant efficiency alteration was observed. Whereas, Al-Shamani et al. [4] observed that heat transfer of the nanofluid decreased with a decrease in size of the nanoparticle. Therefore, there is still a need for further experimentation to conclusively narrate the effects of nanoparticle size on the efficiency of the solar systems.

4.5. Base fluid

Not only the type of nanoparticle affects the performance of the PV/T system but the type of base fluid is also of same significance while predicting the performance of the system. Using base fluids such as ethylene-glycol, polyethylene glycol, cetyltrimethylammonium bromide water mixtures instead of water can considerably elevate the cell efficiency [15]. Addition of surfactant and selection of suitable pH of nanofluid can display pronounced effects [44]. Rajeb et al. [52] examined both numerically and experimentally the effect of variation in concentration (0.1, 0.2 and 0.4 wt%), type of nanoparticle (Al_2O_3 and Cu) and type of base fluid (water and ethylene glycol) on the efficiency of PV/T system being cooled by nanofluid. They observed that increasing the concentration of nanofluid increased the efficiency of the system. The system best performed when water was used as base fluid as compared to ethylene glycol base fluid. According to the drawn results, maximum electrical and thermal efficiency was found to be 13.55% and 77% respectively for Cu/water nanofluid based PV/T system, at 0.4 wt%. Whereas, they found 13.54% electrical and 60% thermal efficiency for Cu/ethylene glycol based PV/T system, at 0.4 wt%. Conclusively, Cu/water nanofluid based system outperformed Al_2O_3/water based system in terms of electrical and thermal efficiency. Hosseinzadeh et al. [53]

found that a PV system being cooled by water only, performed better than the systems cooled by either ethylene glycol only and water-ethylene glycol (50% water and 50% ethylene glycol).

4.6. Nanoparticle type

Maadi et al. [54] stated that for metalloids the viscosity of the nanofluids gets increased and the specific heat capacity is decreased, which is not favorable. This is because, at a given mass fraction the volume of the metalloid nanofluids is increased due to high density. Hasan et al. [48] observed that cooling the PVT by impinging SiC, TiO_2, SiO_2 nanofluids and pure water improved the maximum power output by 62.5%, 57%, 55% and 50% as compared to the conventional PV module. Al-Shamani et al. [55] tested SiO_2, TiO_2 and SiC based nanofluid for the cooling purpose to analyze the efficiency betterment. Following the experimental results, SiC/water nanofluid outperformed rest of the nanofluids. At 1000 W/m^2 irradiance and 0.170 kg/s mass flowrate, SiC/water nanofluid based PV/T system showed 13.529% electrical efficiency whereas, TiO_2/water and SiO_2/water nanofluid based PV/T systems depicted 10.978% and 10.302% electrical efficiency respectively. PV/T system utilizing water solely for cooling, approached 9.608% electrical efficiency.

Kolahan et al. [56] examined the entropy generation in PV/T system due to the addition of nanoparticles both numerically and experimentally. They used $Al_2O_3/water$, $TiO_2/water$ and ZnO/water by 0.2 wt% and SiO_2/water by 1 wt% and 3 wt% nanofluids (along with acetic acid as a surfactant). Following the results, ZnO/water produced least frictional entropy, $SiO_2/water$ produced maximum pressure drop and frictional entropy generation and $Al_2O_3/water$ produced least thermal and total entropy generation. Thermal entropy generation was found to be maximum at inlet, turning points and outlet, due to high temperature differences. For metallic nanofluids, increase of mass fraction caused density and viscosity elevation. Increased mass fraction reduced the velocity which in turns reduced the frictional entropy generation. For metalloid nanofluids, reverse is the case. For ZnO the frictional entropy was decreased by 10.87% at 10 wt%, whereas, for $SiO_2/water$ the frictional entropy was increased by 0.94% compared to pure water. Addition of nanoparticles causes more prominent reduction in thermal entropy generation compared to the frictional entropy generation. Considering the entropy generation view point, metallic nanofluids produce better results than the metalloid nanofluids.

Extensive experimentation has been conducted to examine the effect of magnetic on the performance of nanofluids [66–70]. If the Ferro-nanoparticle is used in the system, employing alternating magnetic field around the channels can improve the efficiency of the system. Experimental results also depicted that the alternating magnetic field improved the system performance whereas, the constant field did not produce significant efficiency enhancement when compared with the no field condition. The system efficiency was found to be 71.91% when there was no field applied, whereas, the efficiency went up to 73.58% in the presence of alternating magnetic field (50 Hz) in case of 1 wt% and $1100\ ^{W}/_{m^2}$ [37]. Shape of nanoparticle and type of magnetic field can influence the performance of nanofluid. Sheikholeslami et al. [66, 67] numerically analyzed the effect of non-uniform magnetic field on Fe_3O_4 -H_2O nanofluid flowing in a porous cavity. Following the results, platelet shape of nanoparticles depicted

Intensity	Cooling Fluid	Electrical Efficiency	Thermal Efficiency
200	SiC-Water	16.90%	8%
1000	SiC-Water	10.90%	48%
200	CuO-Water	16%	6%
1000	CuO-Water	10%	41%
200	Al_2O_3-Water	13.50%	6%
1000	Al_2O_3-Water	9.80%	41%
200	Water	11.90%	4%
1000	Water	8.40%	31%

Table 2. Effect of irradiance on efficiency [38].

highest Nusselt number (i.e. optimum heat transfer) under the influence of non-uniform magnetic field. In the presence of magnetic field, addition of nanoparticles can improve the heat transfer properties of nanofluids [68].

4.7. Channel geometry

Narrow channels offer higher enhancement in the heat transfer coefficient whereas the wide channels depict instabilities in lateral heat transfer. Roughness in the pipes also affects the magnitude of heat transfer. Pipes with greater roughness magnitude offer greater heat transfer due to the increased contact surface. In order to achieve higher performance, the temperature distribution inside the channel should be held uniform, the temperature should be kept low and the pressure drop should also be as minimum as possible [49]. Considering the **Table 3**, helical channel performs best because of greater surface contact of nanofluid with the rear surface of PV unit.

Researcher	Nanoparticle	Base fluid	Concentration	Flowrate	Channel Geometry	Effect on Temperature	Effect on Electrical Efficiency
Karami and Rahimi [34]	Boehmite	Water	0.01 wt%	300 ml/min	Straight	Decreased from 62°C to 32.5°C for flowrate 0–300 ml/min	27.12% increase compared to Conventional PV System
Karami and Rahimi, [57]	Boehmite	Water	0.1 wt%	200 ml/min	Straight	18.33°C Temperature Reduction	20.57% increase compared to conventional PV System
	Boehmite	Water	0.1 wt%	200 ml/min	Helical	24.22°C Temperature Reduction	37.67% increase compared to Conventional PV System

Table 3. Effect of channel geometry on efficiency.

4.8. Circulation method

When cooling the PV module via nanofluid, the circulation method is also of much importance. If the circulation is done via passive method, the increasing intensity of light would cause a reduction in electrical efficiency and enhancement in thermal efficiency because natural convection is not that efficient. Thus, active convection cooling should be employed to obtain optimum results. Whereas, the elevation in thermal efficiency is due to the availability of enough time for the cooling fluid to exchange heat. However, the overall efficiency of the system gets increased if the cooling is employed. Pumping of nanofluid can further improve the efficiency compared to the passive cooling [38].

5. Advantages of nanofluid-based cooling

5.1. Environmental benefits

Fossil fuel based power plants emit tons of noxious gases that detriment the environment. Since the solar power plants are emission free, production of electricity via this method can eliminate the emission of 16,974,57 tons of CO_2 [58]. Hassani et al. [26] evaluated that nanofluid based PV/T systems can omit the emission of 448 kg $CO_2\, m^{-2} yr^{-1}$.

5.2. Economic benefits

PV/T system can provide an economical solution for industrial and domestic power demands. Studies indicate a significant reduction in energy consumption produced from conventional resources due to the use of such system [23, 59]. Taylor et al. [60] also narrated that a solar thermal based power plant of 100 MW capacity can save about $3.5 million per annum if the nanofluid receiver is incorporated with it. Nanofluids need a smaller area for heat transfer thus making the PV system compact and reducing the costs [51]. The economic analysis depicted that the cost of energy produced by nanofluid based PV/T system is 82% less than the current prices in Saudi Arabia [33]. Nanofluid system is predicted to takes only 2 years for pay-back [26]. Sardarabadi et al. [61] evaluated that size reduction by 21, 32,33 and 34 from energy viewpoint and 5,6,7 and 6 from exergy viewpoint for PVT/water, PVT/TiO_2, PVT/ZnO and PVT/Al_2O_3 respectively. By size reduction we mean the amount of material saved for the same required energy and exergy outputs at the same conditions.

6. Conclusion

Cooling of PV module by nanofluids significantly enhances electrical efficiency and thermal energy. Cooling causes the heat removal which in turns halts the development of thermal stresses, making the PV modules to last long and operate more efficiently. Employing nanofluids impedes entropy generation as well. Efficiency of this system escalates with increasing concentration of nanofluid up to a certain limit but as the concentration exceeds this optimum limit, efficiency tends to decline because of the clustering and agglomeration of nanoparticles. Increasing flowrate of nanofluid increases the efficiency but as the flow gets into turbulent regime the instability issues arise and this also requires higher pumping power, in turns reducing overall system's efficiency. Using helical microchannel can increase the heat transfer and thus overall efficiency gets elevated. Using surfactant in the nanofluid can also surge the system's performance. Some of the measures that can refine the performance of these systems include,

1. Glazing can drastically improve the nanofluid based PV/T system's performance [46, 47].
2. Simultaneously using optical filters over the surface and thermal collector at the rear end can also elevate performance.
3. Applying alternating magnetic field around the flow channel can supplement the performance of system if the Ferro-nanoparticles are being used.

The unresolved challenges being faced by the researchers while using nanofluids include instability, agglomeration, high pumping power, and erosions. Stability improvement is the most important need of the hour in order to further proceed towards commercial use of nanofluids, as no perfect method of preparation and processing of stable nanofluid has been determined up-to-date.

Author details

Hafiz Muhammad Ali[1]*, Tayyab Raza Shah[1], Hamza Babar[1] and Zargham Ahmad Khan[2]

*Address all correspondence to: h.m.ali@uetaxila.edu.pk

1 Mechanical Engineering Department, University of Engineering and Technology, Taxila, Pakistan

2 NFC Institute of Engineering and Technology, Multan, Pakistan

References

[1] Sargunanathan S, Elango A, Mohideen ST. Performance Enhancement of Solar Photovol taic Cells Using Effective Cooling Methods: A Review. Renewable and Sustainable Energy Reviews. Vol. 642016 Oct 31. pp. 382-393. DOI: 10.1016/j.rser.2016.06.024

[2] Khanjari Y, Kasaeian AB, Pourfayaz F. Evaluating the environmental parameters affecting the performance of photovoltaic thermal system using nanofluid. Applied Thermal Engineering. 2017 Mar 25;**115**:178-187. DOI: 10.1016/j.applthermaleng.2016.12.104

[3] Gangadevi R, Agarwal S, Roy S. A novel hybrid solar system using nanofluid. Int. J. Engineering Research And Technology. 2013;**6**(6):747-752

[4] Al-Shamani AN, Yazdi MH, Alghoul MA, Abed AM, Ruslan MH, Mat S, Sopian K. Nanofluids for improved efficiency in cooling solar collectors–a review. Renewable and Sustainable Energy Reviews. 2014 Oct 31;**38**:348-367. DOI: 10.1016/j.rser.2014.05.041

[5] Oruc ME, Desai AV, Kenis PJ, Nuzzo RG. Comprehensive energy analysis of a photovoltaic thermal water electrolyzer. Applied Energy. 2016 Feb 15;**164**:294-302. DOI: 10.1016/j.apenergy.2015.11.078

[6] Bashir MA, Ali HM, Ali M, Siddiqui AM. An experimental investigation of performance of photovoltaic modules in Pakistan. Thermal Science. 2013;**19**:134

[7] Ali HM, Mahmood M, Bashir MA, Ali M, Siddiqui AM. Outdoor testing of photovoltaic modules during summer in Taxila, Pakistan. Thermal Science. 2016;**20**(1):165-173. DOI: 10.2298/TSCI131216025A

[8] Bashir MA, Ali HM, Amber KP, Bashir MW, Hassan AL, Imran S, Sajid M. Performance investigation of photovoltaic modules by back surface water cooling. Thermal Science. 2017;**21**(2):290

[9] Ali M, Ali H.M, Moazzam W, Saeed M.B: Performance enhancement of PV cells through micro-channel cooling. WEENTECH Proceedings in Energy GCESD 2015 24th–26th February 2015 Technology Park, Coventry University Coventry, United Kingdom. 2015 Feb;**24**:211. DOI: 10.3934/energy.2015.4.699

[10] Shukla A, Kant K, Sharma A, Biwole PH. Cooling methodologies of photovoltaic module for enhancing electrical efficiency: A review. Solar Energy Materials and Solar Cells. 2017 Feb 28;**160**:275-286. DOI: 10.1016/j.solmat.2016.10.047

[11] Sathe TM, Dhoble AS. A review on recent advancements in photovoltaic thermal techniques. Renewable and Sustainable Energy Reviews. 2017 Sep 30;**76**:645-672. DOI: 10.1016/j.rser.2017.03.075

[12] Michael JJ, Iniyan S, Goic R. Flat plate solar photovoltaic–thermal (PV/T) systems: A reference guide. Renewable and Sustainable Energy Reviews. 2015 Nov 30;**51**:62-88. DOI: 10.1016/j.rser.2015.06.022

[13] Al-Waeli AH, Sopian K, Kazem HA, Chaichan MT. Photovoltaic thermal PV/T systems: A review. International Journal of Computation and Applied Sciences IJOCAAS. 2017;**2**(2)

[14] Lelea D, Calinoiu D.G, Trif-Tordai G, Cioabla A.E, Laza I, Popescu F: The hybrid nanofluid/microchannel cooling solution for concentrated photovoltaic cells. In AIP Conference Proceedings 2015 Feb 17 (Vol. 1646, No. 1, pp. 122-128). AIP. DOI. 10.1063/1.4908592

[15] Yazdanifard F, Ameri M, Ebrahimnia-Bajestan E. Performance of nanofluid-based photovoltaic/thermal systems: A review. Renewable and Sustainable Energy Reviews. 2017 Sep 30;**76**:323-352. DOI: 10.1016/j.rser.2017.03.025

[16] Al-Waeli AH, Sopian K, Chaichan MT, Kazem HA, Hasan HA, Al-Shamani AN. An experimental investigation of SiC nanofluid as a base-fluid for a photovoltaic thermal PV/T system. Energy Conversion and Management. 2017 Jun 15;**142**:547-558. DOI: 10.1016/j.enconman.2017.03.076

[17] Xu Z, Kleinstreuer C. Concentration photovoltaic–thermal energy co-generation system using nanofluids for cooling and heating. Energy Conversion and Management. 2014 Nov 30;**87**:504-512. DOI: 10.1016/j.enconman.2014.07.047

[18] Yousefi T, Veysi F, Shojaeizadeh E, Zinadini S. An experimental investigation on the effect of Al_2O_3–H_2O nanofluid on the efficiency of flat-plate solar collectors. Renewable Energy. 2012 Mar 31;**39**(1):293-298. DOI: 10.1016/j.renene.2011.08.056

[19] Sardarabadi M, Passandideh-Fard M, Maghrebi MJ, Ghazikhani M. Experimental study of using both ZnO/water nanofluid and phase change material (PCM) in photovoltaic

thermal systems. Solar Energy Materials and Solar Cells. 2017 Mar 31;**161**:62-69. DOI: 10.1016/j.solmat.2016.11.032

[20] Radwan A, Ahmed M, Ookawara S. Performance enhancement of concentrated photovoltaic systems using a microchannel heat sink with nanofluids. Energy Conversion and Management. 2016 Jul 1;**119**:289-303. DOI: 10.1016/j.enconman.2016.04.045

[21] Taylor RA, Otanicar T, Rosengarten G. Nanofluid-based optical filter optimization for PV/T systems. Light: Science & Applications. 2012 Oct 1;**1**(10):e34

[22] Crisostomo F, Hjerrild N, Mesgari S, Li Q, Taylor RA. A hybrid PV/T collector using spectrally selective absorbing nanofluids. Applied Energy. 2017 May 1;**193**:1-4. DOI: 10.1016/j.apenergy.2017.02.028

[23] Mittal T, Saroha S, Bhalla V, Khullar V, Tyagi H, Taylor RA, Otanicar T.P: Numerical study of solar photovoltaic/thermal (PV/T) hybrid collector using nanofluids. ASME Paper No. MNHMT2013-22090. 2013 Dec 11

[24] An W, Zhang J, Zhu T, Gao N. Investigation on a spectral splitting photovoltaic/thermal hybrid system based on polypyrrole nanofluid: Preliminary test. Renewable Energy. 2016 Feb 29;**86**:633-642. DOI: 10.1016/j.renene.2015.08.080

[25] Hjerrild NE, Mesgari S, Crisostomo F, Scott JA, Amal R, Taylor RA. Hybrid PV/T enhancement using selectively absorbing Ag–SiO2/carbon nanofluids. Solar Energy Materials and Solar Cells. 2016 Apr 30;**147**:281-287. DOI: 10.1016/j.solmat.2015.12.010

[26] Hassani S, Saidur R, Mekhilef S, Taylor RA. Environmental and exergy benefit of nanofluid-based hybrid PV/T systems. Energy Conversion and Management. 2016 Sep 1;**123**:431-444. DOI: 10.1016/j.enconman.2016.06.061

[27] Saroha S, Mittal T, Modi PJ, Bhalla V, Khullar V, Tyagi H, Taylor RA, Otanicar TP. Theoretical analysis and testing of nanofluids-based solar photovoltaic/thermal hybrid collector. Journal of Heat Transfer. 2015 Sep 1;**137**(9):091015

[28] Jin J, Jing D. A novel liquid optical filter based on magnetic electrolyte nanofluids for hybrid photovoltaic/thermal solar collector application. Solar Energy. 2017 Oct 1;**155**: 51-61. DOI: 10.1016/j.solener.2017.06.030

[29] Hassani S, Taylor RA, Mekhilef S, Saidur R. A cascade nanofluid-based PV/T system with optimized optical and thermal properties. Energy. 2016 Oct 1;**112**:963-975. DOI: 10.1016/j.energy.2016.06.142

[30] Soltani S, Kasaeian A, Sarrafha H, Wen D. An experimental investigation of a hybrid photovoltaic/thermoelectric system with nanofluid application. Solar Energy. 2017 Oct 1;**155**:1033-1043. DOI: 10.1016/j.solener.2017.06.069

[31] Hussien HA, Noman AH, Abdulmunem AR. Indoor investigation for improving the hybrid photovoltaic/thermal system performance using nanofluid (Al2O3-water). Eng Tech J. 2015;**33**(4):889-901

[32] Hussien HA, Hasanuzzaman M, Noman AH, Abdulmunem AR. Enhance photovoltaic/thermal system performance by using nanofluid. 2014

[33] Ebaid MS, Ghrair AM, Al-Busoul M. Experimental investigation of cooling photovoltaic (PV) panels using (TiO2) nanofluid in water-polyethylene glycol mixture and (Al2O3) nanofluid in water-cetyltrimethylammonium bromide mixture. Energy Conversion and Management. 2018 Jan 1;**155**:324-343. DOI: 10.1016/j.enconman.2017.10.074

[34] Karami N, Rahimi M. Heat transfer enhancement in a hybrid microchannel-photovoltaic cell using Boehmite nanofluid. International Communications in Heat and Mass Transfer. 2014 Jul 31;**55**:45-52. DOI: 10.1016/j.icheatmasstransfer.2014.04.009

[35] Gangadevi R, Vinayagam BK, Senthilraja S. Experimental investigations of hybrid PV/spiral flow thermal collector system performance using Al2O3/water nanofluid. In IOP Conference Series: Materials Science and Engineering 2017 May (Vol. 197, No. 1, p. 012041). IOP Publishing. DOI: 10.1088/1757-899X/197/1/012041

[36] Mustafa W, Othman MY, Fudholi A. Numerical investigation for performance study of photovoltaic thermal Nanofluid system. International Journal of Applied Engineering Research. 2017;**12**(24):14596-14602

[37] Ghadiri M, Sardarabadi M, Pasandideh-fard M, Moghadam AJ. Experimental investigation of a PVT system performance using nano ferrofluids. Energy Conversion and Management. 2015 Oct 31;**103**:468-476. DOI: 10.1016/j.enconman.2015.06.077

[38] Al-Waeli AH, Chaichan MT, Kazem HA, Sopian K. Comparative study to use nano-(Al2O3, CuO, and SiC) with water to enhance photovoltaic thermal PV/T collectors. Energy Conversion and Management. 2017 Sep 15;**148**:963-973. DOI: 10.1016/j.enconman.2017.06.072

[39] Manikandan S, Rajan KS. Sand-propylene glycol-water nanofluids for improved solar energy collection. Energy. 2016 Oct 15;**113**:917-929. DOI: 10.1016/j.energy.2016.07.120

[40] Sardarabadi M, Passandideh-Fard M. Experimental and numerical study of metal-oxides/water nanofluids as coolant in photovoltaic thermal systems (PVT). Solar Energy Materials and Solar Cells. 2016 Dec 31;**157**:533-542. DOI: 10.1016/j.solmat.2016.07.008

[41] Khanjari Y, Pourfayaz F, Kasaeian AB. Numerical investigation on using of nanofluid in a water-cooled photovoltaic thermal system. Energy Conversion and Management. 2016 Aug 15;**122**:263-278. DOI: 10.1016/j.enconman.2016.05.083

[42] Abd-Allah SR, Abdellatif OE, El-Kady ES. Performance of cooling photovoltaic cells using Nanofluids

[43] Cieśliński J, Dawidowicz B, Krzyżak J. Performance of the PVT solar collector operated with water–Al_2O_3 nanofluid. Polska Energetyka Słoneczna. **2016**(1-4):5-8

[44] Mahian O, Kianifar A, Kalogirou SA, Pop I, Wongwises S. A review of the applications of nanofluids in solar energy. International Journal of Heat and Mass Transfer. 2013 Feb 28;**57**(2):582-594. DOI: 10.1016/j.ijheatmasstransfer.2012.10.037

[45] Chen M, He Y, Huang J, Zhu J. Synthesis and solar photo-thermal conversion of au, ag, and au-ag blended plasmonic nanoparticles. Energy Conversion and Management. 2016 Nov 1;**127**:293-300. DOI: 10.1016/j.enconman.2016.09.015

[46] Sathieshkumar N, Sureshkumar LN, Balamurugan R. Performance analysis of hybrid solar photovoltaic thermal collector with nanoparticles. Inten. Journal of Current Research in Engineering and Technology. 2017

[47] Yazdanifard F, Ebrahimnia-Bajestan E, Ameri M. Investigating the performance of a water-based photovoltaic/thermal (PV/T) collector in laminar and turbulent flow regime. Renewable Energy. 2016 Dec 31;**99**:295-306. DOI: 10.1016/j.renene.2016.07.004

[48] Hasan HA, Sopian K, Jaaz AH, Al-Shamani AN. Experimental investigation of jet array nanofluids impingement in photovoltaic/thermal collector. Solar Energy. 2017 Mar 1;**144**:321-334. DOI: 10.1016/j.solener.2017.01.036

[49] Hassan I, Phutthavong P, Abdelgawad M. Microchannel heat sinks: An overview of the state-of-the-art. Microscale thermophysical engineering. 2004 Jan 1;**8**(3):183-205. DOI: 10.1080/10893950490477338

[50] Xu Z, Kleinstreuer C. Computational analysis of nanofluid cooling of high concentration photovoltaic cells. Journal of Thermal Science and Engineering Applications. 2014 Sep 1;**6**(3):031009

[51] Elsheikh AH, Sharshir SW, Mostafa ME, Essa FA, Ali MK. Applications of nanofluids in solar energy: A review of recent advances. Renewable and Sustainable Energy Reviews. 2017 Nov 2. DOI: 10.1016/j.rser.2017.10.108

[52] Rejeb O, Sardarabadi M, Ménézo C, Passandideh-Fard M, Dhaou MH, Jemni A. Numerical and model validation of uncovered nanofluid sheet and tube type photovoltaic thermal solar system. Energy Conversion and Management. 2016 Feb 15;**110**: 367-377. DOI: 10.1016/j.enconman.2015.11.063

[53] Hosseinzadeh M, Kazemian A, Sardarabadi M, Passandideh-Fard M. Experimental investigation of using water and ethylene glycol as coolants in a photovoltaic thermal system. Energy Conversion and Management. 2017;**17**:12-20

[54] Maadi S, Kolahan A, Passandideh Fard M, Sardarabadi M. Effects of Nanofluids thermophysical properties on the heat transfer and 1st law of thermodynamic in a serpentine PVT system. In17th Conference On Fluid Dynamics, fd2017 2017 Aug 27

[55] Al-Shamani AN, Sopian K, Mat S, Hasan HA, Abed AM, Ruslan MH. Experimental studies of rectangular tube absorber photovoltaic thermal collector with various types of nanofluids under the tropical climate conditions. Energy Conversion and Management. 2016 Sep 15;**124**:528-542. DOI: 10.1016/j.enconman.2016.07.052

[56] Kolahan A, Maadi S, Passandideh Fard M, Sardarabadi M. Numerical and experimental investigations on the effect of adding nanoparticles on entropy generation in PVT systems. In17th Conference On Fluid Dynamics, fd2017 2017 Aug 27

[57] Karami N, Rahimi M. Heat transfer enhancement in a PV cell using Boehmite nanofluid. Energy Conversion and Management. 2014 Oct 31;**86**:275-285. DOI: 10.1016/j.enconman.2014.05.037

[58] Lari MO, Sahin AZ. Design, performance and economic analysis of a nanofluid-based photovoltaic/thermal system for residential applications. Energy Conversion and Management. 2017 Oct 1;**149**:467-484. DOI: 10.1016/j.enconman.2017.07.045

[59] Vokas G, Christandonis N, Skittides F. Hybrid photovoltaic–thermal systems for domestic heating and cooling—A theoretical approach. Solar Energy. 2006 May 31;**80**(5): 607-615. DOI: 10.1016/j.solener.2005.03.011

[60] Taylor RA, Phelan PE, Otanicar TP, Walker CA, Nguyen M, Trimble S, Prasher R. Applicability of nanofluids in high flux solar collectors. Journal of Renewable and Sustainable Energy. 2011 Mar;**3**(2):023104

[61] Sardarabadi M, Hosseinzadeh M, Kazemian A, Passandideh-Fard M. Experimental investigation of the effects of using metal-oxides/water nanofluids on a photovoltaic thermal system (PVT) from energy and exergy viewpoints. Energy. 2017 Nov 1;**138**: 682-695. DOI: 10.1016/j.energy.2017.07.046

[62] Michael JJ, Iniyan S. Performance analysis of a copper sheet laminated photovoltaic thermal collector using copper oxide–water nanofluid. Solar Energy. 2015 Sep 30;**119**: 439-451. DOI: 10.1016/j.solener.2015.06.028

[63] Al-Waeli AH, Sopian K, Chaichan MT, Kazem HA, Ibrahim A, Mat S, Ruslan MH. Evaluation of the nanofluid and nano-PCM based photovoltaic thermal (PVT) system: An experimental study. Energy Conversion and Management. 2017 Nov 1;**151**:693-708. DOI: 10.1016/j.enconman.2017.09.032

[64] Sardarabadi M, Passandideh-Fard M, Heris SZ. Experimental investigation of the effects of silica/water nanofluid on PV/T (photovoltaic thermal units). Energy. 2014 Mar 1;**66**:264-272. DOI: 10.1016/j.energy.2014.01.102

[65] Hamdan MA, Kardasi KK. Improvement of photovoltaic panel efficiency using nanofluid. Int. J. of Thermal & Environmental Engineering. 2017;**14**(2):143-151. DOI: 10.5383/ijtee.14.02.008

[66] Sheikholeslami M, Shehzad SA. Numerical analysis of Fe_3O_4–H_2O nanofluid flow in permeable media under the effect of external magnetic source. International Journal of Heat and Mass Transfer. 2018 Mar 31;**118**:182-192. DOI: 10.1016/j.ijheatmasstransfer.2017.10.113

[67] Sheikholeslami M. CuO-water nanofluid flow due to magnetic field inside a porous media considering Brownian motion. Journal of Molecular Liquids. 2018 Jan 1;**249**:921-929. DOI: 10.1016/j.molliq.2017.11.118

[68] Sheikholeslami M. Numerical simulation of magnetic nanofluid natural convection in porous media. Physics Letters A. 2017 Feb 5;**381**(5):494-503. DOI: 10.1016/j.physleta.2016.11.042

[69] Sheikholeslami M. Numerical investigation for CuO-H_2O nanofluid flow in a porous channel with magnetic field using mesoscopic method. Journal of Molecular Liquids. 2018 Jan 1;**249**:739-746. DOI: 10.1016/j.molliq.2017.11.069

[70] Sheikholeslami M, Rokni HB. Simulation of nanofluid heat transfer in presence of magnetic field: A review. International Journal of Heat and Mass Transfer. 2017 Dec 31;**115**:1203-1233. DOI: 10.1016/j.ijheatmasstransfer.2017.08.108

Chapter 3

Thermodynamic Mechanism of Nanofluid Minimum Quantity Lubrication Cooling Grinding and Temperature Field Models

Min Yang, Changhe Li, Yanbin Zhang,
Dongzhou Jia, Runze Li and Wenfeng Ding

Additional information is available at the end of the chapter

http://dx.doi.org/10.5772/intechopen.74969

Abstract

Grinding is an indispensable form of machining, in which, a large amount of heat is transferred into workpiece surface, causing surface burn of the workpiece. Flood grinding is easy to cause pollution to the environment while dry grinding and minimum quantity lubrication (MQL) is insufficient of cooling and lubrication effect. The appearance of nanofluid minimum quantity lubrication cooling (NMQLC) technique can effectively solve the problem of heat transfer in grinding zone and also enhance the lubrication characteristics. In this chapter, NMQLC technique, including nanofluid preparation and atomization is summarized first; then a review on the mechanism of grinding thermodynamics under NMQLC condition is presented based on published literatures. Most of the studies, including investigation of grinding forces and temperatures, indicate that NMQLC has realized a lubrication-cooling effect close to that of flood lubrication. According to existing investigations, theoretical models of temperature field are concluded, heat source distribution model, thermal distribution coefficient model, and heat transfer coefficient model under NMQLC condition are developed, and temperature field control equation are determined. This chapter reviews and amasses the current state of the mechanism of grinding thermodynamics and also recommends ways to precision control the grinding temperature field.

Keywords: nanofluid, minimum quantity lubrication, grinding, thermodynamics mechanism, temperature field, thermal models

1. Introduction

The most significant features of grinding process are high grinding wheel speed and high energy consumption. As abrasive grains on grinding wheel surface in grinding process are usually cut at negative rake, the energy consumed for grinding process to remove unit volume of materials is far greater than other machining forms and most of the energy will be converted into heat quantity [1, 2]. And more than 90% of the heat quantity will be transferred inside grinding wheel and workpiece, reducing the life of the grinding wheel and surface quality of the workpiece [3–5].

Cooling and lubrication grinding processing method applied to industrial production at the earliest is flood grinding, namely, injecting a large quantity of grinding fluid in grinding zone in the form of continuous fluid supply. However, the application of a large amount of grinding fluids results in high cost, and moreover, "air barrier" phenomenon generated surrounding the grinding wheel, preventing grinding fluids from entering the wedge-shaped zone between grinding wheel and workpiece [6–8].

As a kind of green and environmentally friendly processing technology appearing at the earliest, dry processing technology has abolished the application of cutting fluids on the precondition that machining precision and tool service life are guaranteed. However, it is difficult for heat quantity generated during the dry grinding process to be taken away by debris and the main heat quantity will still be transferred into grinding wheel and workpiece.

Minimum quantity lubrication (MQL) grinding processing technology is another green machining technology [9–11]. It refers to a grinding processing technology in which minimum quantity lubricant is mixed and atomized with gases under a certain pressure and then sprayed into grinding zone so as to exert cooling and lubricating effects. However, study shows that heat quantity generated by the grinding process and brought away by high-pressure gases are quite limited [12–14].

In view of flood grinding and cooling properties of MQL, researchers have been anxious to seek for a new cooling and lubricating form to apply to the grinding process. Relevant theories about heat transfer enhancement show that the heat transfer capability of solid is greater than that of liquid and gas [15–17]. On this basis, researchers have put forward nanofluid minimum quantity lubrication cooling (NMQLC) technique which adds a certain amount of nanoscale solid particles into degradable minimum quantity lubricant to generate nanofluids, and then nanofluids will be atomized through high-pressure gas and transferred into grinding zone in way of jet flow [18–22]. Sheikholeslami [23, 24] simulated the natural convection of water-based CuO nanofluid considering Brownian motion. Results showed that the Nusselt number decreases with the increase of Hartmann number and increases with the increase of volume fraction and Rayleigh number of nanofluids. NMQLC grinding process has integrated all advantages of MQL grinding process, solved heat transfer problems of MQL grinding, significantly improved workpiece surface quality and burning phenomenon, lengthened service life of grinding wheel and improved working environment. Therefore, it is a green and environmentally friendly, high-efficiency, and low-energy-consumption grinding processing technology [25, 26].

In this chapter, nanofluid preparation and atomization technique are concluded. A review on the mechanism of grinding thermodynamics under NMQLC condition is presented based on published literatures. According to existing investigations, theoretical models of temperature field precision control are obtained. This chapter reviews the current state of the mechanism of grinding thermodynamics and recommends ways to precision control the grinding temperature field.

2. Nanofluid minimum quantity lubrication cooling technique

2.1. Preparation of nanofluids

Preparation of nanofluids is the precondition for NMQLC grinding and high-quality nanofluids can obtain favorable cooling and lubricating effects. Nowadays, preparation method of nanofluids can be divided into two types: single-step method and two-step method. Single-step nanofluid preparation method is to disperse nanoparticles in base fluid while preparing them, which then saves the problems like nanoparticle collection and storage and can effectively avoid the oxidizing reaction of metal nanoparticles in atmosphere. This method is of high cost and small preparation quantity, and it is not suitable for actual batch application in production [6].

Two-step nanofluid preparation method is to add a certain proportion of nanoparticles in base fluid and select corresponding surface dispersant together with supersonic vibration according to physical and chemical properties so that nanoparticles can be distributed uniformly and stably in base fluid and nanofluid with suspension stability will be formed [27]. The two-step nanofluid preparation method is of simple operation and extensive application and it is nearly applicable to the preparation of all kinds of nanofluids.

2.2. Atomization of nanofluids

When nanofluids are applied to NMQLC grinding, studies on their atomization modes are mainly divided into three types: pneumatic atomization, electrostatic atomization, and ultrasonic atomization.

Pneumatic atomization is a process in which the liquid is sprayed in the machining region after small liquid drops are formed through high-pressure gas atomization and it is the most commonly used method for the atomization of nanofluids. Influenced by external forces like aerodynamic force, the liquid will be split while surface tension of the liquid will make the liquid form spherical liquid drops which are also influenced by viscosity. When external force born by the liquid is greater than liquid surface tension and viscous resistance, force balance of the liquid will be broken and the liquid will be split into fine liquid drops which will form nanofluid drop shapes when they are small enough [28].

Besides high-pressure gases are used for atomization of nanofluids, researchers have also put forward electrostatic atomization and ultrasonic atomization. The formers refer to atomizing the liquid drops under electric charge through high-voltage electrostatic field after nanofluids are sprayed out from the nozzle so as to form electrically charged liquid drop flocks, which will be controllably and orderly transported onto workpiece surface under the effect of electric field force. For ultrasonic atomization, micro shock waves are generated during ultrasonic cavitation and bubble closing process in the liquid to damage the interaction between liquid molecules and then liquid particles are dragged out from liquid surface to form liquid drops [29].

3. A review on the law of grinding thermodynamics

3.1. A review on grinding forces under NMQLC condition

More research has been done on the lubrication effect of the NMQLC technique based on published literatures. A good lubrication in the grinding zone can reduce the grinding wheel/workpiece friction, thereby reducing grinding force and heat generation [30–32]. Manojkumar and Ghosh [33] found that nanofluid could substantially outperform soluble oil in terms of grinding force and G ratio (representing wheel life) under NMQLC. Setti et al. [34] showed that NMQLC could reduce normal force (F_n) on an average of 12 and 28% and reduce F_t on an average of 15 and 27% compared to flood and MQL. Sinha et al. [35] reported that grinding forces, coefficient of friction (μ) can be reduced to the maximum by nanofluids. Kalita et al. [36] measured a decline of 45–50% in force-ratio against grinding with flood cooling and MQL. Shen et al. [37] showed that lubricants with MoS_2 nanoparticles significantly reduce the F_t and the overall grinding performance. Jia et al. [38] found that soybean/castor mixed oil obtained the optimal results and lubricating effect compared to castor oil and other mixed base oils. Zhang et al. [39] reported that the energy ratio coefficient of flood lubrication, MQL, and NMQLC was 36.8, 52.1, and 41.4%, respectively, indicating that NMQLC realized a lubrication cooling effect close to that of flood lubrication. Zhang et al. [40, 41] found that NMQLC grinding using mixed nanoparticles obtain lower grinding force ratios and surface roughness (R_a) values, showing that mixed nanoparticles is superior to pure nanoparticles. Lee et al. [42, 43] carried out micro scale grinding experiments under different lubrication conditions and experimental results demonstrated that NMQLC could significantly reduce the grinding forces and enhance the surface quality. Yang et al. [44] investigated the critical maximum undeformed equivalent chip thickness for ductile-brittle transition ($DBh_{\text{max-}e}$) of zirconia ceramics under different lubrication conditions and found that lubrication condition affects the normal force and ultimately influences the resultant force on workpiece, making $DBh_{\text{max-}e}$ decreases with increasing friction coefficient.

Figure 1(a) and **(b)** show the specific tangential grinding force and μ under dry, flood, MQL, and NMQLC conditions in the investigation of Jia et al. [6]. The maximum specific tangential grinding force and μ were obtained in dry grinding. Compared with dry grinding, the specific F_t under MQL, NMQLC, and flood conditions decreased successively by 45.88, 62.34, and 69.33%, respectively. Therefore, flood grinding realized the optimal lubrication effect, followed by NMQLC. It can also be seen from **Figure 1** that the specific tangential grinding force and μ, which NMQLC realized a lubrication cooling effect close to that of flood lubrication.

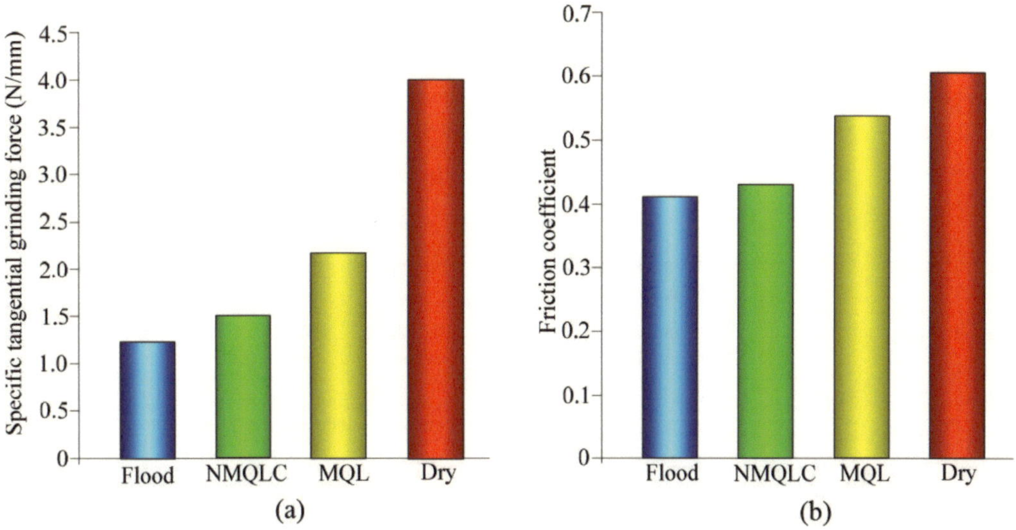

Figure 1. Specific tangential grinding force and frictional coefficient under different lubricating conditions [6]. (a) Specific tangential grinding force. (b) Frictional coefficient.

Table 1 summarizes the details of investigations on nanofluid parameters for the application of NMQLC technique in improving lubrication.

3.2. A review on grinding temperature under NMQLC condition

NMQLC technique was first proposed to solve the problem of inadequate cooling of MQL in grinding [45–48]. Grinding temperature is usually measured by thermocouples. The thermocouples can be placed on the workpiece or on the surface of the grinding wheel. As shown in **Figure 2** [49], a thermocouple forms a junction when the grinding wheel passes over the exposed single pole. The pole is smeared onto the workpiece, thereby forming a junction with the ground surface.

Mao et al. [50] analyzed the effect of nanofluid parameters on grinding performance. It is found that the cooling performance in the grinding zone is improved with the increase of the nanoparticle concentration and nanoparticle diameter. Yang et al. [51] studied the effect of different nanoparticles and concentrations on the temperature field of micro grinding. It was found that different thermal physical properties of nanoparticles have different effects on the temperature field. Yang et al. [52] investigated the dynamic heat flux in micro grinding using different sizes of Al_2O_3 nanoparticles. Results showed that temperatures under NMQLC using nanofluids (30, 50, 70, and 90 nm) are 21.4, 17.6, 16.1, and 8.3% lower, respectively. Li et al. [53] studied the grinding temperature using six types of nanoparticles (MoS_2, ZrO_2,CNT, polycrystalline diamond, Al_2O_3, and SiO_2), and found that CNT nanofluid results in the lowest grinding temperature of 110.7°C with the associated energy proportionality coefficient of 40.1% and the highest heat transfer coefficient of 1.3×10^4 W/(m K).

Authors	Nanoparticle	Particle size (nm)	Base fluid	Concentration
Manojkumar and Ghosh [33]	MWCNT	Not mentioned	Deionized water, soluble oil	0.6, 1 vol.%
Setti et al. [34]	Al_2O_3	40	Water	1 vol.%
Sinha et al. [35]	Ag, ZnO	10 (Ag), 25 (ZnO)	Deionized water	10~30 vol.% (Ag), 0.01~0.5 vol.% (ZnO)
Kalita et al. [36]	MoS_2	70 nm along major axis and 40 nm along minor axis	Paraffin oil	2, 8 wt.%
Shen et al. [37]	MoS_2	70 nm along major axis and 40 nm along minor axis	Paraffin oil, CANMIST oil, soybean oil	5, 20 wt.%
Jia et al. [38]	MoS_2	50	Castor, palm, soybean, peanut, rapeseed, sunflower, maize oil	2, 4, 6, 8, 10 wt.%
Zhang et al. [39]	MoS_2, CNT, ZrO_2	50	Soybean oil	1, 2, 3 vol.%
Zhang et al. [40]	Al_2O_3/SiC	50	Synthetic lipids	6 wt.%
Zhang et al. [41]	MoS_2	50	Palm oil	2 vol.%
Lee et al. [42, 43]	Nanodiamond, Al_2O_3	30, 150	Paraffin oil	2, 4 vol.%
Yang et al. [44]	MoS_2	50	Palm oil	2 vol.%
Jia et al. [6]	MoS_2	50	Soybean oil	6 wt.%

Table 1. Details of investigations on nanofluid parameters for the application of NMQLC in improving lubrication.

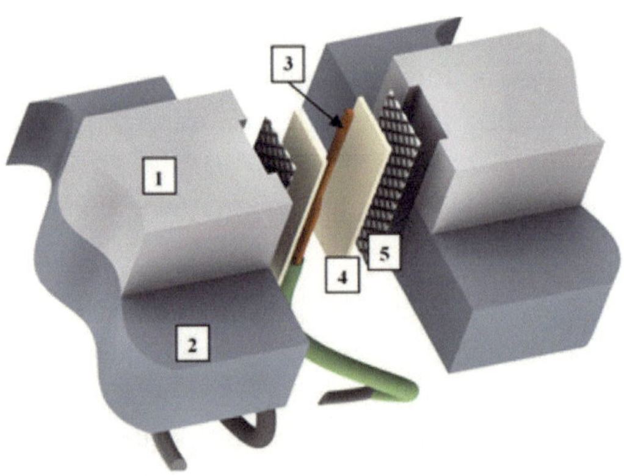

Figure 2. Schematic of single pole configuration thermocouple: where 1-workpiece, 2-workpiece base, 3-thermocouple, 4-mica, 5-varnish layer, 6-grinding wheel [49].

Li et al. [54] analyzed grinding temperature based on the thermal conductivity, viscosity, and contact angle of the nanofluids, and found a lower particle concentration can get a smaller contact angle, thus achieving the optimal heat transfer performance. Lee et al. [55] analyzed the thermal and flow model for the micro-scale grinding process with experiments. Results showed that the grinding temperatures grinding heat flux into the workpiece and grinding energy partition under NMQLC were much lower than those in the cases of compressed air lubrication and pure MQL. Mao et al. [56] investigated the grinding characteristic under different cooling conditions and the results show that NMQLC grinding can significantly reduce the grinding temperature in comparison to pure water MQL grinding as shown in **Figure 3**.

Table 2 summarizes the details of investigations of nanofluid parameters for the application of NMQLC technique in enhanced heat transfer in grinding zone.

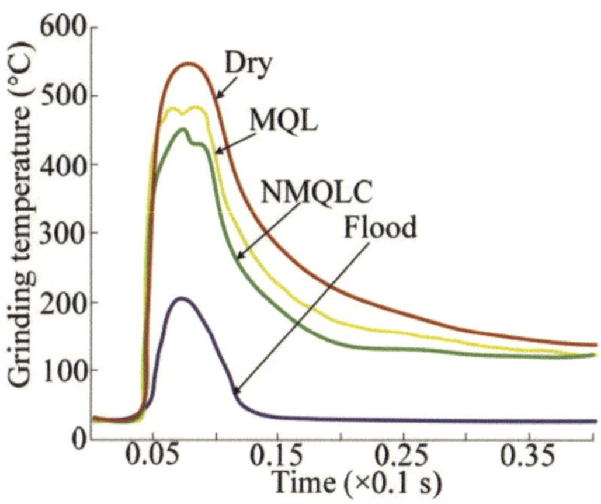

Figure 3. Grinding temperatures under different cooling conditions [56].

Authors	Nanoparticle	Particle size (nm)	Base fluid	Concentration
Mao et al. [50]	Al_2O_3	40, 70, 80	Deionized water, canola oil	1, 3, 5 vol.%,
Yang et al. [51]	Hydroxyapatite, SiO_2, Fe_2O_3, CNTs	Mean particle size is 50 nm, mean length of CNTs is 10–30 μm	Normal saline	2, 4, 6, 8, 10 wt.%,
Yang et al. [52]	Al_2O_3	30, 50, 70, 90	Normal saline	2 vol.%
Li et al. [53]	MoS_2, ZrO_2,CNT, polycrystalline diamond, Al_2O_3, SiO_2	Mean particle size is 50 nm, mean length of CNTs is 10 ~ 30 μm	Palm oil	6 wt.%
Li et al. [54]	CNT	Mean particle size is 50 nm, mean length of CNTs is 10–30 μm	Palm oil	0.5~4 vol.%
Lee et al. [55]	Nanodiamond	30	Paraffin oil	4% vol.%
Mao et al. [56]	Al_2O_3	40	Water	1.2 wt.%

Table 2. Details of investigations for application of NMQLC in enhanced heat transfer in grinding zone.

4. Precise control of temperature field

4.1. Heat source distribution model

In the analysis of temperature fields on workpiece surface, a banding heat source model with continuous equivalent distribution is usually used for heat source in grinding zone to replace the effect of disperse point heat sources to simplify the model [57].

During grinding process, there are three states—scratching, plowing and cutting—between the abrasive grains and workpiece. As shown in **Figure 4** [58], grinding heat sources are under rectangular distribution (heat flux is ξq_w) due to scratching and plowing effects of abrasive grains and triangular distribution (peak heat flux is uq_w) due to cutting effect of abrasive grains, where q_w is the average heat flux transferred into workpiece, and ξ and u are, respectively, heat flux coefficients under rectangular heat source distribution and triangular heat source distribution. Abrasive grains exert scratching and plowing effects in OA segment to generate rectangular heat source distribution, and length is a; they exert cutting effect in AL segment to generate triangular heat source distribution [59]; OL is grinding contact length and its value is l. Point B is the position where peak heat flux of triangular heat sources is located, and OB length is b. On this basis, Zhang and Mahdi [59] established a shape functional equation of triangular heat source distribution:

$$s(x) = \begin{cases} 0 & x \in (-\infty, 0) \\ \xi & x \in (0, a) \\ \dfrac{\xi(b-x) + u(x-a)}{b-a} & x \in (a, b) \\ \dfrac{u(x-1)}{b-l} & x \in (b, l) \\ 0 & x \in (l, +\infty) \end{cases} \tag{1}$$

If abrasive grains are quite pure, there are many abrasive grains exerting scratching and plowing effects under dry grinding state or lubricating performance of grinding fluids is poor, and then comprehensive heat source distribution model is approximate to rectangular heat source

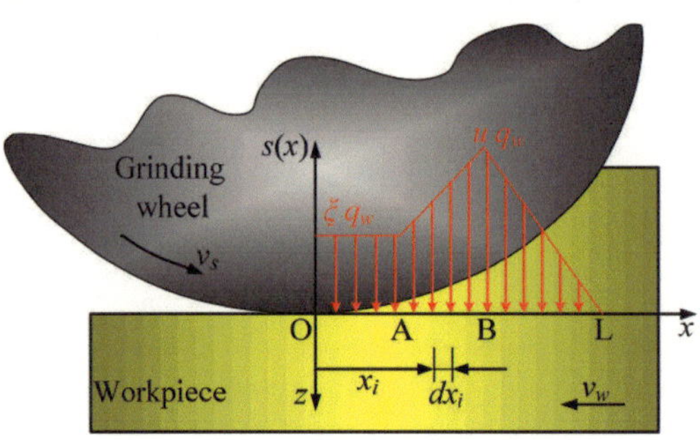

Figure 4. Distribution of heat source [58].

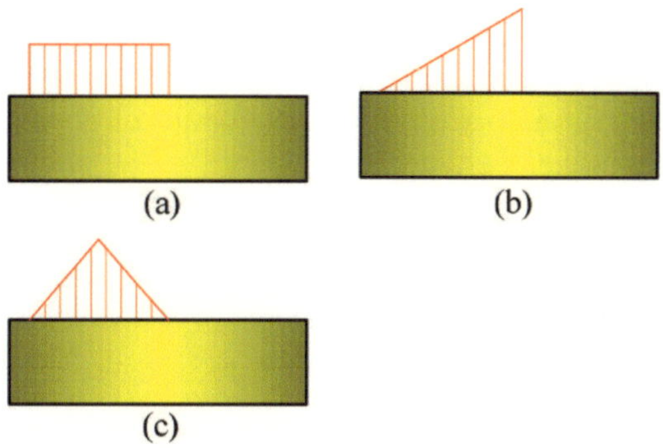

Figure 5. Heat source distribution models in grinding. (a) Rectangular heat source, (b) Right triangular heat source, (c) Isosceles triangular heat source.

distribution model as shown in **Figure 5(a)**. If grinding wheel is dressed very sharp and the lubricating performance of grinding fluids is very good, then abrasive grains mainly exert cutting effect, lengths of scratching and plowing effects are small, and heat source intensity is quite low due to few cutting materials in front end of contact zone. In rear end of contact zone, cutting depth is large, there are many cutting materials and heat source intensity is great, it can be approximate to $a = 0$, $b = l$ and $\xi = 0$, $u = 2$. Comprehensive heat source model can be approximated to right triangular heat source distribution model as shown in **Figure 5(b)**. When $a = 0$, $b = l/2$ and $\xi = 0$, $u = 2$, namely, abrasive grains have strong cutting effect in the middle part of contact zone, generated heat source intensity is great, and comprehensive heat source model can be approximated to isosceles triangular heat source distribution model as shown in **Figure 5(c)**.

4.2. Thermal distribution coefficient

For grinding energies consumed during grinding process, except that a small part of them are consumed on newly generated surface to form needed surface energy, strain energy left on grinding surface layer and kinetic energy for grinding debris to fly out, most part of them are converted into heat energy within contact zone, and these heat energies can be transferred into workpiece, grinding wheel, debris, and grinding fluids in ways of heat conductivity and heat convection. Eqs. (2)–(4) represent the amount of energy transferred into the workpiece (E_w), grinding wheel (E_g), and grinding fluid (E_f), respectively [39]:

$$E_w = \frac{1}{2}\theta_m b' \left[2(\mathrm{k}\rho c)_w v_w l_g\right]^{\frac{1}{2}} \tag{2}$$

$$E_g = \frac{1}{2}\theta_m b' \left[2(k\rho c)_g \left(\frac{A_r}{A}\right)_s v_s l_g\right]^{\frac{1}{2}} \tag{3}$$

$$E_f = \frac{1}{2}\theta_m b' \left[2(k\rho c)_f v_s l_g\right]^{\frac{1}{2}} = \frac{1}{2}\theta_m b' \left\{2\left[\varphi(k\rho c)_n + (1-\varphi)(k\rho c)_f\right] v_s l_g\right\}^{\frac{1}{2}} \tag{4}$$

where θ_m is the maximum temperature rise; k, ρ, c stand for thermal conductivity, density, and specific heat, respectively; the subscript g, w, n, f stand for the properties of grain, workpiece, nanoparticles, and base fluid, respectively; $\frac{A_r}{A}$ is the ratio between the actual and nominal contact areas of grinding wheel and workpiece; v_s is the peripheral speed of grinding wheel; φ is the volume fraction of nanoparticles; b' is the width of grinding wheel; and l_g is the geometric contact arc length.

During grinding, the temperature of the workpiece surface is an important factor to be considered, which is reflected in the thermal distribution coefficient (R) attributed to the workpiece. According to the single abrasive grain model, the amount of heat eliminated by abrasive debris and diffused in convection are too limited to be neglected. During grinding, R can be expressed as follows:

$$R = \frac{E_w}{E_w + E_g + E_f} = \frac{1}{1 + \sqrt{\frac{(k\rho c)_g v_s}{(k\rho c)_w v_w}}\left(\frac{A_r}{A}\right) + \sqrt{\frac{\varphi (k\rho c)_n + (1-\varphi)(k\rho c)_f}{(k\rho c)_w}\frac{v_s}{v_w}}} \tag{5}$$

4.3. Heat transfer coefficient

As NMQLC method can be both heat transfer of normal-temperature gas and boiling heat transfer of nanofluid drop, so this method is the sum of two heat convection methods. According to heat transfer state of single nanofluid drop [60, 61], heat transfer coefficient under NMQLC condition can be solved in three stages: natural convection, nucleate boiling, transition boiling, and film boiling as shown in **Figure 6** [62].

4.3.1. Natural-convection heat transfer stage (I)

When workpiece surface temperature is T_{n1} (lower than boiling point of nanofluid), heat transfer surface will not generate boiling heat transfer and heat transfer enhancement is realized mainly through convective heat transfer of normal-temperature air and convection of nanofluids, mainly being the convective heat transfer of nanofluids [63]. According to Yang's study [28], the nonboiling heat transfer coefficient:

$$h_{n1} = \frac{N_l c_{nf} \rho_{nf} V_l}{\pi {r_{surf}}^2 \cdot t} + h'_a \tag{6}$$

where N_l is the total number of droplets; c_{nf} is the liquid drop specific heat capacity; ρ_{nf} is the density of nanofluids; V_l is the volume of a single droplet; r_{surf} is the spreading radius of a single liquid drop; t is the total time of grinding process; and h'_a is the convective heat transfer coefficient of air at normal temperature.

4.3.2. Nucleate boiling heat transfer and transition boiling heat transfer stages (II and III)

At the end point of nucleate boiling heat transfer and starting point of transition boiling heat transfer, namely, at critical heat flux point, heat transfer coefficient reaches the maximum value h_{n2}. At the end of transition boiling heat transfer stage and in the initial stage of film boiling heat transfer, heat transfer coefficient reaches the minimum value h_{n3} and computational formula of heat transfer coefficient is as follow:

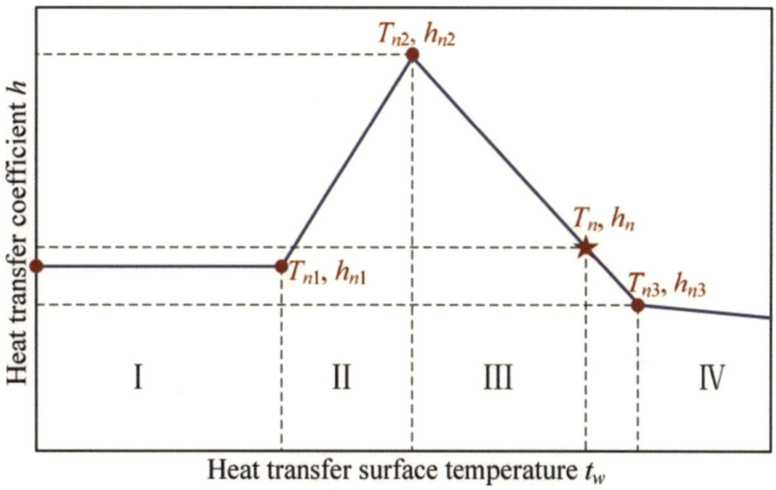

Figure 6. Heat transfer coefficient of grinding surface.

Heat transfer coefficient of critical heat flux density point h_{n2} [64]:

$$h_{n2} = \frac{\left[h_{fa} + c_l\left(T_s - T_{nf}\right)\right]Q'\rho_l}{\pi r_{surf}{}^2\left(T_{n2} - T_{nf}\right)} + h'_a \tag{7}$$

where T_s is the saturation temperature; q'_a is the normal-temperature atmospheric-convection heat transfer quantity; h_{fa} is the latent heat of vaporization; T_{nf} is temperature of nanofluid; c_l is the drop specific heat capacity; and Q' is the nanofluid supply during grinding time.

4.3.3. Transition boiling heat transfer and film boiling heat transfer stages (III and IV)

Computational process of heat transfer coefficient h_{n3} at the end point in transition boiling stage, namely, at starting point in film boiling stage is as follows [65]:

$$h_{n3} = \frac{N_l Q'\rho_l\left[h_{fa} + c_l(T_s - T_l)\right] \cdot \left[0.027e^{\frac{0.08\sqrt{\ln(We/35+1)}}{B^{1.5}}} + 0.21k_d Be^{\frac{-90}{We+1}}\right]}{b' \cdot l \cdot (T_{n3} - T_l)} + h'_a \tag{8}$$

Therefore, the heat transfer coefficient h_n can be obtained by the interpolation calculation when the surface temperature of the workpiece is T_n.

4.4. Temperature field control equation and boundary condition

As shown in **Figure 7**, temperature field model in grinding zone is established and the grinding temperature field can be simplified into 2D heat transfer analysis. Field variable T in transient temperature field meets the equilibrium differential equation of heat conduction [66]:

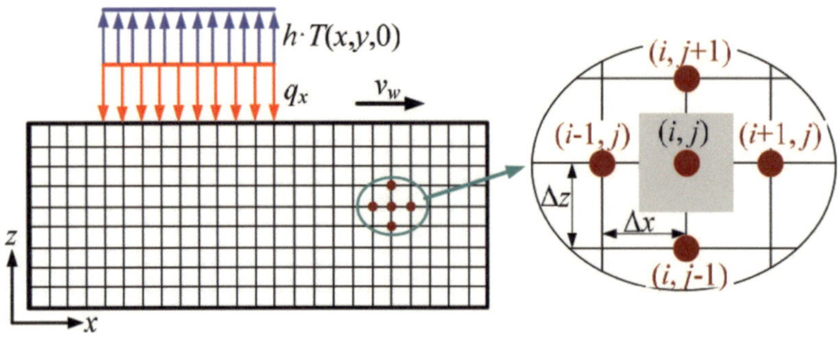

Figure 7. Two-dimensional heat conduction model with nodal network [66].

$$\begin{cases} \frac{\partial^2 T}{\partial x^2} + \frac{\partial^2 T}{\partial z^2} = \frac{1}{\alpha_w}\frac{\partial T}{\partial t} \\ \alpha_w = \frac{k_w}{\rho_w c_w} \end{cases} \tag{9}$$

where α_w is the thermal diffusivity.

Difference in the equation of various nodes in internal grids can be obtained based on (9):

$$T_{t+\Delta t}(i,j) = \left[1 - \frac{2\Delta t(k_x + k_z)}{\rho_w c_w \Delta l^2}\right] T_t(i,j) \cdot \frac{\Delta t\{k_x \cdot [T(i,j+1) + T(i,j-1)] + k_z \cdot [T(i+1,j) + T(i-1,j)]\}}{\rho_w c_w \Delta l^2} \tag{10}$$

As for boundary conditions analysis in grinding zone, coordinate node (i, j) on the workpiece surface is taken as an example. According to energy conservation law [67, 68], the temperature at the node (i, j) after Δt:

$$T_{t+\Delta t}(i,j) = \frac{\Delta t}{\rho_w C_w \Delta l^2}\{k \cdot [T(i-1,j) + T(i+1,j) + T(i+1,j)\text{-}3T(i,j)] + [q - h(T_t - T_a)] \cdot \Delta l\} + T(i,j) \tag{11}$$

4.5. Precise control of temperature field

The temperature field at different times during the steady process can be obtained by solving the difference Eq. (11). **Figure 8** shows the temperature isoline under NMQLC at different times and the corresponding time-space distribution of surface temperature. It can be seen that the grinding process can be divided into three stages, namely, cut-in, steady state, and cut-out [66]:

Cut-in: when abrasive grains start to contact and cut the workpiece, the undeformed chip thickness increases gradually and the heat generated on the grinding interface begins to be transmitted into the workpiece surface.

Steady state: the undeformed chip thickness kept at the mean value and workpiece surface temperature stops increasing. The temperature field reaches the steady state.

Cut-out: the undeformed chip thickness decreases gradually in the cut-out region. According to the theory of heat transfer, the heat conduction in the cut-out region is reduced considering the

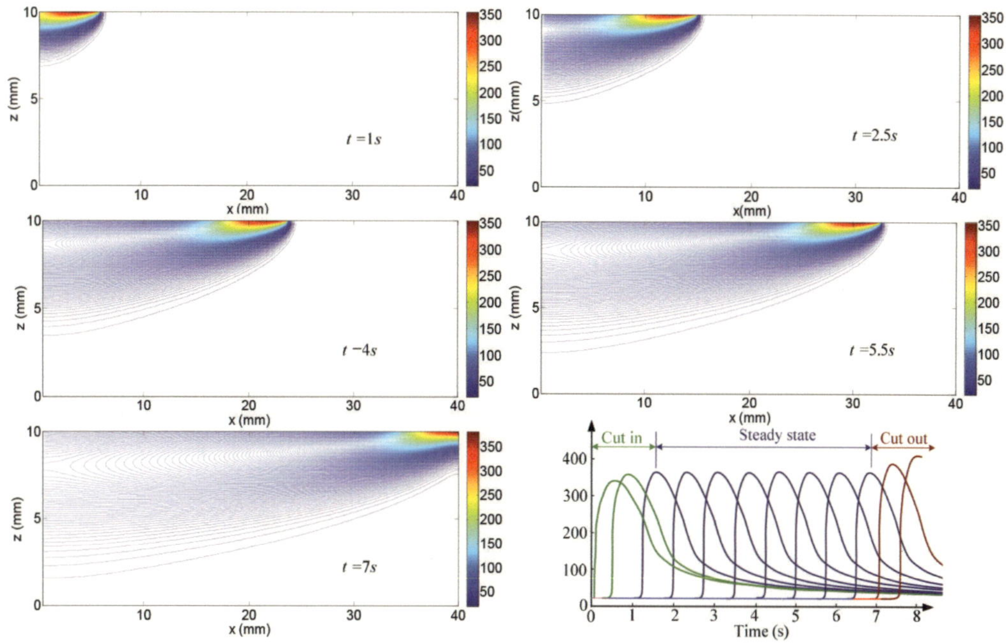

Figure 8. Distributions of the grinding temperature and temperature curves in different states [68].

fixed heat generated on the grinding surface [66]. As the thermal conductivity of the air is very low, more heat is concentrated in the grinding zone, thus increasing the grinding temperature.

5. Conclusions

This chapter has presented a review of published researches in the application of NMQLC during grinding. The following conclusions may be drawn from the present literature review:

1. The amount of nanofluids using NMQLC is very small (7.5–350 mL/h based on published literatures) compared with flood lubrication (usually 60 L/h), so this technique is an environmentally friendly lubrication-cooling method.

2. NMQLC can improve the lubrication condition in grinding area and reduce the friction coefficient effectively, thus reducing the grinding force and specific grinding energy, reducing workpiece surface roughness, and improving the life of grinding wheel.

3. NMQLC can strengthen the heat transfer in the grinding zone that NMQLC could realize a lubrication-cooling effect close to that of flood lubrication.

4. Based on published literatures, nanoparticles that have effective lubrication properties are TiO_2, SiO_2, Al_2O_3, MoS_2, ZnO, and nanodiamond; nanoparticles that have effective cooling properties are CuO, NiO, CNTs, and SiC. A mixed use of nanoparticles with good lubricated properties and nanoparticles with good cooling properties can obtain lower grinding force, grinding temperature, and better surface quality.

5. The lubricating performance of oil-based nanofluid is better than that of water-based nanofluid and the cooling effect is just reverse.

6. Recommendations for future research

The NMQLC technique has been applied by a large number of researchers and can obtain good lubrication and cooling effect. However, there are three points of misgivings based on the summary of available literatures. In order to achieve precision control of the grinding temperature field, more attention is needed to study these misgivings:

1. Most researchers prepare nanofluid using oil or water as base fluid and all can obtain good lubrication or cooling effect close to that of flood lubrication; however, for the addition of nanoparticles into ester, Sridharan and Malkin [69] did not get the ideal result. It is found that when using plain ester oil as base medium, workpiece thermal distortion for NMQLC grinding was reduced by about the same relative amount as the specific energy compared with pure ester oil, which suggests that the addition of nanoparicles to ester oil maybe had no significant effect on workpiece cooling.
2. Li et al. [68] investigated NMQLC grinding temperature using different workpiece materials: 45 steel, Ni based alloy, and cast iron. Experimental results on specific grinding force and grinding temperature revealed that the grinding condition is inapplicable to grinding 45 steel, indicating that the NMQLC method is not applicable to all workpiece materials.
3. Only a few researchers have studied the effect of nanofluid parameters, including the nanoparticle concentration and size, on grinding performance. Some researchers have proved that a higher nanoparticle concentration is more effective for reducing grinding forces [43, 70]. But for the effect of particle size, there is no unified conclusion. The conclusion of Lee [43] and Yang [52] is that the smaller size of nanoparticles is more effective for reducing grinding forces and temperature; Mao [50] found that, when the diameter of nanoparticle increases, tangential grinding force and temperature decreases slightly while the surface finish is deteriorated.

Acknowledgements

This research was financially supported by the National Natural Science Foundation of China (51575290), Major Research Project of Shandong Province (2017GGX30135) and Shandong Provincial Natural Science Foundation, China (ZR2017PEE011 and ZR2017PEE002).

Nomenclature

CNT	carbon nanotubes
EHD	electro-hydro-dynamic

hBN	hexagonal boron nitride
MWCNTs	multi-walled carbon nanotubes
MQL	minimum quantity lubrication
NMQLC	nanofluid minimum quantity lubrication cooling
UAG	ultrasonic-assisted grinding

Author details

Min Yang[1], Changhe Li[1]*, Yanbin Zhang[1], Dongzhou Jia[1], Runze Li[2] and Wenfeng Ding[3]

*Address all correspondence to: sy_lichanghe@163.com

1 School of Mechanical Engineering, Qingdao University of Technology, Qingdao, China

2 Department of Biomedical Engineering, University of Southern California, Los Angeles, CA, USA

3 College of Mechanical and Electrical Engineering, Nanjing University of Aeronautics and Astronautics, Nanjing, China

References

[1] Ding WF, Xu JH, Chen ZZ, et al. Interface characteristics and fracture behavior of brazed polycrystalline CBN grains using Cu-Sn-Ti alloy. Materials Science and Engineering A. 2013;**559**:629-634. DOI: 10.1016/j.msea.2012.09.002

[2] Zhang Y, Li B, Li C, et al. Experimental evaluation of cooling performance by friction coefficient and specific friction energy in nanofluid minimum quantity lubrication grinding with different types of vegetable oil. Journal of Cleaner Production. 2016;**139**:685-705. DOI: 10.1016/j.jclepro.2016.08.073

[3] Guo C, Shi Z, Attia H, et al. Power and wheel wear for grinding nickel alloy with plated CBN wheels. CIRP Annals–Manufacturing Technology. 2007;**56**:343-346. DOI: 10.1016/j.cirp.2007.05.079

[4] Ding W, Zhao B, Xu J, et al. Grinding behavior and surface appearance of (TiCp+TiBw)/Ti-6Al-4V titanium matrix composites. Chinese Journal of Aeronautics. 2014;**27**:1334-1342. DOI: 10.1016/j.cja.2014.08.006

[5] Wang Y, Li C, Zhang Y, et al. Comparative evaluation of the lubricating properties of vegetable-oil-based nanofluids between frictional test and grinding experiment. Journal of Manufacturing Processes. 2017;**26**:94-104. DOI: 10.1016/j.jmapro.2017.02.001

[6] Jia D, Li C, Zhang Y, et al. Experimental research on the influence of the jet parameters of minimum quantity lubrication on the lubricating property of Ni-based alloy grinding. International Journal of Advanced Manufacturing Technology. 2016;**82**:617-630. DOI: 10.1007/s00170-015-7381-y

[7] Ding W, Zhu Y, Zhang L, et al. Stress characteristics and fracture wear of brazed CBN grains in monolayer grinding wheels. Wear. 2015;**332–333**:800-809. DOI: 10.1016/j.wear.2014.12.008

[8] Ding WF, Xu JH, et al. Fabrication and performance of porous metal-bonded CBN grinding wheels;using alumina bubble particles as pore-forming agents. International Journal of Advanced Manufacturing Technology. 2013;**67**:1309-1315. DOI: 10.1007/s00170-012-4567-4

[9] Guo S, Li C, Zhang Y, et al. Experimental evaluation of the lubrication performance of mixtures of castor oil with other vegetable oils in MQL grinding of nickel-based alloy. Journal of Cleaner Production. 2016;**140**:1060-1076. DOI: 10.1016/j.jclepro.2016.10.073

[10] Zhu Y, Ding W, Xu J, et al. An investigation of residual stresses in brazed cubic boron nitride abrasive grains by finite element modelling and raman spectroscopy. Materials & Design. 2015;**87**:342-351. DOI: 10.1016/j.matdes.2015.08.039

[11] Li B, Li C, Zhang Y, et al. Grinding temperature and energy ratio coe cient in MQL grinding of high-temperature nickel-base alloy by using di erent vegetable oils as base oil. Chinese Journal of Aeronautics. 2016;**29**:1084-1095. DOI: 10.1016/j.cja.2015.10.012

[12] Malkin S. Grinding Technology Theory and Applications of Machining with Abrasives. Shenyang: Northeastern University Press; 2002. pp. 9-29

[13] Guo S, Li C, Zhang Y, et al. Analysis of volume ratio of castor/soybean oil mixture on minimum quantity lubrication grinding performance and microstructure evaluation by fractal dimension. Industrial Crops and Products. 2018;**111**:494-505. DOI: 10.1016/j.indcrop.2017.11.024

[14] Li CH. Theory and Key Technology of Nanofluid Minimum Quantity Grinding. Beijing: Science Press; 2018. pp. 91-108

[15] Yang SM, Tao WS. Heat Transfer. Beijing: Higher Education Press; 1998. pp. 33-296

[16] Wang Y, Li C, Zhang Y, et al. Experimental evaluation of the lubrication properties of the wheel/workpiece interface in minimum quantity lubrication (MQL) grinding using different types of vegetable oils. Journal of Cleaner Production. 2016;**127**:487-499. DOI: 10.1016/j.jclepro.2016.03.121

[17] Zhang Y, Li C, Ji H, et al. Analysis of grinding mechanics and improved predictive force model based on material-removal and plastic-stacking mechanisms. International Journal of Machine Tools & Manufacture. 2017;**122**:81-97. DOI: 10.1016/j.ijmachtools.2017.06.002

[18] Sheikholeslami M. Magnetohydrodynamic nanofluid forced convection in a porous lid driven cubic cavity using lattice Boltzmann method. Journal of Molecular Liquids. 2017;**231**:555-565. DOI: 10.1016/j.molliq.2017.02.020

[19] Sheikholeslami M. Magnetic field influence on nanofluid thermal radiation in a cavity with tilted elliptic inner cylinder. Journal of Molecular Liquids. 2017;**229**:137-147. DOI: 10.1016/j.molliq.2016.12.024

[20] Sheikholeslami M. Numerical simulation of magnetic nanofluid natural convection in porous media. Physics Letters A. 2017;**381**:494-503. DOI: 10.1016/j.physleta.2016.11.042

[21] Sheikholeslami M. Influence of Lorentz forces on nanofluid flow in a porous cylinder considering Darcy model. Journal of Molecular Liquids. 2017;**225**:903-912. DOI: 10.1016/j.molliq.2016.11.022

[22] Sheikholeslami M. Influence of Coulomb forces on Fe_3O_4–H_2O nanofluid thermal improvement. International Journal of Hydrogen Energy. 2017;**42**:821-829. DOI: 10.1016/j.ijhydene.2016.09.185

[23] Sheikholeslami M. CuO-water nanofluid free convection in a porous cavity considering Darcy law. The European Physical Journal Plus. 2017;**132**:55. DOI: 10.1140/epjp/i2017-11330-3

[24] Sheikholeslami M, Sadoughi MK. Simulation of CuO-water nanofluid heat transfer enhancement in presence of melting surface. International Journal of Heat and Mass Transfer. 2018;**116**:909-919. DOI: 10.1016/j.ijheatmasstransfer.2017.09.086

[25] Zhang Y, Li C, Jia D, et al. Experimental evaluation of MoS_2 nanoparticles in jet MQL grinding with different types of vegetable oil as base oil. Journal of Cleaner Production. 2015;**87**:930-940. DOI: 10.1016/j.jclepro.2014.10.027

[26] Zhang Y, Li C, Jia D, et al. Experimental study on the effect of nanoparticle concentration on the lubricating property of nanofluids for MQL grinding of Ni-based alloy. Journal of Materials Processing Technology. 2016;**232**:100-115. DOI: 10.1016/j.jmatprotec.2016.01.031

[27] Mao C, Zou H, Zhou X, et al. Analysis of suspension stability for nanofluid applied in minimum quantity lubricant grinding. International Journal of Advanced Manufacturing Technology. 2014;**71**:2073-2081. DOI: 10.1007/s00170-014-5642-9

[28] Yang M, Li C, Zhang Y, et al. Research on microscale skull grinding temperature field under different cooling conditions. Applied Thermal Engineering. 2017;**126**:525-537. DOI: 10.1016/j.applthermaleng.2017.07.183

[29] Mao C, Zou H, Huang X, et al. The influence of spraying parameters on grinding performance for nanofluid minimum quantity lubrication. International Journal of Advanced Manufacturing Technology. 2013;**64**:1791-1799. DOI: 10.1007/s00170-012-4143-y

[30] Wang Y, Li C, Zhang Y, et al. Experimental evaluation on tribological performance of the wheel/workpiece interface in minimum quantity lubrication grinding with different concentrations of Al_2O_3 nanofluids. Journal of Cleaner Production. 2017;**142**:3571-3583. DOI: 10.1016/j.jclepro.2016.10.110

[31] Wang Y, Li C, Zhang Y, et al. Experimental evaluation of the lubrication properties of the wheel/workpiece interface in MQL grinding with different nanofluids. Tribology International. 2016;**99**:198-210. DOI: 10.1016/j.triboint.2016.03.023

[32] Liu G, Li C, Zhang Y, et al. Process parameters optimization and experimental evaluation for nanofluid MQL in grinding Ti-6Al-4V based on grey relational analysis. Materials and Manufacturing Processes. 2017:1-14. DOI: 10.1080/10426914.2017.1388522

[33] Manojkumar K, Ghosh A. Synthesis of MWCNT nanofluid and evaluation of its potential besides soluble oil as micro cooling-lubrication medium in SQL grinding. International Journal of Advanced Manufacturing Technology. 2015;**77**:1955-1964. DOI: 10.1007/s00170-014-6587-8

[34] Setti D, Sinha MK, Ghosh S, et al. An investigation into the application of Al_2O_3 nanofluid-based minimum quantity lubrication technique for grinding of Ti-6Al-4V. International Journal of Precision Technology. 2014;**4**:268-279. DOI: 10.1504/IJPTECH.2014.067742

[35] Sinha MK, Rao PV, Madarkar R, et al. Application of eco-friendly nanofluids during grinding of Inconel 718 through small quantity lubrication. Journal of Cleaner Production. 2017;**141**:1359-1375. DOI: 10.1016/j.jclepro.2016.09.212

[36] Kalita P, Malshe AP, Rajurkar KP. Study of tribo-chemical lubricant film formation during application of nanolubricants in minimum quantity lubrication (MQL) grinding. CIRP Annals - Manufacturing Technology. 2012;**61**:327-330. DOI: 10.1016/j.cirp.2012.03.031

[37] Shen B, Kalita P, Malshe A P, et al. Performance of novel MoS_2 nanoparticles based grinding fluids in minimum quantity lubrication grinding. Transactions of NAMRI/SME. 2008;**36**:357-364

[38] Jia D, Li C, Zhang Y, et al. Specific energy and surface roughness of minimum quantity lubrication grinding Ni-based alloy with mixed vegetable oil-based nanofluids. Precision Engineering. 2017;**50**:248-262. DOI: 10.1016/j.precisioneng.2017.05.012

[39] Zhang D, Li C, Zhang Y, et al. Experimental research on the energy ratio coefficient and specific grinding energy in nanoparticle jet MQL grinding. International Journal of Advanced Manufacturing Technology. 2015;**78**:1275-1288. DOI: 10.1007/s00170-014-6722-6

[40] Zhang X, Li C, Zhang Y, et al. Performances of Al_2O_3/SiC hybrid nanofluids in minimum-quantity lubrication grinding. International Journal of Advanced Manufacturing Technology. 2016;**86**:1-15. DOI: 10.1007/s00170-016-8453-3

[41] Zhang Y, Li C, Jia D, et al. Experimental evaluation of the lubrication performance of MoS_2/CNT nanofluid for minimal quantity lubrication in Ni-based alloy grinding. International Journal of Machine Tools & Manufacture. 2015;**99**:19-33. DOI: 10.1016/j.ijmachtools.2015.09.003

[42] Lee PH, Nam JS, Li C, et al. An experimental study on micro-grinding process with nanofluid minimum quantity lubrication (MQL). International Journal of Precision Engineering and Manufacturing. 2012;**13**:331-338. DOI: 10.1007/s12541-012-0042-2

[43] Lee PH, Nam TS, Li C, et al. Environmentally-friendly nano-fluid minimum quantity lubrication (MQL) meso-scale grinding process using nano-diamond particles. International conference on manufacturing automation. IEEE. 2011:44-49. DOI: 10.1109/ICMA.2010.27

[44] Yang M, Li C, Zhang Y, et al. Maximum undeformed equivalent chip thickness for ductile-brittle transition of zirconia ceramics under different lubrication conditions. International Journal of Machine Tools and Manufacture. 2017;**122**:55-65. DOI: 10.1016/j.ijmachtools.2017.06.003

[45] Sheikholeslami M. CuO-water nanofluid flow due to magnetic field inside a porous media considering Brownian motion. Journal of Molecular Liquids. 2018;**249**:921-929. DOI: 10.1016/j.molliq.2017.11.118

[46] Sheikholeslami M. Numerical investigation of nanofluid free convection under the influence of electric field in a porous enclosure. Journal of Molecular Liquids. 2018;**249**:1212-1221. DOI: 10.1016/j.molliq.2017.11.141

[47] Sheikholeslami M, Rokni HB. Numerical simulation for impact of Coulomb force on nanofluid heat transfer in a porous enclosure in presence of thermal radiation. International Journal of Heat and Mass Transfer. 2018;**118**:823-831. DOI: 10.1016/j.ijheatmasstransfer.2017.11.041

[48] Sheikholeslami M, Shamlooei M, Moradi R. Fe_3O_4-ethylene glycol nanofluid forced convection inside a porous enclosure in existence of Coulomb force. Journal of Molecular Liquids. 2018;**249**:429-437. DOI: 10.1016/j.molliq.2017.11.048

[49] Morgan MN, Barczak L, Batako A. Temperatures in fine grinding with minimum quantity lubrication (MQL). International Journal of Advanced Manufacturing Technology. 2012; **60**:951-958. DOI: 10.1007/s00170-011-3678-7

[50] Mao C, Zhang J, Huang Y, et al. Investigation on the effect of nanofluid parameters on MQL grinding. Materials and Manufacturing Processes. 2013;**28**:436-442. DOI: 10.1080/10426914.2013.763970

[51] Yang M, Li C, Zhang Y, et al. Experimental research on microscale grinding temperature under different nanoparticle jet minimum quantity cooling. Materials and Manufacturing Processes. 2016;**32**:589-597. DOI: 10.1080/10426914.2016.1176198

[52] Yang M, Li C, Zhang Y, et al. Microscale bone grinding temperature by dynamic heat flux in nanoparticle jet mist cooling with different particle sizes. Materials and Manufacturing Processes. 2018;**33**:58-68. DOI: 10.1080/10426914.2016.1244846

[53] Li B, Li C, Zhang Y, et al. Heat transfer performance of MQL grinding with different nanofluids for Ni-based alloys using vegetable oil. Journal of Cleaner Production. 2017; **154**:1-11. DOI: 10.1016/j.jclepro.2017.03.213

[54] Li B, Li C, Zhang Y, et al. Effect of the physical properties of different vegetable oil-based nanofluids on MQLC grinding temperature of Ni-based alloy. International Journal of Advanced Manufacturing Technology. 2017;**89**:3459-3474. DOI: 10.1007/s00170-016-9324-7

[55] Lee PH, Sang WL, Lim SH, et al. A study on thermal characteristics of micro-scale grinding process using nanofluid minimum quantity lubrication (MQL). International Journal of Precision Engineering and Manufacturing. 2015;**16**:1899-1909. DOI: 10.1007/s12541-015-0247-2

[56] Mao C, Tang X, Zou H, et al. Investigation of grinding characteristic using nanofluid minimum quantity lubrication. International Journal of Precision Engineering and Manufacturing. 2012;**13**:1745-1752. DOI: 10.1007/s12541-012-0229-6

[57] Jaeger JC. Moving sources of heat and the temperature of sliding contacts. Journal of the Royal Society of New South Wales. 1942;**76**:203-224

[58] Mao C. Study of Temperature Field and Thermal Damage in Surface Grinding. Hunan University; 2008

[59] Zhang L, Mahdi M. Applied mechanics in grinding-IV. The mechanism of grinding induced phase transformation. International Journal of Machine Tools & Manufacture. 1995;**35**:1397-1409. DOI: 10.1016/0890-6955(95)93590-3

[60] Sheikholeslami M. Numerical investigation for CuO-H_2O nanofluid flow in a porous channel with magnetic field using mesoscopic method. Journal of Molecular Liquids. 2018;**249**:739-746. DOI: 10.1016/j.molliq.2017.11.069

[61] Sheikholeslami M, Shehzad SA. Numerical analysis of Fe_3O_4-H_2O nanofluid flow in permeable media under the effect of external magnetic source. International Journal of Heat and Mass Transfer. 2018;**118**:182-192. DOI: 10.1016/j.ijheatmasstransfer.2017.10.113

[62] Sheikholeslami M, Rokni HB. Simulation of nanofluid heat transfer in presence of magnetic field: A review. International Journal of Heat and Mass Transfer. 2017;**115**:1203-1233. DOI: 10.1016/j.ijheatmasstransfer.2017.08.108

[63] Sheikholeslami M. Influence of magnetic field on nanofluid free convection in an open porous cavity by means of lattice Boltzmann method. Journal of Molecular Liquids. 2017; **234**:364-374. DOI: 10.1016/j.molliq.2017.03.104

[64] Sheikholeslami M. CVFEM for magnetic nanofluid convective heat transfer in a porous curved enclosure. The European Physical Journal Plus. 2016;**131**:413. DOI: 10.1140/epjp/i2016-16413-y

[65] Mao C, Zou H, Huang Y, et al. Analysis of heat transfer coefficient on workpiece surface during minimum quantity lubricant grinding. International Journal of Advanced Manufacturing Technology. 2013;**66**:363-370. DOI: 10.1007/s00170-012-4330-x

[66] Shen B, Shih AJ, Xiao G. A heat transfer model based on finite difference method for grinding. Journal of Manufacturing Science and Engineering. 2011;**133**:031001. DOI: 10.1115/1.4003947

[67] Zhang Y, Li C, Jia D, et al. Experimental evaluation of the workpiece surface quality of MoS_2/CNT nanofluid for minimal quantity lubrication in grinding. Journal of Mechanical Engineering. 2018;**54**:161-170. DOI: 10.3901/JME.2018.01.161

[68] Li B, Li C, Zhang Y, et al. Numerical and experimental research on the grinding temperature of minimum quantity lubrication cooling of different workpiece materials using vegetable oil-based nanofluids. International Journal of Advanced Manufacturing Technology. 2017;**93**:1971-1988. DOI: 10.1007/s00170-017-0643-0

[69] Sridharan U, Malkin S. Effect of minimum quantity lubrication (MQL) with nanofluids on grinding behavior and thermal distortion. Transactions of NAMRI/SME. 2009;**37**:629-636

[70] Molaie MM, Akbari J, Movahhedy MR. Ultrasonic assisted grinding process with minimum quantity lubrication using oil-based nanofluids. Journal of Cleaner Production. 2016;**129**:212-222. DOI: 10.1016/j.jclepro.2016.04.080

Chapter 4

Biological Particle Control and Separation using Active Forces in Microfluidic Environments

Mohd Anuar Md Ali,
Aminuddin Bin Ahmad Kayani and
Burhanuddin Yeop Majlis

Additional information is available at the end of the chapter

http://dx.doi.org/10.5772/intechopen.75714

Abstract

Exploration of active manipulation of bioparticles has been impacted by the development of micro-/nanofluidic technologies, enabling evident observation of particle responses by means of applied tunable external force field, namely, dielectrophoresis (DEP), magnetophoresis (MAG), acoustophoresis (ACT), thermophoresis (THM), and optical tweezing or trapping (OPT). In this chapter, each mechanism is presented in brief yet concise, for broad range of readers, as strong foundation for amateur as well as brainstorming source for experts. The discussion covers the fundamental mechanism that underlying the phenomenon, presenting the theoretical and schematic description; how the response being tuned; and utmost practical, the understanding by specific implementation into bioparticles manipulation engaging from micron-sized material down to molecular level particles.

Keywords: microfluidics, nanofluidics, dielectrophoresis, magnetophoresis, acoustophoresis, thermophoresis, optical tweezing, optical trapping

1. Introduction

Progress in biomedical technologies has emerged into miniaturization of biomedical devices. The main features in miniaturized biomedical devices are establishment of controlled microenvironment that promotes predictive micro-/nanoparticle behaviors and reduction in required sample volume for characterization of scarce materials, e.g., patient-derived samples, which help to reduce the cost and time for diagnosis and therapy [1]. Miniaturization of these biomedical devices demands for implementation of particular procedures in precise approach, which has been extensively studied in micro-/nanofluidic system integrated with active manipulation mechanisms [2–4].

Micro-/nanofluidic system facilitates researchers in creating well-controlled micro-/nanoscale environment and at the same time enables the analysis of micro-/nanoparticles including biological particle (bioparticle) behaviors and responses toward active manipulation mechanisms, in addition to particle-particle reactions and external stimuli [5]. Active manipulation mechanisms make possible the control of bioparticle displacement and motional trajectories in a highly predictable and consistent fashion [2], by introducing tunable external force systems such as dielectrophoresis (DEP) [6, 7], magnetophoresis (MAG) [8], acoustophoresis (ACT) [9], thermophoresis (THM) [10], and/or optical tweezing/trapping (OPT) [11, 12].

In this chapter, description of the fundamental mechanism underlying the phenomenon is presented, covering the theoretical and schematic description, as well as specific implementation into bioparticle manipulation covering from micron-sized material down to molecular-level particles. Conclusion and future perspectives of this multidisciplinary field are provided at the end of this chapter.

2. Bioparticles in biomedical studies

Manipulation of bioparticles has been a major concern in recent development of micro−/nanofluidic studies due to their potential in biomedical application. Those particles, according to their biological structure and physical properties, can be categorized as (1) model organisms; (2) body cells, which include blood cells, tumor and cancer cells, and stem or progenitor cells; (3) bacteria; (4) viruses; (5) nucleic acids; and (6) proteins.

Model organisms, either unicellular (e.g., yeast) or multicellular (e.g., *Danio rerio* zebrafish, *Caenorhabditis elegans* nematode, etc.), are being used for cellular process studies (cell cycle, cell division, metabolism, etc.), genetic and pharmacological studies, as well as pathogenesis and therapy studies. Blood which composed by two components, i.e., (1) blood plasma and (2) the formed elements, including red blood cells (RBCs), white blood cells (WBCs), and platelets, is of high interest as they are being used in various health check and clinical tests, such as blood test to determine whether our organs (e.g., kidneys, liver, thyroid, heart, etc.) are working properly and to diagnose any disease such as HIV/AIDS, cancer, anemia, diabetes, and coronary heart disease. Tumor cells (benign) and cancer cells (malignant) are cells that undergo uncontrolled proliferation, causing tumor development either at the skin, colon, rectum, prostate, breasts, or lungs. They are grouped either as (1) carcinomas (develop in epithelial cells), (2) leukemia and lymphomas (develop at the blood and lymphatic system), or (3) sarcomas (develop at the connective tissue). These cells attempt to develop secondary tumor (metastasis tumor) by spreading through vascular or blood vessel network in the form of circulating tumor cells (CTCs). Stem cells, either adult stem cells or pluripotent stem cells, are cells that can self-renew, generating perfect copies of themselves by division and differentiation, where the produced cells are specific for certain function in the body. These cells are important in research, drug screening test, and cell transplantation therapy. Bacteria, which are unicellular prokaryotic microorganisms, grow and reproduce rapidly through binary fission. They are important in biomedical studies in the development of disease diagnostic tools and in understanding of biological responses under certain stimuli, according to their actions, either as predators, mutualists, or pathogens, in which pathogenic bacteria demonstrate a parasitic association with other organisms to cause infections. Viruses, which are genes enclosed by a protective coat, are infective agents due to lack of metabolic machinery, hence depending on the host for gene expression. They are important in the study of virus detection as well as the study of virus influence to cells. Nucleic acids, which are biomolecules made from nucleotides as monomers, function in encoding, transmitting, and expressing genetic information. There are two types of nucleic acids in living cells, which are deoxyribonucleic acid (DNA) and ribonucleic acid (RNA). They are highly essential in the research of replication, repair, storage and modification of DNA, disease biomarkers, and gene delivery. Proteins are biomolecules constructing a single or a number of long chains of amino acid residues. The discovery of their role within living organisms includes catalyzing metabolic reactions, DNA replication, responding to stimuli and transporting molecules from one location to another, and getting remarkable attention in the biomedical research and application these days.

3. Dielectrophoresis

Dielectrophoresis (DEP) is the motion of polarizable particles under a spatially nonuniform electric field that cause momentary polarization of the particle by dipole establishment within, with an unequal Columbic forces at both ends of the particles, causing the particles to move [2, 6, 7].

3.1. Fundamentals of DEP

Dielectrophoretic force, F_{DEP}, for a stationary alternative current (AC) field is given by

$$F_{DEP} = 2\pi a^3 \varepsilon_m \mathrm{Re}[f_{CM}] \nabla |E|^2 \tag{1}$$

where E is the electric field, ε_m is the permittivity of the suspending medium, a is the particle radius, f_{CM} is the Clausius-Mossotti factor which describes relationship between dielectric constants of two different media, and the $\mathrm{Re}[f_{CM}]$ is the real part of the factor [7].

For a spherical particle, f_{CM} is governed by

$$f_{CM} = \frac{\varepsilon_p^* - \varepsilon_m^*}{\varepsilon_p^* + 2\varepsilon_m^*} \tag{2}$$

where ε^* is the complex permittivity, which is determined by

$$\varepsilon_{p,m}^* = \varepsilon_0 \varepsilon_{p,m} - j\frac{\sigma_{p,m}}{\omega} \tag{3}$$

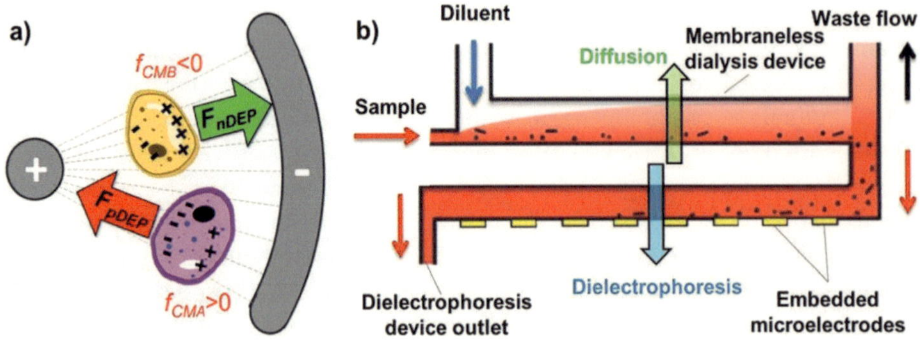

Figure 1. (a) Dielectrophoresis. At a certain frequency, particle A (purple) experiences higher polarizability than the suspending medium ($f_{CMA} > 0$), attracting it to the region with higher electric field gradient, while at the same time particle B (yellow) experiences less polarizability than the suspending medium ($f_{CMB} < 0$), pushing it away to the region with lower electric field gradient (F_{pDEP}, positive dielectrophoretic force; F_{nDEP}, negative dielectrophoretic force) (Reprinted with permission from Md Ali et al. [13]. Copyright 2016 Royal Society of Chemistry). (b) Manipulation of bacteria by dielectrophoresis. Blood specimen mixed with a permeabilizing agent sample is injected into the microfluidic device for desalination process by membraneless dialysis before subsequent enrichment of target bacteria, *E. coli*, at dielectrophoretic electrodes (Adapted with permission from D'Amico et al. [14]. Copyright 2017 Royal Society of Chemistry).

where ε_0 is the vacuum permittivity (8.854 × 10^{-12} F/m, j is the unit imaginary number, i.e., $\sqrt{(-1)}$, σ is the conductivity, and ω is the angular frequency of electric signal supplied. The subscripts "*p*" and "*m*" stand for the particle and the medium, respectively.

The dielectrophoretic force direction is determined by the sign of the f_{CM}, as presented in **Figure 1a**. For f_{CM} with positive sign ($f_{CM} > 0$), the force pulls particles to the region with high electric field gradients, and the situation is termed as positive DEP (pDEP). While for f_{CM} with negative sign ($f_{CM} < 0$), the force pushes particles away from those regions, and this is termed as negative DEP (nDEP). Hence, the dielectrophoretic force is proportional to particle volume and is highly dependent on the electrical properties of the particle, the medium, and the frequency of the AC field [15].

Moreover, for an AC field with spatial variation, the dielectrophoretic force is given by

$$F_{DEP} = 2\pi\varepsilon_m \mathrm{Re}[f_{CM}]\, a^3 \nabla |E|^2 + 2\pi\varepsilon_m \mathrm{Im}[f_{CM}]\, a^3 \times \left(|E_i|^2 \nabla \varphi_i\right) \quad (4)$$

where Re[f_{CM}] and Im[f_{CM}] are the Clausius-Mossotti real part and imaginary part, respectively; φ is the AC field phase, and subscript "*i*" represents each element of the field as well as the phase [15]. AC field with spatial variation is being employed in (1) traveling wave DEP and (2) electrorotation.

In general, the traveling wave DEP is created by application of 90° phase-shifted electric signal, i.e., 0, 90, 180, and 270°, on an array of planar parallel electrodes, causing generation of a traveling wave of electrostatic potential which can vertically suspend a lossy dielectric sphere while at the same time propels it along the array. The Re[f_{CM}] determines whether the bioparticles are levitated (nDEP) or attracted to the electrodes (pDEP), while Im[f_{CM}] determines the translational movement in direction of electrode array. The electrorotation employs quadrupole electrodes with 90° phase-shifted electric signal. The electrodes generate a rotating electric field when excited with this multiphase AC field. Electrorotation is achieved by nDEP ($Re[f_{CM}] < 0$) for bioparticle levitation during the rotational motion which is determined by $Im[f_{CM}]$ [16].

As bioparticles are typically multilayered with the presence of multilayer membranes, Clausius-Mossotti factor calculation needs to consider total permittivity of the bioparticle comprising of the permittivity of all layers. Total permittivity, ε_p, for a spherical multilayered particle is given by

$$\varepsilon_p = \varepsilon_{p_n}\left[\left(\frac{a_n}{a_{n-1}}\right)^3 + 2\left(\frac{\varepsilon_{p_{n-1}} - \varepsilon_{p_n}}{\varepsilon_{p_{n-1}} + 2\,\varepsilon_{p_n}}\right)\right] \Big/ \left[\left(\frac{a_n}{a_{n-1}}\right)^3 - \left(\frac{\varepsilon_{p_{n-1}} - \varepsilon_{p_n}}{\varepsilon_{p_{n-1}} + 2\,\varepsilon_{p_n}}\right)\right] \tag{5}$$

where ε_{p_n} is the permittivity that includes the outermost layer of the bioparticle, while $\varepsilon_{p_{n-1}}$ is permittivity that excludes the outermost layer. a_n is the outermost layer (e.g., cell wall) radius, and a_{n-1} is the second outermost layer (e.g., membrane) radius. The denotation $n = 0, 1, 2, 3\ldots$ is the corresponding layer calculation number that starts from the innermost layer ($n = 0$). The total permittivity, ε_p, is the final layer calculated permittivity, ε_{p_n}, in which n is the final layer number [16].

Transition between the pDEP and nDEP responses of bioparticle happens across the point when the polarization of the particle and the suspending medium are equal, which occurs at a particular frequency known as crossover frequency, f_{xo}. For a spherical structure, f_{xo} is governed by

$$f_{xo} = \frac{\sqrt{2}\,\sigma_m}{2\pi a\, C_m} \tag{6}$$

where σ_m is the conductivity of the surrounding medium, a is the particle radius, and C_m is the capacitance of the bioparticle plasma membrane [17].

3.2. Dielectrophoretic manipulation of bioparticles

Manipulation of model organism has been demonstrated by Chen et al. [18], who perform detection and trapping of *Shewanella oneidensis*, a model organism for electrochemical activity bacteria, using DEP microfluidic chip, which equipped with real-time imaging for trapping process observation. The trapping is enhanced by providing hole arrays on top of the electrode, in which the hole size-bacteria count relation is also studied. Sang et al. [19] demonstrate blood manipulation using DEP. A portable microsystem for separation of living/dead RBCs by DEP which integrated with subsequent evaluation using surface stress sensor for living/dead RBC detection has been developed. This multifunctional portable microsystem is of potential application into diagnostic of hemolytic anemia disease, i.e., the situation where RBCs are damaged and removed from the bloodstream before their normal lifespan is over. Chiu et al. [20] use optically induced DEP (ODEP) to enhance conventional microfluidic isolation and purification of CTCs from whole blood sample. In this system, light image is projected on photoconductive materials which coated on indium tin oxide (ITO) glass substrate to generate dielectrophoretic force field. Hollow circular images are used for targeted CTC trapping, while long rectangular light bar is used to attract other untargeted cells. Adams et al. [21] perform sorting of neural stem and progenitor cells using microfluidic DEP device. The sorting is based on the cell membrane capacitance variation, which specifically determines the future forming cells, either neuron or astrocyte. Bacteria manipulation using DEP microfluidic device is shown by D'Amico et al. [14], in which they isolate and enrich *Escherichia coli* (*E. coli*) which spiked into whole blood sample. The device integrates membraneless dialysis process and dielectrophoretic trapping, presented in **Figure 1b**. Ding et al. [22] demonstrate the capture and enrich of *Sindbis* virus in a gradient insulator-based DEP microfluidic device, i.e., electric signal source is applied across both microchannel ends while insulating polydimethylsiloxane (PDMS) used to distort electric field, hence forming a nonuniform electric field. By tuning the voltage of applied signal, they can capture or release the virus from the saw-shaped electrode tip. Nucleic acid manipulation has been conducted by Jones et al. [23], in which they perform size-based sorting of a wide range of nucleic acid analytes, i.e., 1.0, 10.2, 19.5, and 48.5 kbp double-stranded DNA (dsDNA) analytes, including both plasmid and genomic DNA in a continuous flow microfluidic platform, by using an insulator-based DEP to tune the deflection of the nucleic acids. Viefhues et al. [24] perform dielectrophoretic mobility shift assay of DNA complexes as well as pure DNA in a nanofluidic DEP system to demonstrate new technique in detection of different DNA variants, including protein-DNA complexes. Manipulation of proteins has been achieved by Mohamad et al. [25]. They use DEP to capture and characterize the electrical properties of colloidal protein molecule, i.e., bovine serum albumin (BSA), based on the molecule dispersion impedance exhibited at particular frequencies, which are influenced by the electrical double layer surrounding the molecule. Liao et al. [26] developed a DEP nanofluidic device to perform selective pre-concentration of functional proteins within bio-fluid medium, which in general is a high ionic strength medium. The device is advantageous in the enhancement of DEP trapping forces against electrothermal flow which is challenging in nanoscale device design.

4. Magnetophoresis

Magnetophoresis (MAG) is the motion of particles under the influence of a nonuniform magnetic field, as the particles being magnetized cause them to be attracted toward the regions of high magnetic flux density or repelled away [8]. Magnetic field is generated by either a permanent magnet or an electromagnetic coil.

4.1. Fundamentals of MAG

Magnetophoretic force experienced by a particle is governed by

$$F_{MAG} = \frac{(\chi_p - \chi_m)\, V_p}{\mu_0}(B \cdot \nabla)B \tag{7}$$

where F_{MAG} is the effective magnetic force upon the particle; χ_p and χ_m are the magnetic susceptibility of the particle and the medium, respectively; V_p is the particle volume, B is the magnetic flux density, and μ_0 is the free space permeability [8].

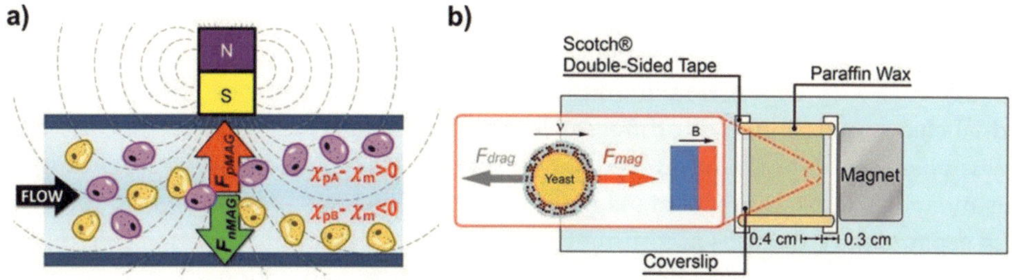

Figure 2. (a) Magnetophoresis. Bioparticles A (purple) are pushed toward the region with maximum magnetic flux density (pMAG) due to positive magnetic susceptibility difference between the bioparticles and the suspending medium (χ_{pA}-χ_m > 0), while bioparticles B (yellow) experience the opposite (χ_{pB}-χ_m < 0), thus being repelled away from the region (F_{pMAG}, positive magnetophoretic force; F_{nMAG}, negative magnetophoretic force) (Reprinted with permission from Md Ali et al. [13]. Copyright 2016 Royal Society of Chemistry). (b) Manipulation of model organism by magnetophoresis. Multilayered magnetic silica film-coated yeast is manipulated in microfluidic device equipped with external permanent magnet (Reprinted with permission from Lee et al. [27]. Copyright 2016 Springer Nature).

The motional direction is controlled by the difference of magnetic susceptibility between the particle and the medium, i.e., χ_p-χ_m. For a positive magnetic susceptibility difference (χ_p-χ_m > 0), the suspended particles are pushed to region with maximum magnetic field flux gradient, and this situation is known as positive MAG (pMAG). In contrast, under negative magnetic susceptibility difference (χ_p-χ_m < 0) situation, the particles are pushed to regions with minimum magnetic field flux gradient, and this is called as negative MAG (nMAG). The conceptual mechanism is shown in **Figure 2a**.

Common practices in magnetophoretic manipulation of bioparticle employ either (1) immunomagnetic manipulation [28] or (2) diamagnetic manipulation [29], in which paramagnetic bounded to bioparticle is exploited in the former, while in the latter, paramagnetic- or ferrofluid-suspending medium is utilized. In immunomagnetic bioparticle manipulation, paramagnetic micro-/nanoparticles, e.g., iron oxide microparticles and streptavidin paramagnetic particles, which have higher susceptibility compared to suspending medium are used. Target bioparticles are bounded to paramagnetic micro-/nanoparticles through antibodies, benefiting from binding affinity with bioparticles. Under influence of magnetic field, the microparticle-bioparticle complexes can be manipulated. While in diamagnetic bioparticle manipulation, a suspending medium with higher magnetic susceptibility compared to target bioparticle is utilized. In this method, the magnetic field manipulates the suspending medium rather than the bioparticles themselves.

4.2. Magnetophoretic manipulation of bioparticles

Lee et al. [27] magnetically functionalize living yeast cells, *Saccharomyces cerevisiae* (*S. cerevisiae*), as micro-magnets, by performing coating of several groups of them with different thicknesses of magnetic silica film (single layer up to seven layers), to control the magnetization degree of the cells as shown in **Figure 2b**. By doing so, multiple subgroups are being formed which can be manipulated independently in the pool of heterogeneous cell mixtures. Magnetophoretic separation of erythrocytes, i.e., maturing RBCs from the mixture with reticulocytes and immature RBCs, in hematopoietic stem cell (HSC) culture, has been performed by Jin et al. [30]. They exploit the paramagnetic property of deoxygenated hemoglobin of maturing erythrocytes compared to reticulocyte which is diamagnetic for the MAG-based separation. Jack et al. [31] demonstrate immunomagnetic isolation of diverse groups of magnetic bead-labeled tumor cells (with different levels of labeling), based on the surface protein expressions. The sample is introduced to multiple levels of magnetic sorter according to the adjustment of external permanent magnet distance to flowing magnetic-labeled cells, resulting in separation of the cells according to their epithelial cell adhesion molecule (EpCAM) levels into low, moderate, and high expression. Bacteria manipulation using MAG has been achieved by Wang et al. [32], who demonstrate the trapping of *Bacillus megaterium*, which are nonmagnetic bioparticles in a ferrofluid suspension while experiencing uniform external magnetic field. The sample continuously flows in a microfluidic channel with nonmagnetic island located in the middle of the channel. Due to the magnetic susceptibility difference between the island and the surrounding ferrofluid, the nonmagnetic bacteria experience magnetophoretic force and, thus attract the bacteria to the island, while the ferrimagnetic particle which comprises the ferrofluid is also attracted to the island; however, the bacteria and the magnetic particles are accumulated at different regions. Wang et al. [33] perform magnetophoretic concentration of avian influenza virus (H5N1) in continuous flow microfluidic system. The H5N1 magnetic nanoparticle complexes are formed using aptamer-biotin-streptavidin binding and injected to three-dimensionally printed magnetophoretic platform which experiences magnetic field from external neodymium magnets, to migrate the complexes from the original sample flow to phosphate-buffered saline carrier flow. Shim et al. [34] demonstrate the magnetophoretic capture of DNA-conjugated magnetic particles in microfluidic device by short-range magnetic field gradient exerted by micro-patterned nickel array on the bottom surface of the separation channel, as well as enhancement with oppositely oriented array of external permanent magnet for a long-range magnetic field gradient at the interfaces between magnets. The DNA is then collected by performing detachment from the captured DNA magnetic particle conjugates by enzymatic reaction with uracil-specific excision reagent enzyme. Magnetophoretic manipulation of protein has been achieved by Lim et al. [35] in which manipulation of Atto-520 biotin-streptavidin-magnetic particle conjugates is demonstrated. The magnetic force is applied by external rotating magnetic field, while soft permalloy ($Ni_{80}Fe_{20}$) magnetic tracks composed of radii and spiral tracks, known as spider web network, are patterned underneath the manipulation plane. The structure facilitates the directional transportation of magnetic particles, which is of potential application as biomolecule cargo, to the desired path, either converging to the spider web center or dispersing away, according to the external rotating magnetic field, i.e., clockwise or counterclockwise.

5. Acoustophoresis

Acoustophoresis (ACT) is the motion of particles when experiencing a surface acoustic wave (SAW) radiation pressure, either by standing surface acoustic wave or traveling surface acoustic wave [36, 37].

5.1. Fundamentals of ACT

Acoustophoretic force, F_{ACT}, experienced by a particle is determined by.

$$F_{ACT} = 4\pi a^3 E_{ACT} k \sin(2kx)\phi \tag{8}$$

where F_{ACT} is the acoustic radiation force, E_{ACT} is the acoustic energy density, a is the particle radius, x is the distance from pressure antinode in the wave propagation axis, k is the wave number ($2\pi f/c_0$), and ϕ is the acoustic contrast factor [9].

The direction of the particle motion, either toward the pressure node or the antinode, is governed by the sign of the acoustic contrast factor, ϕ, given by

$$\phi = \frac{\rho_p + \frac{2}{3}(\rho_p - \rho_m)}{2\rho_p + \rho_m} - \frac{1}{3}\frac{\rho_m c_m^2}{\rho_p c_p^2} \tag{9}$$

where ρ_p and ρ_m are the density of the particle and medium, respectively, while c_p and c_0 are the speed of sound within the particle and medium, respectively. As shown in **Figure 3a**, for a condition with positive acoustic contrast factor (ϕ>0), the particles are pushed toward the pressure node, and the phenomenon is known as positive ACT (pACT). In contrast, for a negative acoustic contrast factor condition (ϕ<0), the particles are pushed toward the pressure antinode, and it is called as negative ACT (nACT).

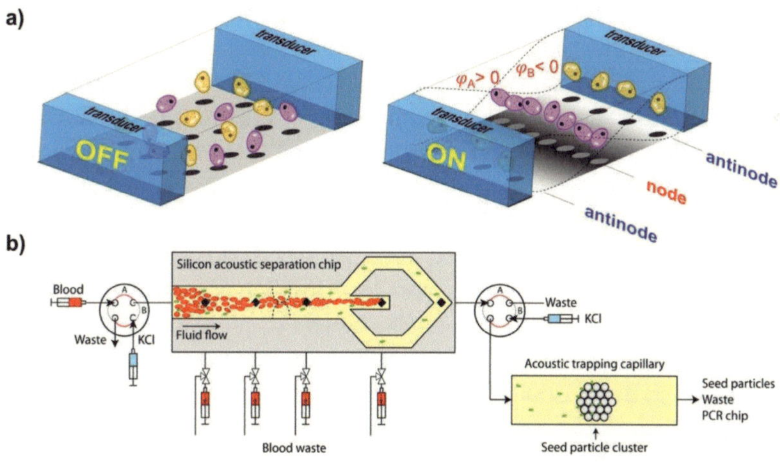

Figure 3. (a) Acoustophoresis. (Left) Bioparticles are randomly dispersed in the suspending medium without the presence of acoustic radiation pressure. (Right) Bioparticles experience acoustic radiation pressure when SAW transducer is turned on. Bioparticles A (purple) are pushed toward the pressure node due to positive acoustic pressure contrast ($\phi_A > 0$). Bioparticles B (yellow) which experience the opposite ($\phi_B < 0$) are pushed toward the pressure antinodes (Reprinted with permission from Md Ali et al. [13]. Copyright 2016 Royal Society of Chemistry). (b) Manipulation of bacteria in blood by acoustophoresis. Separation of bacteria from RBCs is performed by acoustic separation chip. Bacteria enrichment from blood plasma is performed subsequently at acoustic trapping capillary before the release for detection and identification by dry-reagent PCR chips (Reprinted with permission from Ohlsson et al. [38]. Copyright 2016 American Chemical Society).

Mainstream applications of ACT in particle manipulation employ either (1) traveling surface acoustic wave (TSAW) [37] or standing surface acoustic wave (SSAW) [36]. A TSAW is a condition when a surface acoustic wave (SAW) is propagating from interdigitated transducer (IDT) electrodes, while SSAW occurs when two TSAWs constructively interfere and form a standing or stationary SAW. TSAW can be generated by a single IDT electrode, while the SSAW can be generated either by a pair of IDT electrodes or a combination of a single IDT and wave reflectors. In TSAW acoustophoresis, bioparticles move together with the wave propagation, while in SSAW acoustophoresis, they are pushed toward the SAW pressure node or the antinode. Pressure node is the region of constant pressure, while pressure antinodes are regions alternating between maximum and minimum pressure values.

5.2. Acoustophoretic manipulation of bioparticles

Acoustophoretic manipulation of model organism has been demonstrated by Sundvik et al. [39]. They study levitation of zebrafish embryos using acoustic radiation force in a noncontact wall-less platform for a period of less than 2000 s, while the embryos still in development at 2–14 hours postfertilization, and they found that the levitation does not interfere the development, though it might influence mortality rate. Urbansky et al. [40] perform the manipulation of peripheral blood progenitor cells (PBPCs) with the focus on sorting out CD8 lymphocytes (target cells) from the mixture that contains CD4, CD19, CD34, and CD56 lymphocytes as well. They label the CD8 with affinity beads, forming bead-cell complex, to modify the acoustic mobility of the target cells. Furthermore, they modify the medium properties of central buffer, using Ficoll wash buffer, to adjust the acoustic force on different particles, so that bead-CD8 complexes are pushed into central buffer under acoustophoretic force exerted by piezoceramic transducer, while other unbounded cells remain flowing at the side due to lower acoustic mobility. Urbansky et al. [41] further perform separation of mononuclear cells (MNCs) from diluted whole blood using acoustophoretic microfluidic device. They managed to overcome the behavior similarity of MNCs and RBCs in acoustic standing wave by optimizing the buffer conditions to modify the acoustophoretic mobility of the cells. Antfolk et al. [42] accomplish separation of spiked prostate cancer cells (DU145) from whole blood using ACT-DEP-integrated platform consisting of acoustophoretic pre-alignment, separation, and concentration of targeted DU145 cells, prior to single-cell array trapping using DEP microwell. Bacteria manipulation has been accomplished by Ohlsson et al. [38], who developed a microsystem for bacteria separation, enrichment, and detection from blood, as demonstrated in **Figure 3b**. The system is integrated with acoustic separation to remove RBCs from blood sample, with subsequent enrichment of bacteria from plasma by acoustic trapping to polystyrene seed particles, and polymerase chain reaction (PCR) for detection and identification of the bacteria at the final stage. They demonstrate the system using whole blood samples, which, respectively, spiked with *Pseudomonas putida* and *E. coli*. While for virus manipulation, Ness et al. [43] demonstrate extraction and enrichment of MS2 virus from human nasopharyngeal samples by integration of acoustic force to remove host cells, debris, and pollen from the sample and later with electric field force to attract the virus, which is a negatively charged species, and to migrate from sample to co-flowing buffer. Further, Park et al. [44] perform washing and screening of prostate-specific antigen (PSA)-binding aptamer, a single-stranded DNA (ssDNA), using an acoustophoretic separation. The ssDNA pool comprising of ssDNA library is incubated with microbeads which modified with target molecule, before it is introduced to acoustophoretic separation device, where the microbeads with target-bound DNA fragments are focused on central buffer due to acoustophoretic force, while unbound protein and ssDNA are remained in the original buffer at the side flow. The target-bound DNA on the microbeads is collected and amplified by PCR for subsequent round of washing and screening. Recently, Kennedy et al. [45] communicate the process of purifying target biomolecules utilizing acoustic standing wave in a fluidic chamber to partition and maintain solid-phase bead in an acoustically fluidized bed format, for capturing, washing, and elution of target biomolecules, including monoclonal antibody by protein A beads from a crude cell culture system and recombinant green fluorescent protein (GFP) by anion exchange of a crude cell lysate.

6. Thermophoresis

Thermophoresis (THM) is the motion of particles driven by thermal gradients in the suspending medium. Thermal gradients are commonly generated by local absorption of infrared (IR) laser. The thermal gradients induce diffusional motion of the particles, either toward higher or lower temperature regions [10].

6.1. Fundamentals of THM

Thermophilic particles diffused to the region with higher temperature, while thermophobic particles move to the opposite direction, as shown in **Figure 4a**.

Liquid flow density, J, driven by thermophoretic field, is governed by

$$J = D\left[\nabla C + S_T C(1 - C)\nabla T\right] \tag{10}$$

where D is the diffusion coefficient, C is the concentration, T is the temperature, and S_T is the Soret coefficient, defined as the ratio of thermal diffusion coefficient, D_T, over diffusion coefficient, D, which is given by

$$S_T = \frac{D_T}{D} \tag{11}$$

Steady-state concentration changes for a given spatial temperature difference, ΔT, which is given by

$$\frac{C_{hot}}{C_{cold}} = exp(-S_T \Delta T) \tag{12}$$

where C_{hot} is the molecule concentration in the hot area, while C_{cold} is in the cold area.

Studies prove that a temperature difference between 2 and 8 K in the beam center with a $1/e^2$ diameter of 25 μm managed to induce thermophoretic motion, while the "$1/e^2$ diameter" (e = 2.71828) indicates the beam diameter where intensity drops to 13.5% of the maximum value. The temperature rise of the suspending medium must be kept low to avoid bioparticle damage, such as in the case of DNA, which is from 23 to 31°C [47].

6.2. Thermophoretic manipulation of bioparticles

Thermophoretic manipulation of yeast cells has been demonstrated by Lin et al. [46] using low-power and flexible all-optical manipulation method, which presented in **Figure 4b**. They generate light-controlled temperature gradient field thus to trap the suspended cells due to permittivity gradient in the electric double layer of the cell membrane-charged surface. In fact, they manage to realize arbitrary spatial arrangement, as well as precise rotation of single-cell assemblies, with resolution down to 100 nm. J. Chen et al. [48] demonstrate thermophoretic manipulation of *E. coli* by inducing thermal gradient by microscale electric thermal heater. The electric thermal heater is fabricated by gold thin film by means of direct writing with a femtosecond laser. A thin SiO_2 is coated over the gold thin film to electrically isolate the sample as well as to ensure that the trapping is entirely due to thermal effect. Osterman et al. [49] use microscale thermophoresis to study all potential intraviral protein-protein interactions of hepatitis E virus in order to understand the viral replication cycle. The thermal gradient is generated using 100 V-powered red LED for 25 s radiation. Remarkable thermophoretic manipulation of nucleic acids has been demonstrated by He et al. [50], in which translocation of DNA through nanopore utilizing cross-pore thermal gradient has been achieved. Heating of *cis* chamber is performed by exterior heater to maintain the environment at the melting temperature of the double-strand DNA to transform into single-strand DNA. Extensive temperature drop across the pore is caused by thermal-insulating membrane that separate the *cis* and *trans* chambers, resulting in the thermophoretic translocation of the DNA from *cis* to *trans* chamber. Wolff et al. [51] perform microscale thermophoresis upon different forms of the protein α-synuclein, which is associated with Parkinson's disease, to quantitatively characterize

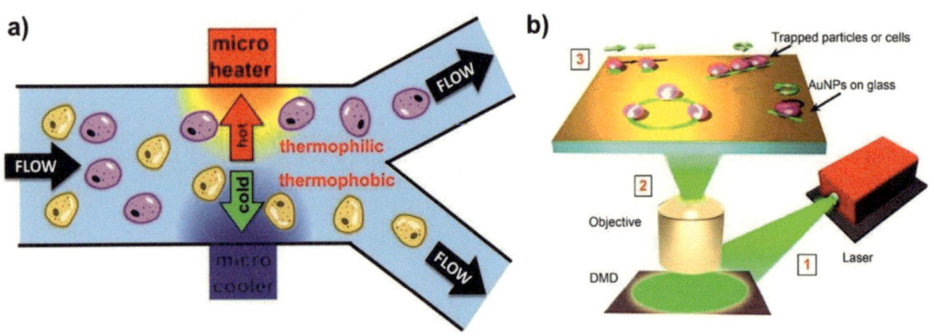

Figure 4. (a) Thermophoresis. Thermophilic bioparticles A (purple) diffuse to regions with higher temperature, while thermophobic bioparticles B (yellow) diffuse to lower temperature regions (Reprinted with permission from Md Ali et al. [13]. Copyright 2016 Royal Society of Chemistry). (b) Manipulation of model organisms by thermophoresis. Optothermal thermophoretic tweezer manipulation is performed by the following: (1) Laser is directed to the digital micromirror device (DMD). (2) Resultant image is focused on gold nanoparticle (AuNP) substrate for surface plasmon excitation. (3) Plasmon-enhanced optothermal potentials defined by the DMD-controlled optical images are exploited to trap and arbitrarily manipulate colloidal particles or biological cells (Adapted with permission from Lin et al. [47]. Copyright 2017 American Chemical Society).

the thermophoretic behavior of the monomeric, oligomeric, and fibrillary forms of the protein. An infrared laser is focused inside a capillary borosilicate where homogenous protein suspension is placed, to generate localized heating, resulting in the motion of the molecules along the temperature gradient until the steady state is established.

7. Optical tweezing/trapping

Optical tweezing or trapping (OPT) indicates the manipulation of particles using optical forces, referring to the exploitation of light radiation pressure to displace and demobilize target particles [11, 12].

7.1. Fundamentals of OPT

Emission of light by a light source induces scattering and gradient forces, which affect particle in the light propagation axis. Scattering force, F_{scat}, affects in the direction of propagation, pushing the particle away from the light source. Gradient force, F_{grad}, on the other hand, affects the direction of the optical field gradient, attracting the particle to the region with peak spatial light intensity [11] as shown in **Figure 5a**.

Optical tweezing or trapping depends on the dimension range of the particle under manipulation, which is governed by two physical principles, i.e., (1) *Mie* scattering and (2) *Rayleigh* scattering [12]. *Mie* regime condition governs for the condition where the particle dimension range is greater than the wavelength of light ($d >> \lambda$) which can be explained by ray optics. Cell-type bioparticle (micron-sized) manipulation lies on this regime. Rays of light carry momentum and

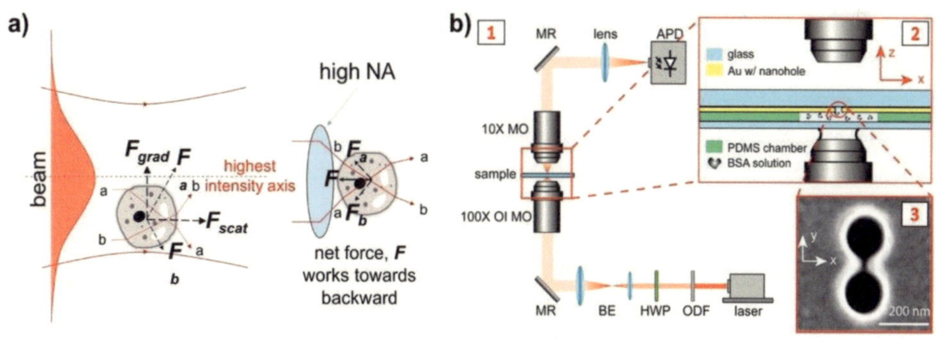

Figure 5. (a) Optical tweezing or trapping. (Left) Scattering force, F_{scat}, pushes bioparticle away from the light source following the direction of light propagation. Gradient force, F_{grad}, attracts bioparticle to the spatial light intensity peak according to the direction of the optical field gradient. (Right) Bioparticle is trapped at the highest intensity region, i.e., light focus, as emission of light through a high numerical aperture number lens generates $F_{grad} > F_{scat}$ condition (Reprinted with permission from Md Ali et al. [13]. Copyright 2016 Royal Society of Chemistry). (b) Manipulation of protein by optical tweezing or trapping: (1) Double-nanohole optical tweezer setup. (2) Details of manipulation of microfluidic device construct. Proteins are trapped at the focus of optical trap at the vicinity of the gold nanohole surface while at the same time unfolding between the double nanoholes. (3) SEM image of the double nanohole (ODF, optical density filter; HWP, half-wave plate; BE, beam expander; MR, mirror; MO, microscope objective; OI MO, oil immersion microscope objective; APD, avalanche photodiode) (Reprinted with permission from Pang and Gordon [52]. Copyright 2011 American Chemical Society).

is refracted when passing through a particle with a refractive index, n_2, which is greater than the surrounding medium, n_1. The rate of the momentum change in the detected rays develops an equal and opposite rate of momentum change of the particle, producing a force by Newton's second law due to conservation of momentum. When a particle is placed in a light gradient, the sum of all rays passing through it creates an imbalance in force, pushing the particle toward the region with higher intensity of light. A focus constructs a trap as the strong light gradient points to the center. *Rayleigh* regime governs for bioparticles with dimension far less than the wavelength ($d \ll \lambda$), such as protein and nucleic acids. In this regime, the particles are treated as extremely small point dipoles polarized by a uniform electric field, which then interacts with light field. The dipole moment, p_{dipole}, induced by a uniform electric field E, is given by

$$p_{dipole} = 4\pi n_{\mathrm{m}}^2 \varepsilon_0 a^3 \left(\frac{m^2-1}{m^2+2}\right) E \tag{13}$$

where n_{1m} and n_{2p} are refractive indices of the suspending medium and the particle, respectively, a is the particle radius, ε_0 is the vacuum permittivity, and m is the contrast ratio of the indices, i.e., $m = n_p/n_m$. The magnitudes of scattering force, F_{scat}, and gradient force, F_{grad}, based on point dipole interaction with light field method are given by

$$F_{scat} = \frac{8\pi n_{\mathrm{m}} k^4 a^6}{3c} \left(\frac{m^2-1}{m^2+2}\right) I \tag{14}$$

$$F_{grad} = \frac{2\pi n_{\mathrm{m}} a^3}{c} \left(\frac{m^2-1}{m^2+2}\right) \nabla I \tag{15}$$

where F_{scat} is the scattering force, F_{grad} is the gradient force, I is the light intensity, c is the speed of light, k is the wave number ($k = 2\pi/\lambda$), and λ is the wavelength. Trapping is achieved at the highest intensity axis when $F_{grad} > F_{scat}$.

OPT can be created by light emission through a high numerical aperture number (NA) of microscope objective (MO), which focuses light tightly and results in a force along the highest intensity axis, but in the backward direction, which causes the bioparticle to be demobilized, as presented in **Figure 5a**.

7.2. Optical tweezing or trapping of bioparticles

Favre-Bulle et al. [53] use optical trapping in vivo to manipulate otoliths in larval zebrafish to stimulate the vestibular system. Lateral and medial forces upon the otolith cause complementary corrective tail motion, while lateral force on either otolith causes a rolling correction in both eyes. Fascinating manipulation of blood has been demonstrated by Zhong et al. [54], who perform manipulation of RBCs in vivo, i.e., within subdermal capillaries in living mice, using infrared optical tweezers. They demonstrate the optical trapping and three-dimensional manipulation of single RBC in the capillary, as well as multiple RBC trapping, forming capillary blockage and clearance, by turning on and off the optical tweezer. Pradhan et al. [55] use optical trapping to bring a single cell of neural tumor cell lines into close proximity of another and measure the time required for cell-cell adhesion to form, as this method can be used to assess the differentiation status of cancerous cells. They perform the measurement for human neuroblastoma SK-N-SH and rat C6 glioma cells. Stem cell manipulation has been performed by Kirkham et al. [56]. They develop remarkable holographic optical tweezer and demonstrate the micromanipulation of several stem cells, including mouse embryonic stem cells, mouse mesenchymal stem cells, and mouse primary calvarea cells, as well as microstructures, such as poly(DL-lactic-co-glycolic acid) microparticles and electrospun fiber fragments. They succeeded in accurately construct three-dimensional architecture with varying geometries from cocultured cells and microstructure and then stabilized them using hydrogels and cell-cell adhesion methods. Zakrisson et al. [57] perform the optical trapping of nonpiliated strain of *E. coli* bacteria, as well as transformed strain HB101/pHMG93 (expressed as P pili) and the HB101/pAZZ50 strain (expressed as S_{II} pili) in the development of method to determine the presence of pili on a single bacterium. The method comprises of measurement of the bacterium by imaging, then estimation of the fluid drag using an analytical model based on the size, and measurement of the effective fluid drag by oscillating the sample, while the single bacterium is trapped by an optical tweezer. Variation between estimation and measured fluid drag determines the existence of pili. Progress in virus manipulation has been shown by Pang et al. [58]. They conduct optical trapping of individual human immunodeficiency virus (HIV-1) in culture fluid under native conditions and then subsequently perform multiparameter analysis of individual virions, including diameter measurement, concentration-dependent aggregation, and monitoring of viral protein using two-photon fluorescence. Nucleic acid optical manipulation has been achieved by Ngo et al. [59] who use single-molecule assay which integrates fluorescence and optical tweezers generated by infrared laser to manipulate a single nucleosome under force and simultaneously analyze its local conformational transitions. The nucleosome is affixed to a polyethylene glycol (PEG)-coated glass surface at one end of the DNA and pulled via a λ-DNA tethered to the other end by an optical trap to apply force. Pang and Gordon [52], as shown in **Figure 5b**, perform optical trapping and stable unfolding of an individual BSA molecule using a double nanohole in a gold film. The double nanohole with 15 nm separated sharp tips is milled on Au/Ti/glass substrate by a focused ion beam, while the optical trap is generated by 820 nm laser focused onto the sample using a 100× oil immersion microscope objective in a polarization with electric field aligned with the tips of the double nanohole.

8. Conclusion and future perspectives

Manipulation of bioparticles in micro-/nanofluidic, integrated with active manipulation mechanisms, i.e., dielectrophoresis, magnetophoresis, acoustophoresis, thermophoresis, and optical tweezing/trapping, has been discussed in this chapter. Description of the underlying fundamental theory is provided at the beginning, and state-of-the-art implementations into a wide range of bioparticles are carefully introduced. DEP has shown rapid progress into exploration of extremely small bioparticle manipulation, i.e., virus, nucleic acids, and protein. In particular, demonstration of DNA sorting [23] and impedance-based protein capturing [25] prove the potential for nanoscale application, as well as genetic and molecular biology studies. MAG is advantageous in selective manipulation benefited by biofunctionalization of magnetic micro-/nanoparticle for affinity binding to target bioparticles. Novel achievement in customization of target bioparticle magnetization using predefined multiple layers of magnetic particles [27] is promising for application into heterogeneous suspension manipulation, such as whole blood, progenitor, and cancerous cell detection and sorting. ACT's greatest progress is in the integration with other mechanisms, e.g., DEP, to establish a complete biomedical device [42], showing that transformation from conventional devices to microfluidic biomedical devices which are superfast, precise, and portable is soon to be realized. THM technology particularly is matured enough in extremely small bioparticle manipulation, i.e., protein and nucleic acids, specifically as a measurement tool for binding interaction between molecules. In fact, the recent exploration of thermal gradient-based DNA translocation [50], as well as cell arbitrary manipulation benefited from permittivity gradient in the electric double layer of cell membrane [47], potentially open for new path in THM research. OPT has emerged into in vivo studies [53, 54], indicating that the clinical application is promising and soon to be achieved. In addition, OPT capability in application to genetic and stem cell studies is of high potential, as demonstrated in construction of three-dimensional bioparticle assemblies [55, 56]. Rapid progress of studies on these micro-/nanofluidic active manipulation mechanisms toward bioparticles has a significant impact to biomedical research and technology development. Evolution into high-precision, superfast, and portable miniaturized biomedical devices is pretty soon to be achieved.

Author details

Mohd Anuar Md Ali[1], Aminuddin Bin Ahmad Kayani[2,3]* and Burhanuddin Yeop Majlis[1]

*Address all correspondence to: aminuddin.kayani@mmu.edu.my

1 Institute of Microengineering and Nanoelectronics (IMEN), Universiti Kebangsaan (UKM), Bangi, Selangor, Malaysia

2 Center for Advanced Materials and Green Technology, Multimedia University (MMU), Cyberjaya, Selangor, Malaysia

3 Functional Materials and Microsystems Research Group, School of Engineering, RMIT University, Melbourne, Australia

References

[1] Alberts B, Johnson A, Lewis J, et al. Molecular Biology of the Cell. New York: Garland Science; 2008

[2] Kayani AA, Khoshmanesh K, Ward SA, Mitchell A, Kalantar-zadeh K. Optofluidics incorporating actively controlled micro- and nano-particles. Biomicrofluidics. 2012;**6**:31501

[3] Çetin B, Özer MB, Solmaz ME. Microfluidic bio-particle manipulation for biotechnology. Biochemical Engineering Journal. 2014;**92**:63-82

[4] Sheikholeslami M. Numerical investigation of nanofluid free convection under the influence of electric field in a porous enclosure. Journal of Molecular Liquids. 2018;**249**:1212-1221

[5] Whitesides GM. The origins and the future of microfluidics. Nature. 2006;**442**:368-373

[6] Pohl HA. Dielectrophoresis: The Behavior of Neutral Matter in Nonuniform Electric Fields. Cambridge/New York: Cambridge University Press; 1978

[7] Voldman J. Electrical forces for microscale cell manipulation. Annual Review of Biomedical Engineering. 2006;**8**:425-454

[8] Zborowski M, Chalmers JJ. Magnetophoresis: Fundamentals and applications. In: Webster JG, editor. Wiley Encyclopedia of Electrical and Electronics Engineering. Chichester: Wiley; 2015. pp. 1-23

[9] Lenshof A, Laurell T. Acoustophoresis. In: Bhushan B, editor. Encyclopedia of Nanotechnology. Springer: Dordrecht; 2012. pp. 45-50

[10] Reineck P, Wienken CJ, Braun D. Thermophoresis of single stranded DNA. Electrophoresis. 2010;**31**:279-286

[11] Ashkin A. Acceleration and trapping of particles by radiation pressure. Physical Review Letters. 1970;**24**:156-159

[12] Woerdemann M. Structured Light Fields: Applications in Optical Trapping, Manipulation, and Organisation. Berlin/Heidelberg: Springer; 2012

[13] Md Ali MA, Ostrikov K (Ken), Khalid FA, Majlis BY, Kayani AA. Active bioparticle manipulation in microfluidic systems. RSC Advances. 2016;**6**:113066-113094

[14] D'Amico L, Ajami NJ, Adachi JA, Gascoyne PRC, Petrosino JF. Isolation and concentration of bacteria from blood using microfluidic membraneless dialysis and dielectrophoresis. Lab on a Chip. 2017;**17**:1340-1348

[15] Çetin B, Li D. Dielectrophoresis in microfluidics technology. Electrophoresis. 2011;**32**:2410-2427

[16] Jones TB. Basic theory of dielectrophoresis and electrorotation. IEEE Engineering in Medicine and Biology Magazine. 2003;**22**:33-42

[17] Muratore M, Mitchell S, Waterfall M. Plasma membrane characterization, by scanning electron microscopy, of multipotent myoblasts-derived populations sorted using dielectrophoresis. Biochemical and Biophysical Research Communications. 2013;**438**:666-672

[18] Chen X, Liang Z, Li D, et al. Microfluidic dielectrophoresis device for trapping, counting and detecting Shewanella oneidensis at the cell level. Biosensors & Bioelectronics. 2018;**99**:416-423

[19] Sang S, Feng Q, Jian A, et al. Portable microsystem integrates multifunctional dielectrophoresis manipulations and a surface stress biosensor to detect red blood cells for hemolytic anemia. Scientific Reports. 2016;**6**:33626

[20] Chiu T-K, Chou W-P, Huang S-B, et al. Application of optically-induced-dielectrophoresis in microfluidic system for purification of circulating tumour cells for gene expression analysis- Cancer cell line model. Scientific Reports. 2016;**6**:32851

[21] Adams TNG, Jiang AYL, Vyas PD, Flanagan LA. Separation of neural stem cells by whole cell membrane capacitance using dielectrophoresis. Methods. 2018;**133**:91-103

[22] Ding J, Lawrence RM, Jones PV, Hogue BG, Hayes MA. Concentration of Sindbis virus with optimized gradient insulator-based dielectrophoresis. Analyst. 2016;**141**:1997-2008

[23] Jones PV, Salmon GL, Ros A. Continuous separation of DNA molecules by size using insulator-based Dielectrophoresis. Analytical Chemistry. 2017;**89**:1531-1539

[24] Viefhues M, Regtmeier J, Anselmetti D. Nanofluidic devices for dielectrophoretic mobility shift assays by soft lithography. Journal Micromechanics Microengineering. 2012;**22**:115024

[25] Mohamad AS, Hamzah R, Hoettges KF, Hughes MP. A dielectrophoresis-impedance method for protein detection and analysis. AIP Advances. 2017;**7**:15202

[26] Liao K, Chaurey V, Tsegaye M, Chou C, Swami NS. Nanofluidics for selective protein trapping in bio-fluids. In: Landers JP, editor. Fifteenth International Conference on Miniaturized Systems for Chemistry and Life Sciences. San Diego: Chemical and Biological Microsystems Society; 2011. pp. 1071-1073

[27] Lee H, Hong D, Cho H, et al. Turning diamagnetic microbes into multinary micromagnets: Magnetophoresis and spatio-temporal manipulation of individual living cells. Scientific Reports. 2016;**6**:38517

[28] Schneider T, Moore LR, Jing Y, et al. Continuous flow magnetic cell fractionation based on antigen expression level. Journal of Biochemical and Biophysical Methods. 2006;**68**:1-21

[29] Zeng J, Deng Y, Vedantam P, Tzeng T-R, Xuan X. Magnetic separation of particles and cells in ferrofluid flow through a straight microchannel using two offset magnets. Journal of Magnetism and Magnetic Materials. 2013;**346**:118-123

[30] Jin X, Abbot S, Zhang X, et al. Erythrocyte enrichment in hematopoietic progenitor cell cultures based on magnetic susceptibility of the hemoglobin. PLoS One. 2012;**7**:e39491

[31] Jack R, Hussain K, Rodrigues D, et al. Microfluidic continuum sorting of sub-populations of tumor cells via surface antibody expression levels. Lab on a Chip. 2017;**17**:1349-1358

[32] Wang ZM, Wu RG, Wang ZP, Ramanujan RV. Magnetic trapping of Bacteria at low magnetic fields. Scientific Reports. 2016;**6**:26945

[33] Wang Y, Li Y, Wang R, Wang M, Lin J. Three-dimensional printed magnetophoretic system for the continuous flow separation of avian influenza H5N1 viruses. Journal of Separation Science. 2017;**40**:1540-1547

[34] Shim S, Shim J, Taylor WR, et al. Magnetophoretic-based microfluidic device for DNA concentration. Biomedical Microdevices. 2016;**18**:28

[35] Lim B, Torati SR, Kim KW, Hu X, Reddy V, Kim C. Concentric manipulation and monitoring of protein-loaded superparamagnetic cargo using magnetophoretic spider web. NPG Asia Materials. 2017;**9**:e369

[36] Antfolk M, Antfolk C, Lilja H, Laurell T, Augustsson P. A single inlet two-stage acoustophoresis chip enabling tumor cell enrichment from white blood cells. Lab on a Chip. 2015;**15**:2102-2109

[37] Destgeer G, Alazzam A, Sung HJ. High frequency travelling surface acoustic waves for microparticle separation. Journal of Mechanical Science and Technology. 2016;**30**:3945-3952

[38] Ohlsson P, Evander M, Petersson K, et al. Integrated acoustic separation, enrichment, and microchip polymerase chain reaction detection of bacteria from blood for rapid sepsis diagnostics. Analytical Chemistry. 2016;**88**:9403-9411

[39] Sundvik M, Nieminen HJ, Salmi A, Panula P, Hæggström E. Effects of acoustic levitation on the development of zebrafish, Danio rerio, embryos. Scientific Reports. 2015;**5**:13596

[40] Urbansky A, Lenshof A, Dykes J, Laurell T, Scheding S. Affinity-bead-mediated enrichment of CD8+ lymphocytes from peripheral blood progenitor cell products using acoustophoresis. Micromachines. 2016;**7**:101

[41] Urbansky A, Ohlsson P, Lenshof A, Garofalo F, Scheding S, Laurell T. Rapid and effective enrichment of mononuclear cells from blood using acoustophoresis. Scientific Reports. 2017;**7**:17161

[42] Antfolk M, Kim SH, Koizumi S, Fujii T, Laurell T. Label-free single-cell separation and imaging of cancer cells using an integrated microfluidic system. Scientific Reports. 2017;**7**:46507

[43] Ness K, Rose KA, Jung B, Fisher K, Mariella RP Jr. Improved bacterial and viral recoveries from complex samples using electrophoretically assisted acoustic focusing. In: Locascio LE, editor. Twelfth International Conference on Miniaturized Systems for Chemistry and Life Sciences. San Diego: Chemical and Biological Microsystems Society; 2008. pp. 802-804

[44] Park JW, Lee SJ, Ren S, Lee S, Kim S, Laurell T. Acousto-microfluidics for screening of ssDNA aptamer. Scientific Reports. 2016;**6**:27121

[45] Kennedy T, Pluskal M, Gilmanshin R, Lipkens B. Acoustophoresis mediated chromatography processing: Capture of proteins from cell cultures. The Journal of the Acoustical Society of America. 2017;**141**:3505

[46] Yu L-H, Chen Y-F. Concentration-dependent thermophoretic accumulation for the detection of DNA using DNA-functionalized nanoparticles. Analytical Chemistry. 2015;**87**:2845-2851

[47] Lin L, Peng X, Wei X, Mao Z, Xie C, Zheng Y. Thermophoretic tweezers for low-power and versatile manipulation of biological cells. ACS Nano. 2017;**11**:3147-3154

[48] Chen J, Cong H, Loo FC, et al. Thermal gradient induced tweezers for the manipulation of particles and cells. Scientific Reports. 2016;**6**:35814

[49] Osterman A, Stellberger T, Gebhardt A, et al. The hepatitis E virus intraviral interactome. Scientific Reports. 2015;**5**:13872

[50] He Y, Tsutsui M, Scheicher RH, Bai F, Taniguchi M, Kawai T. Thermophoretic manipulation of DNA translocation through nanopores. ACS Nano. 2013;**7**:538-546

[51] Wolff M, Mittag JJ, Herling TW, et al. Quantitative thermophoretic study of disease-related protein aggregates. Scientific Reports. 2016;**6**:22829

[52] Pang Y, Gordon R. Optical trapping of a single protein. Nano Letters. 2012;**12**:402-406

[53] Favre-Bulle IA, Stilgoe AB, Rubinsztein-Dunlop H, Scott EK. Optical trapping of otoliths drives vestibular behaviours in larval zebrafish. Nature Communications. 2017;**8**:630

[54] Zhong M-C, Wei X-B, Zhou J-H, Wang Z-Q, Li Y-M. Trapping red blood cells in living animals using optical tweezers. Nature Communications. 2013;**4**:1768

[55] Pradhan M, Pathak S, Mathur D, Ladiwala U. Optically trapping tumor cells to assess differentiation and prognosis of cancers. Biomedical Optics Express. 2016;**7**:943-948

[56] Kirkham GR, Britchford E, Upton T, et al. Precision assembly of complex cellular microenvironments using holographic optical tweezers. Scientific Reports. 2015;**5**:8577

[57] Zakrisson J, Singh B, Svenmarker P, et al. Detecting bacterial surface organelles on single cells using optical tweezers. Langmuir. 2016;**32**:4521-4529

[58] Pang Y, Song H, Kim JH, Hou X, Cheng W. Optical trapping of individual human immunodeficiency viruses in culture fluid reveals heterogeneity with single-molecule resolution. Nature Nanotechnology. 2014;**9**:624-630

[59] Ngo TTM, Zhang Q, Zhou R, Yodh JG, Ha T. Asymmetric unwrapping of nucleosomes under tension directed by DNA local flexibility article asymmetric unwrapping of nucleosomes under tension directed by DNA local flexibility. Cell. 2015;**160**:1135-1144

Chapter 5

Particle Deposition in Microfluidic Devices at Elevated Temperatures

Zhibin Yan, Xiaoyang Huang, Chun Yang,
Mingliang Jin and Lingling Shui

Additional information is available at the end of the chapter

http://dx.doi.org/10.5772/intechopen.78240

Abstract

In microchannels, interaction and transport of micro-/nanoparticles and biomolecules are crucial phenomena for many microfluidic applications, such as nanomedicine, portable food processing devices, microchannel heat exchangers, etc. The phenomenon that particles suspended in liquid are captured by a solid surface (e.g., microchannel wall) is referred to as *particle deposition*. Particle deposition is of importance in numerous practical applications and is also of fundamental interest to the field of colloid science. This chapter presents researches on fouling and particle deposition in microchannels, especially the effects of temperature and temperature gradient, which have been frequently 'ignored' but are important factors for thermal-driven particle deposition and fouling processes at elevated temperatures.

Keywords: particle deposition, microchannel, temperature, temperature gradient

1. Fouling

In the oil and gas industry, people first studied fouling and then it begun to be often adopted to describe any undesirable deposit that led to an increase of flow resistance in fluid pipes or thermal resistance in heat exchanger [1]. Most of the published researches on fouling are focused on macro-scale phenomena and parameters, such as changes of hydraulic performances and thermal resistances. According to the causes of formation, fouling can be categorised into different types, which include crystallisation fouling, or particulate fouling, scale formation, chemical reaction fouling, corrosion fouling, biological fouling, solidification fouling and mixed fouling [2]. In this chapter, the authors will concentrate on the studies on particulate fouling in microfluidic applications. Particulate fouling refers to that finely suspended solid particles accumulate onto solid surfaces. The diameters of particles are usually less than microns. Without other mechanisms having particles firmly attached onto solid surface, the deposit formed by particulate fouling is normally thin and easily removed.

2. Particle deposition

Particulate fouling is caused by particle deposition onto the solid surfaces of collectors. This is a complex process affected by diffusion, convection, colloidal, and external interaction forces. To better understand the physical process of particle deposition, it can be conceptually divided into three steps [3]:

1. Particle transport/diffusion

   When particles are at large distances from the collector surface, the particles are transported from the bulk fluid to the collector by diffusion, flow convection, and migration due to external forces. Particle concentration gradients usually exist between the bulk fluids to the collector surfaces and drive the suspended particles approaching the collector surfaces. Flow convection, especially in turbulent flow, could give particles momentum to move across the bulk fluid toward the collector surface. Besides, particles would migrate to the collector when they are affected by external fields, such as gravity filed and electric field.

2. Particle-wall hydrodynamic interaction

   When particles get close to the collector within a distance comparable to the particle radius, the motion of the fluid between the particle surface and the collector surface becomes much more difficult compared to the scenario at large distances. It is because that the particles need to experience additional hydrodynamic drag caused by the presence of the collector surface. Thus, the reduction of particle mobility is commonly attributed to the particle-wall hydrodynamic interaction.

3. Particle adhesion

   As particles approach the collector at even closer distances, typically less than 100 nm, the motion of the particles is influenced by colloidal forces besides the hydrodynamic interaction. Within such extremely short distance, the influences of the colloidal forces are overwhelmingly stronger than those of external forces. Thus, the particle adhesion onto the collector is mainly determined by the competition of the attractive and repulsive colloidal forces. The colloidal forces consist of the universal van der Waals (VDW) force and the electric double layer (EDL) force. These two interaction forces form the basis of the Derjaguin-Landau-Verwey-Overbeek (DLVO) theory of colloidal chemistry. The van der Waals force originates from spontaneous electrical and magnetic polarizations, which generate a fluctuating electromagnetic field within the particle and the collector as well as in the gap between them. Solid surfaces in aqueous media (electrolyte solution) are always charged because of the dissociation of ionisable surface sites or the adsorption of ionic surface active sites. As a result, electric double layers are formed in the vicinity of both the particle and the collector surfaces. When these two charged surfaces approach each other in the electrolyte solution, two electric double layers overlap and a repulsive interaction is developed in this region if the particle and the collector carry the same sign of charge. This repulsive interaction is known as the electric double layer force. In addition, other non-DLVO colloidal interactions might affect the particle deposition in short ranges (0.5–5 nm) under certain physicochemical conditions.

3. Theory of surface forces

The behaviours of particles in aqueous media are significantly influenced by the physicochemical characteristics of the interaction forces between particles and solid surfaces [4]. Thus, the interaction forces exerted on colloidal particles determine particulate fouling or particle deposition. In this section, the colloidal forces between particles and surfaces are briefly reviewed, including colloidal interaction forces, van der Waals attraction force and electrical double layer repulsion force. These two forces form the basis of the Derjaguin-Landau-Verwey-Overbeek (DLVO) theory, which were independently developed by Derjaguin and Landau [5] from Soviet/Russian and Verwey and Overbeek [6] from Netherlands. Besides the DLVO forces, non-DLVO forces are also involved in particle deposition and aggregation, such as polymer bridging, solvation forces, steric forces or hydrophobic forces.

3.1. Derjaguin-Landau-Verwey-Overbeek (DLVO) theory

In the DLVO theory, van der Waals attraction force and electrostatic repulsion force are suggested as the dominant interactions between two charged hydrophobic or lyophobic particles/surfaces in electrolyte solution. Moreover, the total interaction between particles and solid surfaces in a liquid is assumed as the sum of the two interactions. This is the first theory enabling to explain and predict the experimental observations of particle deposition and aggregation in a quantitative way. The van der Waals interaction arises from the electromagnetic effects of the molecules composing the particles while the electric double layer interaction is caused by the overlapping of the electric double layers of two particles/surfaces in an aqueous medium. Normally, the former is attractive and the latter is repulsive, which could be changed depending on the material properties in some specific cases [4].

3.1.1. van der Waals force

The van der Walls force, also known as London-van der Waals force, originates from a fluctuating electromagnetic field in particles and between particle and solid surface which is induced by the spontaneous magnetic and electrical polarisation. The van der Waals force can be either attractive or repulsive depending on the material property and is always attractive between identical materials. A number of methods have been proposed to calculate the van der Waals interaction energy [7–11]. Basically, there are two computation methods: the microscopic and the macroscopic.

For the microscopic methods, perturbation theory was initially adopted to solve the Schrödinger equation for the interactions between two hydrogen atoms at a large separation distance by Wang [12] and London [13], and they considered the interactions between the protons and electrons of the two atoms in the calculation. Their study provides a basis of quantum-mechanical analysis of the interaction between two non-polar molecules. Margenau [14] improved the analysis with consideration of higher moments. The retardation effect for the interactions was further investigated by Casimir and Polder [15] when the separation distance was shorter than the characteristic wavelength of radiation. Subsequently, Hamaker [11] proposed a simplified microscopic approximation in which the interaction between two solids is pair-wise additive. In another words, the total interaction force can be obtained by simply summing up the forces over all pairs of atoms in both solids. It is worth mentioning that Hamker's microscopic method neglects the retardation effect and many-body interactions. However, the influence of neighbouring atoms cannot be ignored, especially for condensed medium such as liquid. As a result, the pair additivity is difficult to be implemented for interacting objects in aqueous medium.

A more rigorous approach, macroscopic theory, was proposed in order to account for the aforementioned challenges. Dzyaloshinskii et al. [9] developed a new theory to avoid the problem of additivity encountered in the microscopic methods, known as Lifshitz theory. In this theory, large subjects are treated as a continuous medium without considering the atomic structure. The interaction forces between the subjects are calculated based on the bulk material properties including dielectric constants and refractive indices. The retardation effect is implicitly considered in the full Lifshitz treatment, but it is readily take account of the effect via modifying the Hamaker constant. In terms of calculation of Hamaker constant, various approaches have been developed and details can be found in the literatures [16–19].

Hamaker constants are most accurately calculated by Lifshitz theory, which determines the magnitude of the interaction through the frequency dependent dielectric properties of the intervening media [9, 18]. The Hamaker constant is estimated from the frequency dependent dielectric properties of the individual materials comprising the system as

$$A_{132} = \frac{3}{2}kT\left(\begin{array}{l}\frac{1}{2}\left(\frac{\varepsilon_1(iv_0)-\varepsilon_3(iv_0)}{\varepsilon_1(iv_0)+\varepsilon_3(iv_0)}\right)\left(\frac{\varepsilon_2(iv_0)-\varepsilon_1(iv_0)}{\varepsilon_2(iv_0)+\varepsilon_1(iv_0)}\right)\\ +\sum_{j=1}^{\infty}\left(\frac{\varepsilon_1(iv_j)-\varepsilon_3(iv_j)}{\varepsilon_1(iv_j)+\varepsilon_3(iv_j)}\right)\left(\frac{\varepsilon_2(iv_j)-\varepsilon_1(iv_j)}{\varepsilon_2(iv_j)+\varepsilon_1(iv_j)}\right)\end{array}\right) \tag{1}$$

where A_{132} is the Hamaker constant between particles '1' and the plate '3' in medium '2' and is a measure of the magnitude of the interaction between two objects. Israelachvili [20] proposed a simplified expression for the function $\varepsilon(iv_j)$ based on the refractive index and the absorption frequency of materials.

$$\varepsilon(iv_j) = 1 + \frac{n^2-1}{v^2} \tag{2}$$

In practice, it is difficult to obtain all the parameters in Eq. (2), mainly the absorption frequency. Assuming the absorption frequencies of the three media are the same, the Tabor-Winterton (TW) expression was developed to overcome the difficulties, shown as [21]

http://dx.doi.org/10.5772/intechopen.78240

$$A_{132} \approx \frac{3}{4} k_B T \left(\frac{\varepsilon_1 - \varepsilon_3}{\varepsilon_1 + \varepsilon_3}\right)\left(\frac{\varepsilon_2 - \varepsilon_3}{\varepsilon_2 + \varepsilon_3}\right) + \frac{3hv_e}{8\sqrt{2}} \frac{\left(n_1^2 - n_3^2\right)\left(n_2^2 - n_3^2\right)}{\left(n_1^2 + n_3^2\right)^{0.5}\left(n_2^2 + n_3^2\right)^{0.5}\left[\left(n_1^2 + n_3^2\right)^{0.5} + \left(n_2^2 + n_3^2\right)^{0.5}\right]} \quad (3)$$

where the refractive index n_i and zero frequency term $\varepsilon_{0,i}$ are the temperature-dependent factors for van der Waals interaction, k_B is the Boltzmann constant, and T is the absolute temperature of the system. The zero frequency term for most of the aqueous colloids is about equal to 3kT/4 or around 3×10^{-21} J. Water has a dielectric constant of about 80 at room temperature and non-polar media have values in the range 2–3.

In consideration of the retardation effect, Suzuki and Higuchi [22] proposed an approximated expression for the van der Waals interaction potential between the sphere and the plate as

$$F_{vdw} = \frac{-A_{132}}{6kT} \frac{\overline{\lambda}\left(\overline{\lambda} + 22.232H\right)}{H^2\left(\overline{\lambda} + 11.116H\right)^2} \quad (4)$$

where a is the radius of the sphere, h is the minimum surface-to-surface separation between the sphere and the plate, and H (h/a) and $\overline{\lambda}$ (λ/a) are the separation distance between the sphere and plate surface and the dimensionless characteristic wavelength, respectively, as illustrated in **Figure 1**. λ is the characteristic wavelength of the interaction which has a value of about 100 nm for most materials.

3.1.2. *Electrostatic double layer force*

Because of the electric double layer (EDL) force, particles can be well dispersed in liquids other than forming aggregation. The EDL force originates from the repulsion between the charged surfaces of the particles and solid surfaces immersed in liquids of high dielectric constants. The charges form the so-called electric double layer in the vicinity of the particles and the solid surfaces. The charging mechanism of a solid surface in a liquid medium can be categorised into two: (1) ionisation or dissociation of surface groups on the solid surface and (2) adsorption or binding of ions from electrolyte solutions onto a surface with oppositely charged sites or an

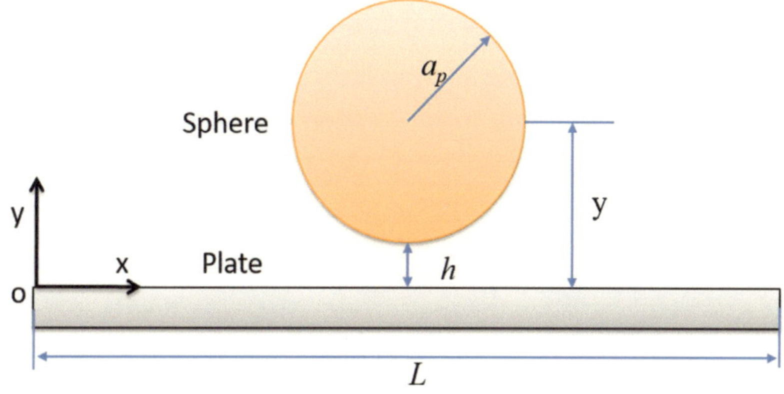

Figure 1. Interaction between the sphere and the plate.

originally uncharged solid surface. For a single particle suspended in a liquid medium, the particle is covered by the electric double layer (**Figure 2**). With consideration of the finite size of ions, Stern [23] developed an electric double layer model in which one immobilised layer of ions is absorbed onto the particle surface and the other layer is filled with diffusive space charges from the liquid medium (**Figure 2**). The former layer is termed as Stern layer and the latter layer is called as diffuse or Gouy layer.

Due to the nonuniform distribution of charges around the charged surface, electric potential reduces gradually with the separation distance from the solid surface to the bulk liquid phase. In the electric double layer model, several potentials are defined including surface potential on the solid surface, Stern potential at the Stern layer and zeta potential (ζ) at slipping plane. Assuming ions of identical property and average surface charge over the whole solid surface, the electric potential (ψ) and the average charge distribution in the diffuse layer of the electric double layer can be computed based on the non-linear Poisson-Boltzmann equation (PBE) as [24–26]

$$\nabla^2\psi = \frac{-1}{\varepsilon_0\varepsilon}\sum_i n_i^0 z_i e \exp\left(\frac{-z_i e\psi}{k_B T}\right) \tag{5}$$

where n^0 is the number density of ions in bulk, i represents the component i, z and e are the valence and the elementary electric charge, ε_0 is the permittivity of vacuum, and ε is the static dielectric constant.

As illustrated in **Figure 2**, a particle approaches a solid surface in an electrolyte solution or two charged particles approach each other, and their diffuse layers would overlap with each other. EDL force is repulsive for two surfaces with charges of same sign, while it becomes attractive for two particles with charges of opposite sign. The accuracy of calculating the EDL interaction is influenced by various factors. To simplify the calculation, two important assumptions are

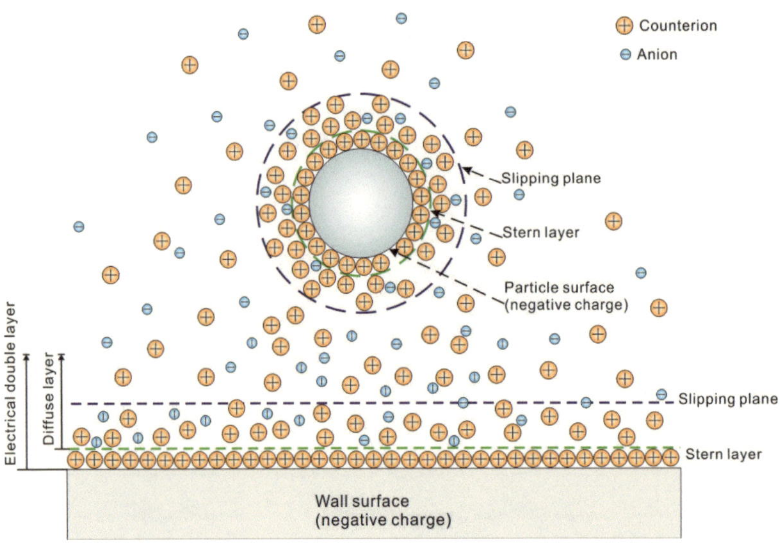

Figure 2. Schematic of a diffuse double layer of a charged particle in the vicinity of a charged solid/wall surface.

made: interactions with constant surface potential and constant surface charge density. For the constant surface potential cases, surface-chemical equilibrium is maintained while two particles/surfaces are approaching in a very short time. This may not be realistic for some practical cases [27]. For the constant surface charge cases, two particles/surfaces have fixed surface charge densities in the approaching process. These two assumptions are applied to the potential and charge on the particle/solid surfaces, whereas the interaction between electrical double layers is determined by the potential at the Stern plane. The charges at the Stern layer may behave differently from those on the particle/solid surface during the approach process. Recently, Barisik et al. [28] and Zhao et al. [29] have applied a complex charge regulation as boundary conditions to calculate the EDL interactions in nanoscale.

Generally, the EDL interaction energy can be computed based on two methods. One method is to directly solve the Poisson-Boltzmann equation for systems of particle/solid surfaces. Normally, it is difficult to obtain simple analytical solutions by this method. The other method is to construct the formula based on known expressions for each of the surfaces involved without consideration of influences of the other surfaces. The approximations of EDL interaction energy obtained in this way are often more attractive for practical applications which require fairy accuracy and simplicity [30].

3.1.3. *Sphere-plate double layer interactions*

In 1934, Derjaguin [31] developed an integration method to calculate the electric double layer interactions between two spheres in a dilute suspension. It has become a widely adopted method in colloidal chemistry since then. The EDL interaction energy between two spheres with overlapping electric double layers can be calculated as

$$V_{edl} = \frac{2\pi a_1 a_2}{a_1 + a_2} \int_h^\infty \upsilon_E dh \tag{6}$$

where h denotes the minimum separation distance between two sphere surfaces, and a_1 and a_2 are the radii of two spheres. The EDL interaction force can be obtained by differentiating the interaction energy, V_{edl}, with the separation distance, h, as

$$F_{edl} = \frac{2\pi a_1 a_2}{a_1 + a_2} \upsilon_E(h) \tag{7}$$

It should be noted that the above expressions are only applicable for cases in which $\kappa a_p > 5$ and $h \ll a_\mathrm{p}$ are valid. By allowing one of the radii to approach infinity, the sphere-plate interactions can be derived from the sphere-sphere interactions. Assuming constant surface potential, a theoretical expression of EDL interaction energy between a sphere and a plane was developed by Hogg et al. [32] as

$$V_{edl} = \pi\varepsilon_0\varepsilon a\left(\xi_p^2 + \xi_w^2\right)\left[\frac{2\xi_p^2\xi_w^2}{\xi_p^2 + \xi_w^2}\ln\left[\frac{1 + \exp(-\kappa h)}{1 - \exp(-\kappa h)}\right] + \ln\left[1 - \exp(-2\kappa h)\right]\right]. \tag{8}$$

The electrical double layer (EDL) interaction force can be obtained by differentiating electrical double layer interaction energy with separation between two surfaces.

$$F_{edl} = -\frac{\partial V_{edl}}{\partial h} = 2\pi\varepsilon_0\varepsilon a\kappa\left(\xi_p^2 + \xi_w^2\right)\left[\frac{2\xi_p^2\xi_w^2}{\xi_p^2 + \xi_w^2}\frac{\exp\left(-\kappa h\right)}{1 - \exp\left(-2\kappa h\right)} - \frac{\exp\left(-2\kappa h\right)}{1 - \exp\left(-2\kappa h\right)}\right] \tag{9}$$

where ζ_p and ζ_w denote the zeta potentials of colloid particle and channel wall, respectively. ε_0 and ε_r represent the permittivity of vacuum and relative permittivity, respectively. The EDL thickness, also known as Debye length, κ^{-1} is defined as

$$\kappa^{-1} = \left(\frac{2e^2z^2n_\infty}{\varepsilon k_B T}\right)^{-0.5} \tag{10}$$

where e represents the electron charge, z is the valance of ions and n_∞ denotes the bulk number density of ions. When the conditions of $h \ll a_p$ and $\kappa a_p \gg 1$ are satisfied, the expressions above can work well with cases of small potential. Alternatively, the EDL interaction energy can be computed with either the linear superposition method or the complete numerical solution of the nonlinear Poisson-Boltzmann equation [33]. Considering ion-ion interactions, a complex statistical mechanical model was developed for calculating the EDL interaction based on the thermodynamic entropy and Helmholtz free energy approach. Different boundary conditions have been studied, such as charge regulation and constant surface charge density [28, 34, 35].

The total interaction energy, V, in the DLVO theory is obtained by the summation of the electrostatic and van der Waals contributions as illustrated in **Figure 3**. With the electronic double layer potential and van der Waals potential described under previous mentioned assumptions, the total interaction is calculated as

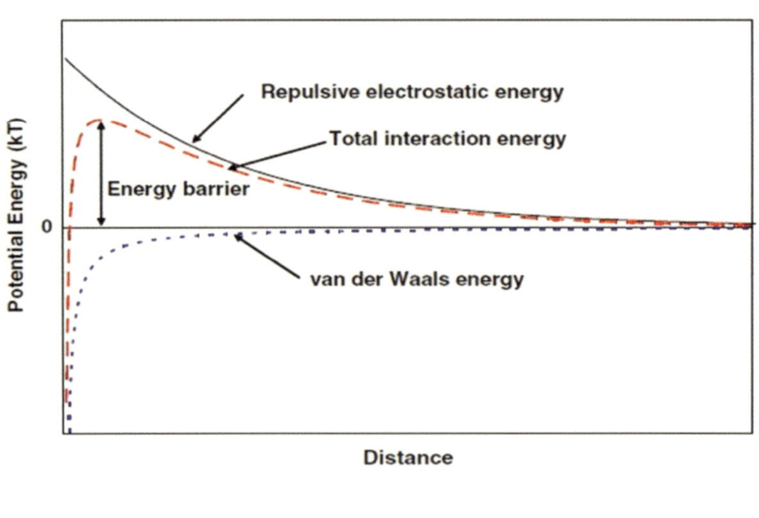

Figure 3. Example diagram of potential energy vs. separation distance [36].

$$V = V_{vdw} + V_{edl} \tag{11}$$

Eq. (11) gives both a theoretical framework to predict and compare experimentally measured colloidal interactions, and the knowledge of how surface interactions can be controlled.

3.2. Non-DLVO forces

In the classical DLVO theory, van der Waals and electrical double layer interactions play vital roles in colloidal particle interactions. This theory has been successfully utilised to explain many experimental observations. Whereas, there are situations in which theoretical predications based on interactions of electrical double layer and van der Waals force cannot provide reasonable agreement with experimental results [6, 37]. For instance, the classic DLVO theory fails to explain the interactions with ultra-short separation distance (i.e., shorter than a few nanometres). The continuum theories are not valid in such short distances, and bulk material properties (e.g., refractive index, density, and dielectric constant) cannot be used to describe such interactions. For these cases, some additional non-DLVO forces can be introduced into the DLVO theory, such as Born repulsion [38], polymer bridging [39], hydration forces [40], hydrophobic interaction [41], and steric interaction [42]. In this section, a brief introduction for non-DLVO forces is provided and more details can be found in a comprehensive review by Liang [4].

Polymer bridging theory applies to polymer flocculation. It is postulated that polymer bridges are built between neighbouring solid particles in a suspension in order to form a loose porous 3D network of solid particles (i.e., floc). When the detailed spatial variation of the short-range forces are not crucial, Elimelech et al. [30] reported that the microscopically averaged Born repulsion could be a convenient approach to consider effects of non-DLVO interactions. As particles interact with adsorbed fluid layers, solvation or hydration forces begin to take effects. Grabbe and Horn [43] suggested that the repulsive hydration force plays a dominant role for two interacting silica surfaces in a short range immersed in an electrolyte solution (NaCl). Unlike the electric double layer force, the hydration force was found to be independent on the electrolyte concentration over the range in their experiments. However, the physical mechanism of the hydration force is still unclear. The anomalous polarisation of water near the interfaces could generate the hydration force. It also could originate from the entropic repulsion of thermally activated molecular groups from protrusions on the surfaces [44–46]. Water molecules between two hydrophobic surfaces tend to migrate from the narrow gap to the bulk liquid at extremely short separation distance. It is because that the opportunities for hydrogen bonding are unlimited in the bulk liquid and free energy is lower than in the gap. As a result, an attractive force, hydrophobic force, would be generated between the two surfaces. This attractive force works in much greater range (up to 80 nm) than the van der Waals force and is one to two orders of magnitude stronger [41, 47, 48]. When two polymer-covered particles approach to each other, steric or osmotic forces would be developed between the particles. The steric force is related to the repulsive entropic force caused by the entropy of confining these chains for overlapping polymer molecules. So far, theories of steric forces are not well-established. Many components can affect the magnitude of the steric forces, such as bonding stress between the polymer molecules, the quantity or coverage of polymer molecules on each solid surface and solid surfaces (i.e., reversible process or not) [49–52].

4. Temperature control in microfluidic systems

Temperature is a crucial parameter in many microfluidic applications, for example, microscale milk pasteurisation unit [53], polymerase chain reaction (PCR) [54, 55], mixing [56], and temperature gradient focusing [57] or separation [58]. Lab-on-a-chip devices and systems are compact and multi-functional platforms which can be integrated into a small chip. Temperature control is one of the important functions in various microfluidic applications. In this section, different techniques for temperature control in microfluidic systems will be summarised into two categories: bulk temperature control and temperature gradient control [59].

4.1. Bulk temperature

The bulk temperatures of liquids in microfluidic systems are uniformly distributed and it can be changed by either external heating or internal heating approaches. A number of techniques have been implemented for various applications.

To control the bulk temperature of the liquids, the external heating approaches usually use either commercial heaters (e.g., Peltier elements) or preheat/cool liquids prior to being injected into microchannels. Velve Casquillas et al. [60] designed an external temperature control system with two Peltier elements that is able to readily vary the temperature of yeast channel underneath the temperature control channel (shown in **Figure 4a**). By using this system, they can rapidly regulate the yeast channel temperature in a wide range (5–45°C within 10 s). Similarly, Khandurina et al. [61] utilised two Peltier elements as a sandwich assembly to directly heat up a microfluidic chip for the polymerase chain reaction (PCR) (**Figure 4b**). This

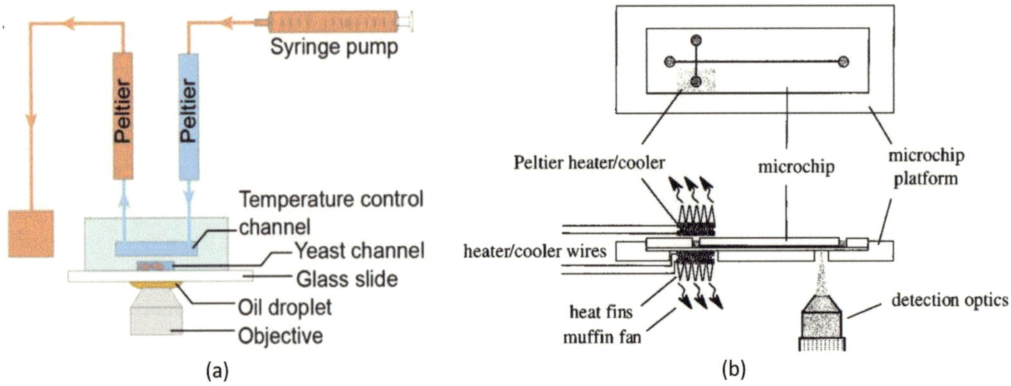

Figure 4. (a) A schematic of the temperature control device by an external Peltier element; the yeast channel is placed below the temperature control channel [60] and (b) schematic of the dual Peltier assembly for rapid thermal cycling followed by electrophoretic analysis on-chip [61].

http://dx.doi.org/10.5772/intechopen.78240

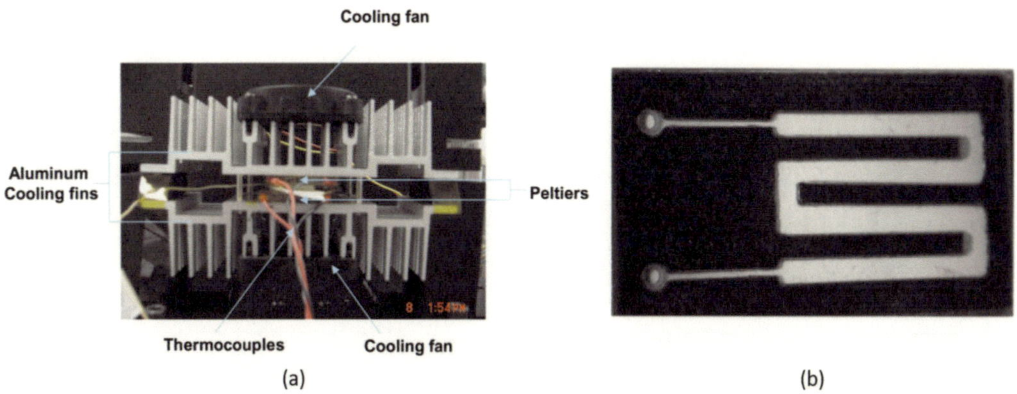

Figure 5. (a) Peltier thermocycler for the PCR microreactor and (b) a photograph of the serpentine polycarbonate microchannel [55].

compact setup is able to complete 10 thermal cycles within 20 min. With the similar configuration, Yang et al. [55] fabricated a serpentine microchannel on a thin polycarbonate plate (thickness: 0.75 mm) as a PCR microreactor (**Figure 5**). They used thermocouples to measure the surface temperatures of the intrachamber temperature and Peltier elements. With the developed device, they performed 30 thermal cycles in 30 min and the heating rate and cooling rate are 7–8 and 5–6°C/s, respectively.

Furthermore, Maltezos et al. [62] chose using liquid metal in combination with Peltiers to explore the limits of PCR speed in a microchip. Sample tubes were first immerged in a liquid medium (gallium eutectic) and then sandwiched between two high-powered Peltier elements. Heat can be rapidly transferred between the DNA/RNA in the sample tubes and the Peltier elements via the liquid interface. High heating rate and cooling rate can be up to 106 and 89°C/s in their experiments, respectively.

On the other side, the direct integration of heating or cooling elements into microfluidic systems is a popular method. Integrated heating approaches are able to reduce the whole system size and improve its portability.

Joule heating has been the most widely-used technique as an integrated heating solution for the temperature control in microfluidic systems. Lao et al. [63] filled platinum into silicon-based microchannels as in-chip heaters and sensors. A well-controlled thermal environment (±1°C) was achieved for gas and liquid phase reactions (**Figure 6**). The silicon microchannel was thermally isolated by a thin silicon nitride membrane to save power consumption. Precise and prompt temperature control was facilitated with a digital feedback system developed by them (heating rate: 20°C/s, cooling rate: 10°C/s, and response time: 5 s). By using similar serpentine channel configurations, Wu et al. [64] injected silver paint into the PDMS microchannels as integrated microheaters and took an advantage of compressed air for rapid cooling in parallel channels. Heating rate was 20°C/s and error of temperature was about ±0.5°C in the steady state. With this device, DNA can be successfully amplified in 25 thermal cycles with 1 min per cycle. Instead of applying metal, de Mello et al. [65] made use of ionic

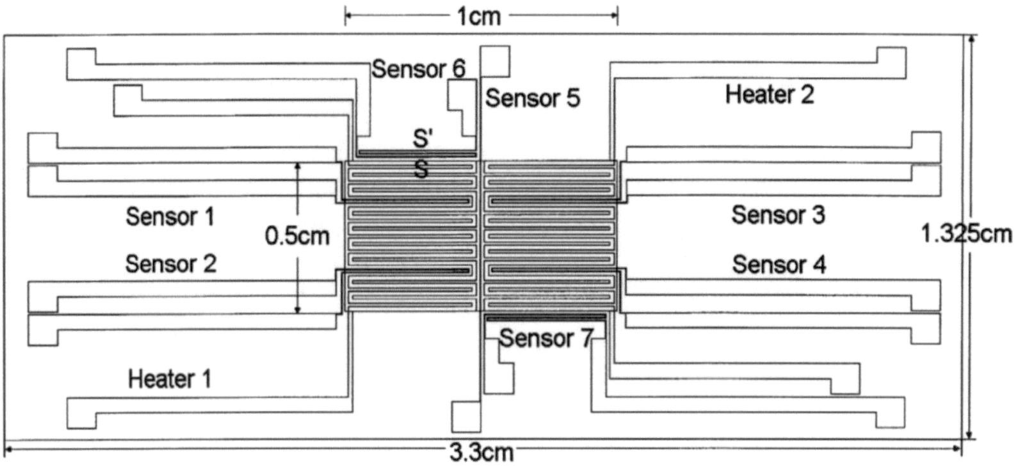

Figure 6. Top view of the microchip showing the patterned heaters and temperature sensors [63].

liquids ([BMIM][PF_6] and [BMIM][Tf_2N]) as heating elements. The ionic liquid was Joule heated by an AC current (power: 1 W, frequency: 50 Hz, and voltage: ≤ 3.75 kV). By controlling the applied voltage of the ionic liquids, the bulk temperature was regulated from 50 to 90°C with an accuracy of 0.2°C.

Other than the commonly used physical heating approach (i.e., Joule heating), chemical approach can be another integrated option. To locally control the temperature in a microchannel, Guijt et al. [66] made use of heat adsorption and dissipation via endothermic and exothermic chemical reactions, respectively (**Figure 7a**). Heating/cooling can be achieved in a temperature control channel (TCC) where two reagents from separate reactant channels (RC_1 and RC_2) merged and had chemical reaction. Heating was achieved with a dissolution of 97 wt% H_2SO_4 (Reagent 1) in water (Reagent 2), while cooling was realised with the evaporation process acetone (Reagent 1) in the air (Reagent 2). A wide temperature range from −3 to 76°C with a ramp rate of 1°C/s can be realised in the central channel by manipulating the flow rate ratio between two reagents (shown in **Figure 7b**).

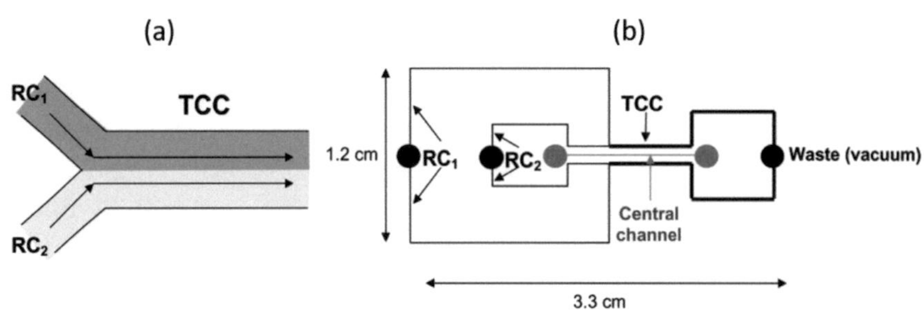

Figure 7. (a) Two reactant channels (RC) merging into a temperature control channel (TCC) and (b) layout of the device used for demonstration experiments [66].

http://dx.doi.org/10.5772/intechopen.78240

4.2. Temperature gradient

For several applications such as Soret effect and droplet actuation, temperature gradients are demanded in microfluidic systems. Temperature gradients have been generated in either a controlled temperature profile or an arbitrary way by various techniques.

As illustrated in **Figure 8a**, Mao et al. [67] generated a linear temperature gradient along the horizontal direction of the PDMS microchannel by using the pre-heated liquid approach. Hot and cold fluids were channelled into two side channels separately, and a stable temperature gradient has been established in the central channel (**Figure 8b**). Zhao et al. [68] implemented this approach for studies on thermophoresis using a microchip made of stainless steel. Due to a higher conductivity of stainless steel than PDMS, a wider range of variations for both bulk temperature and temperature gradient (1.5×10^4 K/m) can be formed in the microchannel. In order to fabricate similar structures in PDMS microchip, Yan et al. [69] proposed a fast prototyping method for single-layer PDMS microfluidic devices with abrupt depth variations by combining the laser ablation and NOA81 moulding. The whole fabrication process can be completed within 2 h. This method can readily produce PDMS microfluidic devices with micrometre and millimetre structures in one step. Moreover, this method can be applied in a non-clean-room environment and does not require complicated and expensive soft lithography equipment or etching processes.

Instead of preheating liquids, Vigolo et al. [70] chose to use silver-filled epoxy with a similar parallel channel configuration to generate temperature gradients in a microchannel. This microscale heating element can be powered by two ordinary AAA batteries. Thus, this design shows its potential as a cost effective and portable solution for thermal control of microfluidic devices. Alternatively, the parallel-channel configuration has also been implemented for control of bulk temperatures in microchannels. It can be readily realised by changing the directions of preheated liquid and DC electric current [71–73].

In addition to regulating bulk temperatures, Peltier elements can be utilised to establish a constant temperature gradient in microfluidic systems. Matsui et al. [74] designed and fabricated a hybrid temperature gradient focusing (TGF) chip (materials: PDMS/glass) by applying two Peltier elements. A range of temperature gradient can be formed along the horizontal

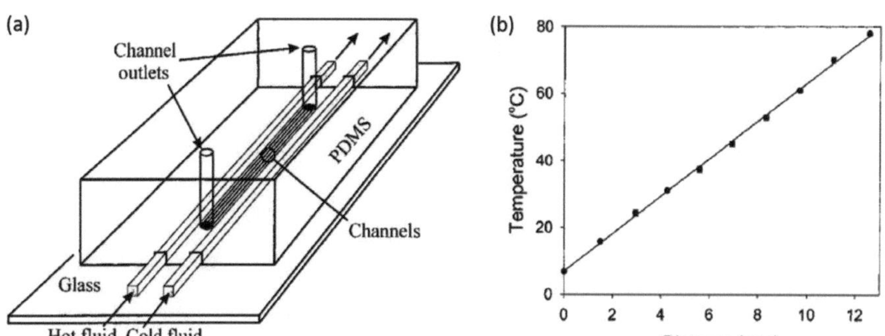

Figure 8. (a) Schematic diagram of a device with an on-chip linear temperature gradient and (b) a plot of temperature vs. position of the temperature gradient device [67].

direction of the microchannel (**Figure 9**). The maximum temperature gradient generated in their device was 13.75°C/mm.

Jiao et al. [75] designed a microfluidic droplet manipulation system with integration of four microheaters (materials: titanium and platinum) generating planar temperature gradients in the square region (10 mm × 10 mm). By controlling the frequency and amplitude of square wave signals, they successfully manipulated the trajectory of a microdroplet based on the periodic thermocapillary actuation caused by temperature gradients. Yap et al. [76] deposited a thin film platinum heater in a microchannel to control the thermal field for droplet formation in a bifurcated microchannel. The trajectory and splitting of the droplet can be controlled by modulating viscosity and interfacial tension of liquids under different heating conditions. Because of the temperature gradients generated by metal microheaters, the PDMS microchannel would undergo a slight dilation. The temperature-induced dilation changed the height of the microchannel so that the bubble would be driven away from the constricted region. Selva et al. [77] successfully controlled the bubble motion by utilising this thermomechanical actuation under temperature variation. Miralles et al. [78] from the same research group optimised their previous design by using localised heating resistor whose size was smaller than the droplet size.

The aforementioned designs of temperature-gradient devices are not suitable for researches on the microscale mechanism of particle deposition in microchannels due to the obstruction of heating or cooling elements along the optical path of microscope for observation or inconsistency of the directions of temperature gradient and particle deposition. The authors' group has developed a novel design of temperature-gradient microchip to investigate particle deposition in microchannels [79].

There are three major parts for the temperature-gradient microchip (**Figure 10**), including a microchannel, a thermoelectric cooler (TEC) unit as cold end, and a thin glass slide coated with indium tin oxide (ITO) film as hot end. A temperature gradient can be established along the vertical direction inside the microchannel, by cooling the top surface of the microchip with the TEC unit and heating the bottom surface with the ITO film heater.

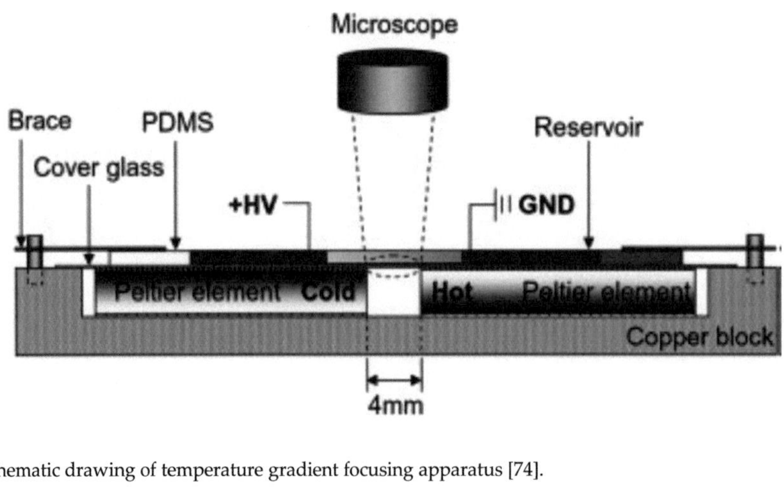

Figure 9. Schematic drawing of temperature gradient focusing apparatus [74].

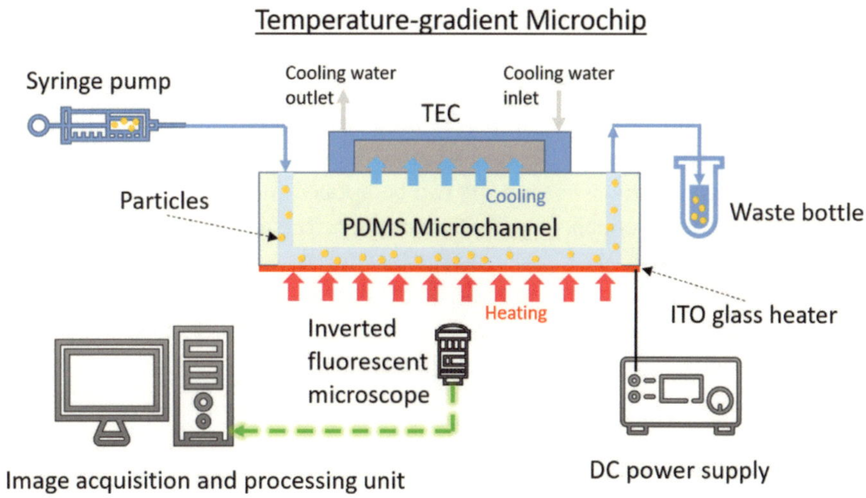

Figure 10. Schematic of the temperature-gradient microchip, consisting of the PDMS microchannel, the glass slide coated with the ITO film and the TEC unit. Various temperature gradients can be achieved by cooling the top surface of the microchip with the TEC unit and heating the bottom surface with the ITO film heater. (figure is for not drawn to scale) [79].

The microchannel can be fabricated with polydimethylsiloxane (PDMS) using standard soft lithography techniques. The PDMS monomer and the curing agent are fully mixed in a mass ratio of 10:1, and then they are vacuumed for 45 min to evacuate air bubbles remaining in the PDMS mixture. The mixture is applied onto the master mould constructed by SU8 on a silicon wafer. PDMS is cured after being heated with the wafer in an oven at 80°C for about 1 h. A thin layer of polymerised PDMS (e.g., 1 mm in thickness) is peeled off from the mould. Two cylindrical openings are then punched at both ends of the microchannel as the inlet and outlet for sample fluids.

For the ITO glass slide, a thin indium tin oxide film (e.g., 200 nm in thickness) is deposited on one side of a glass slide (e.g., 0.71 mm in thickness) as a heater for the microchip. Indium tin oxide is a solid mixture including 90% In_2O_3 and 10% SnO_2 by mass. On the other side of the glass slide, a thin PDMS film was coated on the bare glass surface. The ITO glass heater has three major advantages for the particle deposition study. (1) The ITO glass (10 ohms/sq) can easily generate heat by Joule heating when being connected into an electrical circuit. The heat dissipation rate of the ITO glass can be well controlled by regulating the applied electrical current and voltage. (2) The ITO glass has excellent optical transparency allowing direct observation on particle deposition onto the bottom surface of the microchannel along the direction of the applied temperature gradient. The bottom view of the particle deposition can be readily captured via using an inverted microscope equipped with a CCD camera. (3) A wide range of customised dimensions and patterns can be precisely achieved for ITO heaters by implementing the standard photolithography techniques.

A closed microchannel is formed via irreversibly oxygen plasma bonding the treated ITO glass slide and the PDMS block with microchannel structure. The bonded microchannel is heated in the oven to reinforce the plasma-bonding strength. The thermoelectric cooler (TEC) unit is mounted on the top of the bonded microchannel to provide a stable cold end over the microchannel. The temperatures of hot/cold ends can be readily adjusted via the DC power supplies for the ITO film heater and the TEC unit, respectively. Moreover, the temperature gradient can be well controlled in the microchannel along the same direction as the particle deposition. Thermal conductive silicon paste can be applied in the gap between the TEC unit and the PDMS microchip to enhance the heat conduction. This specially designed temperature-gradient microfluidic system provides a useful tool for researches on dynamics of particle deposition under different thermal conditions. To the best knowledge of the authors, it could be the first microfluidic device allowing to directly observe the dynamic process of particle deposition along the direction of applied temperature gradient.

5. Particle deposition in microchannels

Particle deposition onto solid surfaces has been intensively studied in both experimental and theoretical approaches. In this section, major works on particle deposition in microchannels are reviewed.

5.1. Experimental studies

Experiments are usually conducted in three types of setup, including cylindrical, parallel-plate channels, and impingement jet chamber. Various parameters have been evaluated experimentally in terms of their effects on particle deposition, such as pH, electrolyte concentration, particle, and solid surface.

5.1.1. Effect of pH on particle deposition

The repulsive interaction would exist between charged particles of same like sign and hinder fouling behaviour. The zeta potentials of the particle and microchannel surface determine the magnitude of the repulsion and are closely related to the pH value of working fluid. The pH value can control the charge signs of the particle and channel surface. Newson et al. [80] investigated the mechanisms of deposition, removal and sticking in a haematite/water system (particle diameter: 0.2 μm and concentration: 100 ppm). They found that the sticking coefficient of particles from turbulent water suspension (Re: 11,000–14,000) was significantly dependent upon the pH value of suspension. Perry and Kandlikar [36] also adjusted the pH value of the nanofluids and were able to effectively mitigate particulate fouling. The reduction of the deposition rate was mainly attributed to the augmented EDL repulsion as pH was increased.

5.1.2. Effect of the ion concentration on particle deposition

Gu and Li studied the influence of the electrolyte concentration on deposition of silicon oil microdrops onto cylindrical surfaces via both experimental and numerical approaches [81, 82]. They found that the increase of the Sherwood number (dimensionless mass transport rate) resulted from the neutralisation effect of the electrolyte concentration (10^{-6}–10^{-3} M) which crucially determined the zeta potentials of bare glass tube surfaces and silicon oil drops. Moreover, they found that the Sherwood number was increased significantly with the increase of the cationic surfactant concentration (CTAB, 10^{-6}–10^{-4} M) but was reduced monotonically with adding anionic surfactant (SDS) into the oil droplet emulsion. Kar et al. [83] performed experiments with $CaCO_3$ microparticles and hollow fibre membrane to study the effect of salt concentration gradients on particle deposition. They reported that the diffusiophoretic particle transport has crucial influence on particle deposition when different electrolyte ions of salts in solution have different diffusion coefficients. Furthermore, Guha et al. [84] found the diffusiophoresis has significant influence on the colloidal fouling in a low salinity reverse osmosis system.

5.1.3. *Effect of properties of particle and wall on particle deposition*

Salim et al. [85] investigated the effects of protein (fibrinogen and lysozyme) adsorption on the electroosmotic flow (EOF) behaviour of the plasma-polymerised glass microchannel surfaces. Three types of plasma-polymerised surfaces (pp.TG, pp.AAm, and pp.AAc) were tested which had different surface charges and charge densities. They observed a non-fouling phenomenon with tetraglyme coating in the presence of protein, and this coating provided stable EOF performance in the glass microchannel.

Mustin and Stoeber [86] conducted experiments with polystyrene microsphere suspensions in a PDMS microchannel. They found that the dynamics of channel blockage was influenced by particle size distribution besides the particle size alone. Recently, they performed another experiment in a mini impingement jet flow cell made of PDMS for particle deposition (**Figure 11**) [87]. They noticed discrepancies between the experimental measurements and numerical simulation results based on both DLVO and extended DLVO theories, and proposed the surface roughness and electrostatic charge heterogeneity of the PDMS surface could be the most plausible reason for such discrepancies.

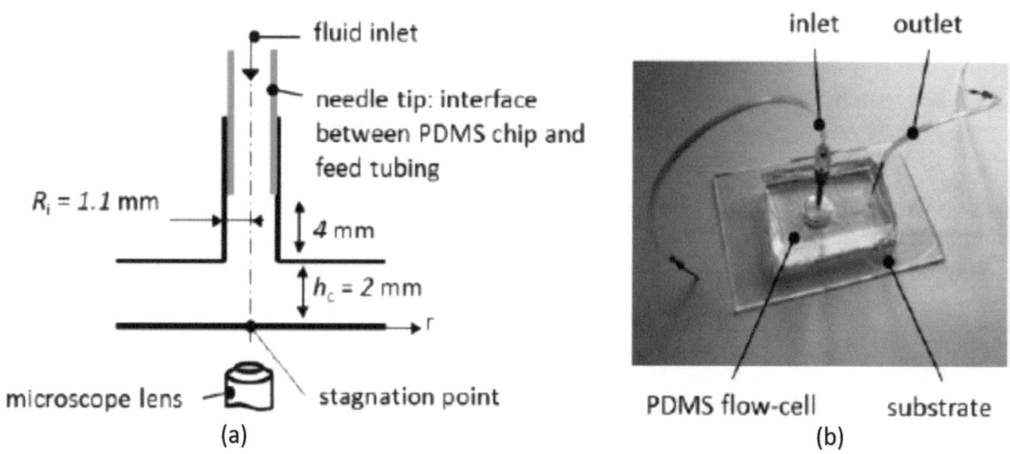

Figure 11. (a) Cross-section view of the deposition chamber and (b) flow cell on substrate without clamping [87].

5.1.4. *Effect of external field on particle deposition*

In the previous literature, transport of micron-size particles is normally simplified as a mass transfer process that is majorly affected by Brownian diffusivity. However, gravity or a constant body force exerted on the particles could have significant influence on the deposition process, even for such tiny colloidal particle size (less than 1 μm). Yiantsios and Karabelas [88] reported that gravity played a key role for the deposition rate of spherical Ballotini glass particles (diameter: 1.8 μm). Gravity could control the particle transport boundary layer in a horizontal narrow channel under laminar flow over a fairly wide range of flow rate. They [89] conducted further experiments on the effects of physicochemical and hydrodynamic conditions by using dilute microsized glass particle suspensions (diameter: 1.5 μm) in a parallel-plate channel. They concluded that gravity was a determining factor for deposition at low wall shear stresses. While the hydrodynamic wall shear stress was increased, particle deposition rates were noticeably decreased because of hydrodynamic lift or drag forces hindering transport or attachment.

Stamm et al. [90] experimentally examined the initial stage of cluster growth in a particle-laden flow in a microchannel and investigated the parametric effects of a void fraction, flow shear strain rate and channel height to particle diameter ratio. Thereafter, Gudipaty et al. [91] studied the cluster formation of colloidal particles in a PDMS microchannel and found that the clusters were initiated by the attachment of individual flowing particle onto the bottom surface. However, they have not either addressed the physical mechanism of the initial particle attachment to the surface or observed the adherence process in the experiments.

Unni and Yang [92] experimentally investigated the dynamics of particle deposition in an electroosmotic flow using video microscope, and reported that the increased surface coverage at higher salt concentrations resulted from weakened EDL repulsion with particles being adsorbed onto the channel surface. Hydrodynamic blocking became relatively weaker with lower electric field strengths because the surface blocking was majorly caused by electrical interactions.

5.1.5. *Effect of temperature on particle deposition*

Most of the researches about deposition of micro-/nanoparticles in microchannel are conducted in room temperature environment, seldom with high bulk temperature or temperature gradient, which is a crucial factor for heat exchangers in reality. Yan et al. [73] investigated the effect of bulk solution temperature on particle deposition in a microchannel under well-controlled temperature conditions using a microfluidic temperature control device. To the best knowledge of the authors, this was the first attempt to study the thermal effect on the deposition of colloidal particles in an aqueous dispersion onto a microchannel wall. It is found that the temperature of solution has a considerable effect on the particle deposition in microchannels. The static particle deposition rate (Sherwood number) has been measured over a range of temperatures between 20 and 70°C. It is found that the Sherwood number is monotonically increased up to 265%, with the solution temperature within the test range. They developed a deterministic model based on the Derjaguin-Landau-Verwey-Overbeek theory with consideration of temperature dependence, and found that by increasing the solution temperature, the attraction energy (van der Waals force) between the particles and the solid surface is increased while the repulsive energy (electric double layer force) is decreased. Moreover, they further studied the hydrodynamic effects on particle deposition in microchannels at elevated temperatures, including steady flow and pulsatile flow [71, 72]. The dimensionless particle deposition rate (Sherwood number) was found to be reduced with the Reynolds number and changed insignificantly for the Reynolds number beyond 0.091 (0.5 mL/h) within the tested range with a given solution temperature (324.85 K) and an electrolyte concentration (5×10^{-4} M). Under the pulsatile flow condition, the normalised particle deposition rate was found to be reduced significantly as the flow oscillation frequency was increased from 0 Hz to 1 Hz, while keeping the steady flow component and the amplitude of the flow oscillation unchanged.

5.2. Theoretical modelling/studies

Spielman and Friedlander [93] theoretically analysed the effect of the electrical double layer on particle deposition based on the equation of convective diffusion in an external force field. They reported that the deposition process of Brownian particles was equivalent to ordinary convective diffusion in the bulk with a first-order surface reaction at the collector. With respect to the net interaction potential, a formula can be derived for the surface reaction coefficient.

Another analytical model was developed by Adamczyk and Van De Ven [94] for particle deposition kinetics onto the surfaces of parallel-plate and cylindrical channels (**Figure 12**). As governing equation, the mass transport equation was formulated with consideration of electrical double layer force, van der Waals force, and external forces such as gravity. The 'perfect sink' boundary condition was applied to solve the mass transport equation. Different dimensionless parameters (Ad, Pe, Dl, and Gr) were proposed to account for dispersion, convection/diffusion, electrical double layer, and gravity, respectively.

Studies on particle deposition onto permeable surfaces are practically meaningful for membrane filtration industry. Song and Elimelech [95, 96] theoretically studied this phenomenon in

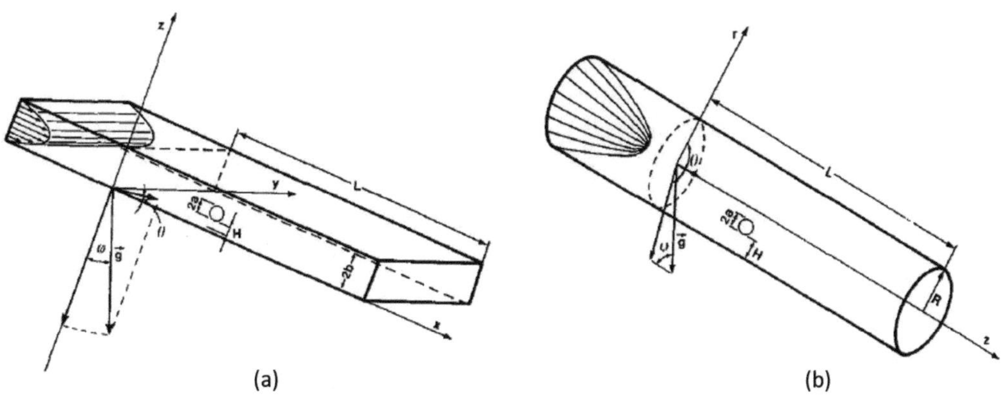

Figure 12. Schematics of parallel-plate and cylindrical channels [94].

a system of parallel-plate channel. The convection-diffusion equation was solved numerically with consideration of lateral transport. The lateral transport would be induced by inertial lift, permeation drag, and transport that are determined by the collective effect of gravity force and surface forces. Parametric studies were systematically carried out regarding the initial particle deposition rate, including the effects of permeation velocity, cross flow velocity, particle size, and solution ionic strength.

Since the distribution of surface charge is not uniform in practice, Nazemifard et al. [97, 98] performed a trajectory analysis of particles close to a micropatterned charged plate based on the radial impinging jet setup for the influences of surface charge heterogeneity on deposition efficiency and particle trajectory. The surface charge heterogeneity was controlled by concentric bands with varied properties, such as geometric dimension and types of surface charges. In their analysis, the van der Waals, electrostatic double layer force, hydrodynamic force, and gravity, have been taken into consideration. Due to the coupled effects of colloidal and hydrodynamic forces, the deposition efficiencies and particle trajectories were remarkably influenced by surface charge heterogeneity when a particle flowed radially away from the stagnation point in the radial jet impingement setup. This analysis demonstrates how the existing particle transport models could possibly be modified in consideration of chemical heterogeneity and additional surface interactions. Similarly, Chatterjee et al. [99] applied the convection-diffusion-migration equation (Eulerian model) with fully developed Poiseuille flow velocity profile to investigate the transport of particles in patchy heterogeneous cylindrical microchannels (**Figure 13**). They evaluated the effects of surface chemical heterogeneity on particle transport and deposition, and found that particles tend to travel further along the microchannel in the heterogeneous channels compared to the homogeneously favourable channels.

Using a soft-sphere discrete element method, Marshall [100] studied the caption of particles by wall and particle aggregation in a microchannel. According to their results, the particle lift-off from the wall was caused by adhesion and collision with particle aggregate or a passing particle when a single particle with a large size was attached to the wall. The fluid forces were not the direct reason for the particle lift-off.

Unni and Yang [92] developed an electrokinetic particle transport model in a parallel-plate microchannel (**Figure 14**) based on the Stochastic Langevin equation. They incorporated random Brownian motion of colloidal particles and the hydrodynamic, electrical, DLVO colloidal interactions into the equation. Based on the developed model, particle trajectories can be stochastically simulated using Brownian dynamics simulation and the surface coverage was calculated under a range of electrical and physicochemical conditions.

In the absence of repulsive energy barrier, Jin et al. proposed concurrent modelling for the effects of interaction forces and hydrodynamics on particle deposition on rough spherical surfaces [101]. The model considered the hydrodynamic retardation functions and flow field profiles. Their works showed that the hydrodynamic effects remarkably affect the particle deposition behaviours which were different from the predictions based on DLVO forces alone. In addition, surface roughness played an important role in particle deposition experiment/simulation. They conducted another study on deposition of colloidal particles and reported a non-monotonic, non-linear effect of nanoscale roughness on particle deposition without

http://dx.doi.org/10.5772/intechopen.78240

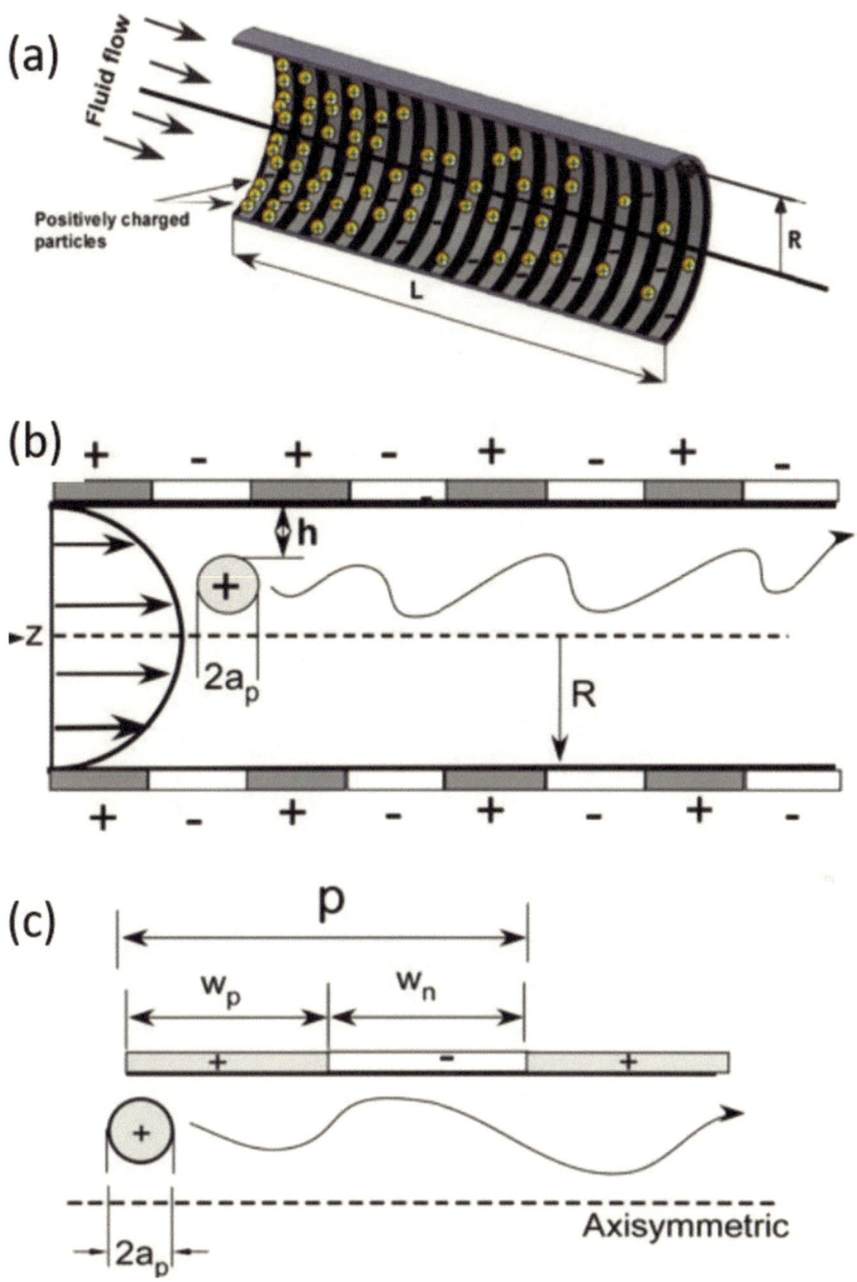

Figure 13. Schematic of patchy heterogeneous cylindrical microchannels. (a) 3D schematic representation of positively charged particles deposition along the the microchannel. (b) 2D axisymmetric view of the microchannel geometry. (c) The zoomed view of one pitch length [99].

energy barrier using both the convection-diffusion model and parallel-plate chamber experimental system [102]. Their results showed particle deposition flux could reach the minimum value when a critical roughness size was provided.

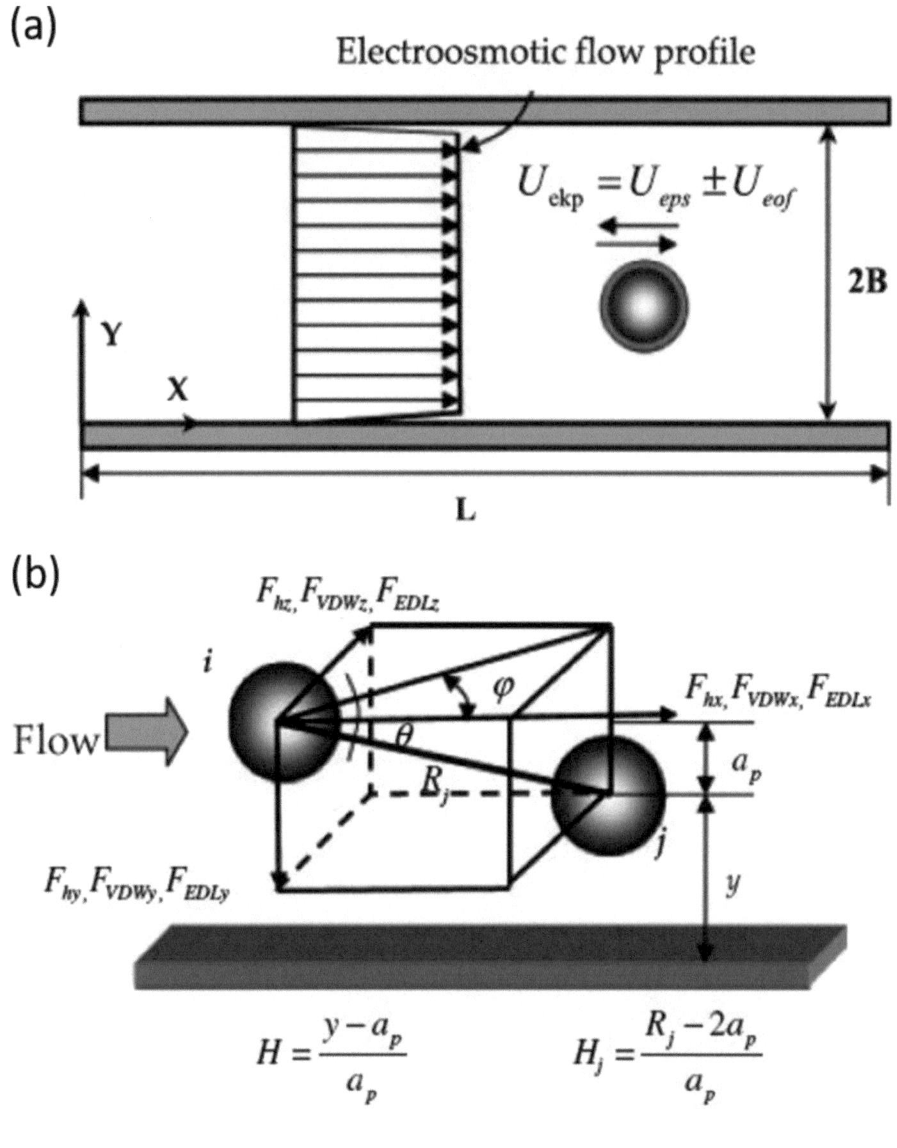

Figure 14. (a) Schematic of electrokinetic transport of a particle in a parallel-plate channel. (b) Forces acting on a moving particle in the vicinity of an attached particle and the channel wall [92].

Based on the Derjaguin-Landau-Verwey-Overbeek theory, Yan et al. [73] developed a qualitative theoretical model with consideration of the temperature effect for the first time. All the driving forces during the whole particle deposition process are temperature dependent, so the particle deposition is determined by the collective effects of the variations of the forces caused by temperature changes. By plotting interaction potential curves and energy barriers, the effect of bulk solution temperature can be clearly seen in **Figure 15** that the interaction potential

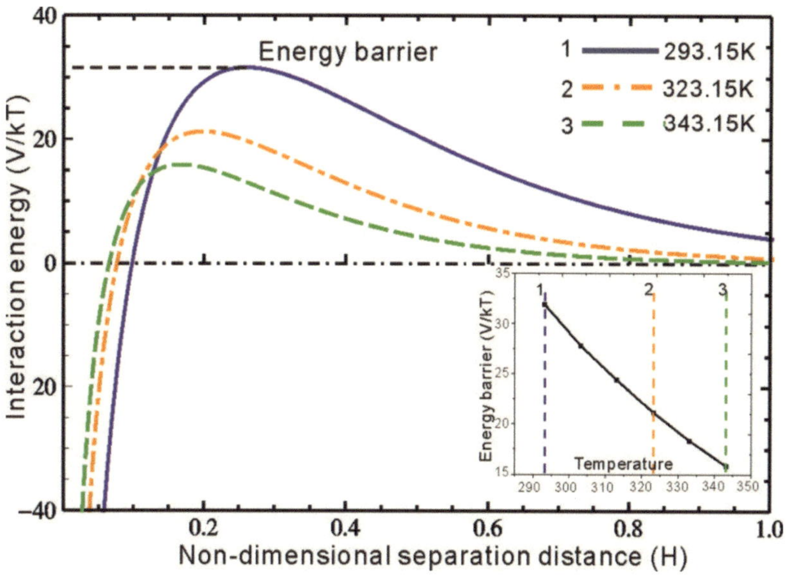

Figure 15. Dimensionless particle-wall interaction potential curves for three different solution temperatures (T = 293.15, 323.15, and 343.15 K). Curves are obtained with the temperature dependent Hamaker constant, zeta potential and the thickness of electric double layer. Inset shows the energy barrier that polystyrene particles should overcome to achieve deposition onto the PMMA surface at different temperatures for the DI water case [73].

curves are varied with the bulk solution temperature and the energy barrier reduces considerably with increasing the bulk solution temperature as shown in the inset of **Figure 15**.

Furthermore, Yan et al. [71] developed a simplified one-dimensional mass transport model (i.e., Eulerian model) for calculating the particle deposition rate in microchannel flows at elevated temperatures. A schematic of microparticles flowing through a microchannel at an elevated temperature (T) is shown in **Figure 16**. For a dilute spherical particle monodispersion in the absence of chemical reactions, the interactions between particles are neglected. The deposition rate of particles from solution onto the microchannel surface at the steady state can be described by the general convection-diffusion equation as

$$\nabla \cdot \mathbf{j} = -\nabla \cdot (\mathbf{D} \cdot \nabla n) + \nabla \cdot (\mathbf{u}n) + \nabla \left(\frac{D\mathbf{F}}{k_B T_0} n \right) = 0, \tag{12}$$

where **j** is the particle flux vector (the number of particles per unit area per second), the particle flux comprises three components as shown in Eq. (12): Brownian diffusion, fluid convection, and migration under external forces. Besides, the particle flux **j** can be decomposed into two portions along the x-direction and y-direction, $\mathbf{j} = j_x\vec{x} + j_y\vec{y}$, n is the local particle number concentration, **D** is the diffusion coefficient tensor, **u** is the particle velocity, **F** indicates external forces exerted on the particles, k_B is the Boltzmann constant, and T_0 is the reference temperature. By an appropriate scaling with the dimensionless parameters given in **Table 1**, the mass transport equation can be expressed in a dimensionless form as

Dimensionless parameters	Expression
Scaled particle flux	$J_x = \frac{a_p j_x}{n_{\infty,p} D_\infty}, J_y = \frac{a_p j_y}{n_{\infty,p} D_\infty}$
Scaled particle concentration	$\overline{n} = \frac{n}{n_{\infty,p}}$
Scaled external force	$\overline{F} = \frac{a_p F}{k_B T_0}$
Scaled interaction energy	$\overline{V} = \frac{V}{k_B T_0}$
Scaled particle-wall separation distance	$H = \frac{h}{a_p} = \frac{y - a_p}{a_p}$
Scaled distance from channel entrance	$X = \frac{x}{w}$
Scaled particle ratio	$R = \frac{a_p}{w}$
Peclet number	$Pe = \frac{3U_{avg} a_p^3}{2w^2 D_\infty}$
Hydrodynamic retardation functions [101]	$f_1(H) = 1 - 0.399\exp(-0.14869H) - 0.601\exp\left(-1.2015H^{0.92667}\right)$
	$f_3(H) = 1 - 0.3752\exp(-3.906H) - 0.625\exp\left(-3.105H^{0.15}\right)$
	$f_4(H) = 1 - 1.23122\exp(-0.2734H) + 0.8189\exp\left(-0.175H^{1.2643}\right)$

Table 1. Dimensionless parameters utilised for the mass transport equation [71].

$$-f_4(H)R^2\frac{\partial^2 \overline{n}}{\partial X^2} + \{Pef_3(H)(H+1)[2-(H+1)R]\}\frac{\partial \overline{n}}{\partial X} + \frac{\partial}{\partial H}\left(-f_1(H)\frac{\partial \overline{n}}{\partial H} + f_1(H)\overline{F}\overline{n}\right) = 0. \quad (13)$$

All the forces acting on particles are along the vertical direction (y-axis in **Figure 16**), and they are colloidal forces (van der Waals force $\overline{\mathbf{F}}_{vdw}$ and electric double layer force $\overline{\mathbf{F}}_{edl}$) and external forces (gravity $\overline{\mathbf{F}}_G$, hydrodynamic lift force $\overline{\mathbf{F}}_L$, and thermophoretic force $\overline{\mathbf{F}}_T$). The thermophoretic force is neglected because of its low magnitude compared to other forces in the study [73]. The non-DLVO forces are excluded for simplicity. Hence, the total force ($\mathbf{F} = \mathbf{F}_y$) can be treated as a scalar in the following sections. The dimensionless total force ($\overline{F}$) is calculated as

$$\overline{F} = \overline{F}_y = \overline{F}_{edl} + \overline{F}_{vdw} + \overline{F}_G + \overline{F}_L. \quad (14)$$

The potential of each force is computed by integrating the force over the separation distance between a particle and the channel surface (H). The total potential ($\overline{V}$) can be obtained by using superposition of individual potentials: the van der Waals potential ($\overline{V}_{vdw}$), the EDL potential ($\overline{V}_{edl}$), and the potentials contributed by gravity and lift force ($\overline{V}_G$, $\overline{V}_L$),

$$\overline{V} = \overline{V}_{edl} + \overline{V}_{vdw} + \overline{V}_G + \overline{V}_L = \int \overline{F}_{edl} dH + \int \overline{F}_{vdw} dH + \int \overline{F}_G dH + \int \overline{F}_L dH \quad (15)$$

As the scaled particle ratio ($R = a_p/w$) in Eq. (13) is about the order of 10^{-3} ($a_p \ll w$) for their experiment system, the particle diffusion term (first term in Eq. (13)) and the particle convection term (second term in Eq. (5)) in x-direction can be neglected for the low Peclet number (<1) in the present study. Besides, the variation of the deposition rate for a dilute particle solution

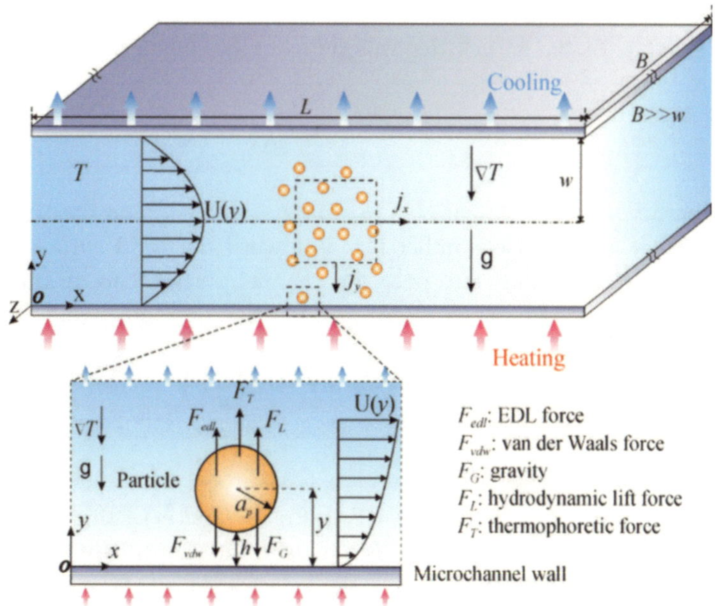

Figure 16. Schematic of microparticle transport in a microchannel. The forces on the particle are van der Waals force (F_{vdw}), gravity force (F_G), electric double layer force (F_{edl}), thermophoretic force (F_T), and hydrodynamic lift force (F_L). The radius of the particle is a_p, the minimum separation distance between the particle surface and the bottom surface of the microchannel is h, the flow velocity distribution is U(y), and the applied temperature gradient in the microchannel is ∇T (the figure is not drawn to scale) [71].

along the x-direction in downstream was found to be insignificant [96, 103]. Consequently, the mass transport equation can be further simplified to one-dimensional as

$$\frac{\partial}{\partial H}\left(-f_1(H)\frac{\partial \overline{n}}{\partial H}+f_1(H)\overline{F}\overline{n}\right)=0. \tag{16}$$

Eq. (16) can be solved with the boundary conditions as

$$\overline{n}=0, \text{at } H=H_0 \tag{17}$$

$$\overline{n}=1, \text{at } H=H_\infty. \tag{18}$$

Eq. (17) refers to the 'Perfect sink' boundary condition. H_0 indicates the minimum dimensionless particle-wall distance (h_0/a_p). This boundary condition has been widely used in particle deposition studies, and it assumes that all particles are irreversibly adhered to the solid surface when they move into the primary energy minimum (PEM) region. It can be explained by that the attractive van der Waals force in the PEM region becomes much stronger than the repulsive electric double layer force. Thus, the particles would deposit onto the solid surfaces. The second boundary condition, given by Eq. (18), states a natural boundary condition for the particle concentration. The particle concentration gets close to that in the bulk phase when the particle-wall distance becomes an 'infinite' distance.

Having obtained dimensionless concentration distribution ($\overline{n}$) along the dimensionless separation distance (H) in Eq. (6), the particle deposition flux to the channel surface can be found as

$$J_0 = -f_1(H_0)\left(\frac{d\overline{n}}{dH}\right)_{H=H_0}. \tag{19}$$

Here, J_0 is the particle number flux at $H = H_0$. The negative sign on the right hand side of Eq. (9) indicates that the particle number flux is toward the solid surface. Moreover, the dimensionless particle deposition rate onto the channel surface can be quantified by the Sherwood number

$$\text{Sh}_{num} = -\frac{j_0}{(D_\infty n_{\infty,p}/a_p)} = f_1(H_0)\left(\frac{d\overline{n}}{dH}\right)_{H=H_0}. \tag{20}$$

The thermal effects on the particle deposition rate (i.e., Sherwood number) are influenced by the temperature dependences of all the forces ($\overline{\mathbf{F}}_{vdw}$, $\overline{\mathbf{F}}_{edl}$, $\overline{\mathbf{F}}_G$, and $\overline{\mathbf{F}}_L$) acting on the particles. For van der Waals force, the Hamaker constant is a temperature dependent parameter. For EDL force, EDL thickness, zeta potential, and relative dielectric constants of materials are varied with the temperature. For gravity and hydrodynamic lift forces, the magnitude of forces is changed due to the variations of density and viscosity of liquid. Details of the calculations for the temperature dependences can be found in [71, 73].

For the first time, based on the DLVO theory with considering the temperature-dependent interactions, a simplified one-dimensional mass transport model was developed and it can serve as a semi-quantitative approach for describing particle deposition phenomena in microchannel flows at elevated temperatures.

6. Summary and future prospects

Particle deposition and particulate fouling have been ubiquitous phenomena in natural and industry processes. Thermal effects (i.e., temperature and temperature gradient) on particle deposition are important but always 'ignored' in literatures. Most of the published research works about micro-/nanoparticles deposition in a microchannel were conducted in the room temperature environment, seldom with consideration of elevated bulk temperature or temperature gradient, which is a crucial factor for thermal driven fouling phenomena in reality. Especially, the microscale mechanism of particle deposition in microchannel at elevated temperature was still in its infancy. In this chapter, researches on particle deposition and particulate fouling on surfaces have been extensively reviewed both theoretically and experimentally from the published works. This chapter has summarised relevant concepts of particle deposition, key parameters, and experimental techniques (e.g., device design) as well as theoretical methodologies (e.g., modelling). The physics of particle deposition phenomena under different parametric influences has been discussed in detail. The authors have presented a new microfluidic temperature-gradient device that can be used to directly observe particle deposition

http://dx.doi.org/10.5772/intechopen.78240

along the direction of temperature gradient with a single-particle resolution. Moreover, a simplified mass transport model (Eulerian model) with consideration of thermal effects has been presented to describe the particle deposition phenomena in microchannels at elevated temperatures based on the Derjaguin-Landau-Verwey-Overbeek theory. Both the theoretical modelling and experimental measurements have shown that the thermal effects have profound the impact on particle deposition in microchannels.

Future research in this field lies in the development of investigations on coupling effects of thermal field and other external fields, such as optical, acoustic, and magnetic fields [104–106]. Microfluidic devices and systems offer ideal experimental platforms which provide well-controlled external fields applied to the particle deposition process. Especially, the dynamic behaviour of particle deposition under complex external fields can be observed directly with a resolution of micrometre even nanometre (e.g., single molecule detection). With such unique information, trans-scale theoretical modelling which can bridge the gap between particle kinetics in microscale and fouling phenomena in macroscale will be highly appreciated. The current investigations at elevated temperatures rely on either ensemble average value of a population of particles (i.e., average particle deposition rate) or bulk property of fouling (i.e., fouling resistance). When the deposition of single nanoparticle at elevated temperatures can be reliably investigated, researchers will be able to understand the dynamic processes of nanoparticle deposition in microchannels which elucidates particulate fouling of nanofluids in heat exchangers and nanomaterial drug delivery *in vivo* in details.

Acknowledgements

This work has been partially supported by the National Key Research & Development Program of China (2016YFB0401502), the National Natural Science Foundation of China (61574065, 51561135014, and U1501244), Science and Technology Planning Project of Guangdong Province (2016B090906004), Special Fund Project of Science and Technology Application in Guangdong (2017B020240002), PCSIRT Project No. IRT_17R40, the National 111 Project, the MOE International Laboratory for Optical Information Technologies, and the Cultivation Project of National Engineering Technology Center of Optofluidic Materials and Devices (2017B090903008). Zhibin Yan also acknowledges the financial support by the Foundation for Distinguished Young Talents in Higher Education of Guangdong, China and the Research and Cultivation Fund for Young Faculty supported by South China Normal University (SCNU). Xiaoyang Huang and Chun Yang appreciate the financial support from the Ministry of Education of Singapore under Grant No. RG80/15 and RG97/13, respectively.

Nomenclature

a_p	particle radius (m)
a_1	radius of interacting particle 1 (m)

a_2	radius of interacting particle 2 (m)
A_{132}	Hamaker constant for the interaction between substances 1 and 2 in medium 3 (J)
B	width of the microchannel (m)
w	half channel height (m)
D_x	diffusivity in X direction (m^2/s)
D_y	diffusivity in Y direction (m^2/s)
D_T	thermal diffusion coefficient (m^2/s)
D_∞	Stokes-Einstein diffusivity (m^2/s)
Dl	double layer number in the HHF expression for the EDL interaction
Da	double layer asymmetry number in the HHF expression for the EDL interaction
$f_1(H)$	UHCC correction function for diffusivity perpendicular to the solid surface
$f_3(H)$	UHCC correction function for diffusivity parallel to the solid surface
$f_4(H)$	UHCC correction function for diffusivity along the flow direction (X)
F_x	X component of the total force exerted on the particle (N)
F_y	Y component of the total force exerted on the particle (N)
$\overline{F}_{edl}$	dimensionless EDL interaction force
$\overline{F}_{vdw}$	dimensionless van der Waals interaction force
$\overline{F}_G$	dimensionless gravity force
$\overline{F}_L$	dimensionless hydrodynamic lift force
$\overline{F}_T$	dimensionless thermophoretic force
h	minimum separation distance between particle and wall (m)
H	dimensionless particle-wall separation
H_0	dimensionless primary energy minimum separation
H_∞	dimensionless particle-wall separation in bulk fluid
j_x	particle flux in X direction ($m^{-2}\ s^{-1}$)
j_y	particle flux in Y direction ($m^{-2}\ s^{-1}$)
$\overline{J}_x$	dimensionless mass flux in X direction

$\bar{J}_y$	dimensionless mass flux in Y direction
k_B	Boltzmann's constant ($1.38064852 \times 10^{-23}$ J/K)
L	length of the microchannel (m)
n	particle number concentration (/m^3)
$\overline{n}$	dimensionless particle number concentration
$n_{\infty,p}$	particle number concentration in bulk liquid (/m^3)
Pe	Peclet number based on average flow velocity
Q	volume flow rate of liquid (m^3/s)
Re	Reynolds number based on average flow velocity
Sh	Sherwood number
t	time (s)
T	absolute temperature (K)
T_0	reference temperature (293.15 K)
U_{avg}	average flow velocity in microchannels (m/s)
$\overline{V}$	dimensionless total interaction potential
$\overline{V}_{edl}$	dimensionless EDL interaction potential
$\overline{V}_{vdw}$	dimensionless van der Waals interaction potential
$\overline{V}_G$	dimensionless gravity potential
$\overline{V}_L$	dimensionless lift potential
$\overline{V}_T$	dimensionless thermophoretic potential
z	ionic valence
e	elementary electric charge ($1.602176620898 \times 10^{-19}$ C)

Greek symbols

ε	dielectric constant
ε_0	absolute dielectric constant of vacuum (8.845×10^{-12} C^2/N m^2)
ξ	zeta potential (V)
λ	London retardation wavelength (100 nm)
κ	Debye-Hückel parameter (m)
μ	dynamic viscosity of liquid (N s/m^2)

ν	kinematic viscosity of liquid (m^2/s)
ρ	density (kg/m^3)
ψ	electric potential (V)
τ	ratio of particle radius (a_p) to the Debye length (κ^{-1})
ω	mobility of particle

Author details

Zhibin Yan[1,2,3]*, Xiaoyang Huang[3], Chun Yang[3], Mingliang Jin[1,2] and Lingling Shui[1,2]

*Address all correspondence to: zhibin.yan@m.scnu.edu.cn

1 Guangdong Provincial Key Laboratory of Optical Information Materials and Technology and Institute of Electronic Paper Displays, South China Academy of Advanced Optoelectronics, South China Normal University, Guangzhou, P. R. China

2 National Center for International Research on Green Optoelectronics, South China Normal University, Guangzhou, P. R. China

3 School of Mechanical and Aerospace Engineering, Nanyang Technological University, Singapore

References

[1] Liu R. A Study of Fouling in a Heat Exchanger with an Application of an Electronic Antifouling Technology. Pennsylvania, United States: Drexel University; 1999. pp. 286-286

[2] Bott TR. Fouling of Heat Exchangers. Amsterdam, Netherlands: Elsevier Science B.V.; 1995

[3] Yang C, Dabros T, Li D, Czarnecki J, Masliyah JH. Kinetics of particle transport to a solid surface from an impinging jet under surface and external force fields. Journal of Colloid and Interface Science. 1998;**208**(1):226-240

[4] Liang Y, Hilal N, Langston P, Starov V. Interaction forces between colloidal particles in liquid: Theory and experiment. Advances in Colloid and Interface Science. 2007;**134–135**: 151-166

[5] Derjaguin BV, Landau EM. Theory of the stability of strongly charged lyophobic sols and of the adhesion of strongly charged particles in solutions of electrolytes. Acta Physicochimica. 1941;**14**:633-662

[6] Verwey EJW, Overbeek JTG. Theory of the stability of lyophobic colloids. Nature. 1948; **162**:315-316

[7] Ninham BW, Parsegian VA. van der Waals Forces. Biophysical Journal. 1970;**10**(7):646-663

[8] Hough DB, White LR. The calculation of Hamaker constants from liftshitz theory with applications to wetting phenomena. Advances in Colloid and Interface Science. 1980;**14** (1):3-41

[9] Dzyaloshinskii IE, Lifshitz EM, Pitaevskii LP. General theory of van der Waals forces. Soviet Physics Uspekhi. 1961;**4**(2):153

[10] Bowen WR, Jenner F. Theoretical descriptions of membrane filtration of colloids and fine particles: An assessment and review. Advances in Colloid and Interface Science. 1995;**56**: 141-200

[11] Hamaker HC. The London-van der Waals attraction between spherical particles. Physica. 1937;**4**(10):1058-1072

[12] Wang S. The mutual influence between the two atoms of hydrogen. Zeitschrift für Physik. 1927;**28**:663-666

[13] London F. Zur Theorie und Systematik der Molekularkräfte. Zeitschrift für Physik. 1930; **63**(3):245-279

[14] Margenau H. The role of quadrupole forces in van der Waals attractions. Physical Review. 1931;**38**(4):747-756

[15] Casimir HBG, Polder D. The influence of retardation on the London-van der Waals forces. Physical Review. 1948;**73**(4):360-372

[16] Prieve DC, Russel WB. Simplified predictions of Hamaker constants from Lifshitz theory. Journal of Colloid and Interface Science. 1988;**125**(1):1-13

[17] Israelachvili JN. The calculation of van der Waals dispersion forces between macroscopic bodies. Proceedings of the Royal Society of London A: Mathematical, Physical and Engineering Sciences. 1972;**331**(1584):39-55

[18] Horn RG, Israelachvili JN. Direct measurement of structural forces between two surfaces in a nonpolar liquid. The Journal of Chemical Physics. 1981;**75**(3):1400-1411

[19] Gregory J. The calculation of Hamaker constants. Advances in Colloid and Interface Science. 1970;**2**(4):396-417

[20] Israelachvili JN. Intermolecular and Surface Forces. 1st ed. Burlington, MA: Academic Press; 1992

[21] Tabor D, Winterton RHS. The direct measurement of normal and retarded van der Waals forces. In: Proceedings of the Royal Society of London. A. Mathematical and Physical Sciences. 1969;**312**(1511):435-450

[22] Suzuki A, Ho NFH, Higuchi WI. Predictions of the particle size distribution changes in emulsions and suspensions by digital computation. Journal of Colloid and Interface Science. 1969;**29**(3):552-564

[23] Stern O. The theory of the electrolytic double shift. Zeitscrift für Elektrochemie. 1924;**30**: 508-516

[24] Russel WB, Saville DA, Schowalter WR. Colloidal Dispersions. Cambridge, UK: Cambridge University Press; 1989

[25] van de Ven TGM. Colloidal Hydrodynamics. Boston USA: Academic Press; 1989

[26] Anderson JL. Journal of Fluid Mechanics. 1991;**222**:692-694

[27] Frens G, Overbeek JTG. Repeptization and the theory of electrocratic colloids. Journal of Colloid and Interface Science. 1972;**38**(2):376-387

[28] Barisik M, Atalay S, Beskok A, Qian S. Size dependent surface charge properties of silica nanoparticles. The Journal of Physical Chemistry C. 2014;**118**(4):1836-1842

[29] Zhao C, Ebeling D, Siretanu I, van den Ende D, Mugele F. Extracting local surface charges and charge regulation behavior from atomic force microscopy measurements at heterogeneous solid-electrolyte interfaces. Nanoscale. 2015;**7**(39):16298-16311

[30] Elimelech M, Gregory J, Jia X, Williams RA. Particle Deposition & Aggregation: Measurement, Modelling and Simulation. Oxford: Butterworth-Heinemann; 1995

[31] Derjaguin BV. Analysis of friction and adhesion, IV. The theory of the adhesion of small particles. Kolloid Zeitschrift. 1934;**69**:155-164

[32] Hogg R, Healy TW, Fuerstenau DW. Mutual coagulation of colloidal dispersions. Transactions of the Faraday Society. 1966;**62**:1638-1651

[33] Adamczyk Z, Warszyński P. Role of electrostatic interactions in particle adsorption. Advances in Colloid and Interface Science. 1996;**63**(0):41-149

[34] Kralj-Iglič V, Iglič A. A simple statistical mechanical approach to the free energy of the electric double layer including the excluded volume effect. Journal de Physique II. 1996;**6** (4):477-491

[35] Briscoe WH, Attard P. Counterion-only electrical double layer: A constrained entropy approach. The Journal of Chemical Physics. 2002;**117**(11):5452-5464

[36] Perry J, Kandlikar S. Fouling and its mitigation in silicon microchannels used for IC chip cooling. Microfluidics and Nanofluidics. 2008;**5**(3):357-371

[37] Elimelech M. Particle deposition on ideal collectors from dilute flowing suspensions: Mathematical formulation, numerical solution, and simulations. Separations Technology. 1994;**4**(4):186-212

[38] Ruckenstein E, Prieve DC. Adsorption and desorption of particles and their chromatographic separation. AICHE Journal. 1976;**22**(2):276-283

[39] Healy TW, La Mer VK. The energetics of flocculation and redispersion by polymers. Journal of Colloid Science. 1964;**19**(4):323-332

[40] Pashley RM, Israelachvili JN. Dlvo and hydration forces between mica surfaces in Mg^{2+}, Ca^{2+}, Sr^{2+}, and Ba^{2+} chloride solutions. Journal of Colloid and Interface Science. 1984;**97**(2):446-455

[41] Claesson PM, Christenson HK. Very long range attractive forces between uncharged hydrocarbon and fluorocarbon surfaces in water. The Journal of Physical Chemistry. 1988;**92**(6):1650-1655

[42] Tipping E. Colloids in the aquatic environment. Chemistry and Industry (Chichester). 1988;**15**:485-490

[43] Grabbe A, Horn RG. Double-layer and hydration forces measured between silica sheets subjected to various surface treatments. Journal of Colloid and Interface Science. 1993; **157**(2):375-383

[44] Leckband D, Israelachvili J. Intermolecular forces in biology. Quarterly Reviews of Biophysics. 2001;**34**(02):105-267

[45] Ruckenstein E, Manciu M. The coupling between the hydration and double layer interactions. Langmuir. 2002;**18**(20):7584-7593

[46] Israelachvili JN, Wennerstrom H. Hydration in electrical double layers. Nature. 1997;**385** (6618):690-690

[47] Rabinovich YI, Derjaguin BV. Interaction of hydrophobized filaments in aqueous electrolyte solutions. Colloids and Surfaces. 1988;**30**(3–4):243-251

[48] Israelachvili JN, Pashley RM. Measurement of the hydrophobic interaction between two hydrophobic surfaces in aqueous electrolyte solutions. Journal of Colloid and Interface Science. 1984;**98**(2):500-514

[49] Milner ST, Witten TA, Cates ME. Theory of the grafted polymer brush. Macromolecules. 1988;**21**(8):2610-2619

[50] Vrij A. Polymers at interfaces and the interactions in colloidal dispersions. Pure and Applied Chemistry. 1976;**48**(4):471-483

[51] Napper DH. The steric stabilization of hydrosols by nonionic macromolecules. Journal of Colloid and Interface Science. 1969;**29**(1):168-170

[52] de Gennes PG. Polymers at an interface; a simplified view. Advances in Colloid and Interface Science. 1987;**27**(3):189-209

[53] Tuckerman DB, Pease RFW, Guo Z, Hu JE, Yildirim O, Deane G, Wood L. Microchannel heat transfer: Early history, commercial applications, and emerging opportunities. ASME Conference Proceedings. 2011;**2011**(44649):739-756

[54] Mavraki E, Moschou D, Kokkoris G, Vourdas N, Chatzandroulis S, Tserepi A. A continuous flow μPCR device with integrated microheaters on a flexible polyimide substrate. Procedia Engineering. 2011;**25**(0):1245-1248

[55] Yang J, Liu Y, Rauch CB, Stevens RL, Liu RH, Lenigk R, Grodzinski P. High sensitivity PCR assay in plastic micro reactors. Lab on a Chip. 2002;**2**(4):179-187

[56] Kim S-J, Wang F, Burns MA, Kurabayashi K. Temperature-programmed natural convection for micromixing and biochemical reaction in a single microfluidic chamber. Analytical Chemistry. 2009;**81**(11):4510-4516

[57] Becker M, Mansouri A, Beilein C, Janasek D. Temperature gradient focusing in miniaturized free-flow electrophoresis devices. Electrophoresis. 2009;**30**(24):4206-4212

[58] Edwards TL, Gale BK, Frazier AB. A microfabricated thermal field-flow fractionation system. Analytical Chemistry. 2002;**74**(6):1211-1216

[59] Miralles V, Huerre A, Malloggi F, Jullien M-C. A review of heating and temperature control in microfluidic systems: Techniques and applications. Diagnostics. 2013;**3**(1):33-67

[60] Velve Casquillas G, Fu C, Le Berre M, Cramer J, Meance S, Plecis A, Baigl D, Greffet J-J, Chen Y, Piel M, Tran PT. Fast microfluidic temperature control for high resolution live cell imaging. Lab on a Chip. 2011;**11**(3):484-489

[61] Khandurina J, McKnight TE, Jacobson SC, Waters LC, Foote RS, Ramsey JM. Integrated system for rapid PCR-based DNA analysis in microfluidic devices. Analytical Chemistry. 2000;**72**(13):2995-3000

[62] Maltezos G, Johnston M, Taganov K, Srichantaratsamee C, Gorman J, Baltimore D, Chantratita W, Scherer A. Exploring the limits of ultrafast polymerase chain reaction using liquid for thermal heat exchange: A proof of principle. Applied Physics Letters. 2010;**97**(26):264101

[63] Lao AIK, Lee TMH, Hsing IM, Ip NY. Precise temperature control of microfluidic chamber for gas and liquid phase reactions. Sensors and Actuators A: Physical. 2000;**84**(1–2):11-17

[64] Wu J, Cao W, Wen W, Chang DC, Sheng P. Polydimethylsiloxane microfluidic chip with integrated microheater and thermal sensor. Biomicrofluidics. 2009;**3**(1):012005

[65] de Mello AJ, Habgood M, Lancaster NL, Welton T, Wootton RCR. Precise temperature control in microfluidic devices using joule heating of ionic liquids. Lab on a Chip. 2004;**4**(5):417-419

[66] Guijt RM, Dodge A, van Dedem GWK, de Rooij NF, Verpoorte E. Chemical and physical processes for integrated temperature control in microfluidic devices. Lab on a Chip. 2003;**3**(1):1-4

[67] Mao H, Yang T, Cremer PS. A microfluidic device with a linear temperature gradient for parallel and combinatorial measurements. Journal of the American Chemical Society. 2002;**124**(16):4432-4435

[68] Zhao Y, Zhao C, He J, Zhou Y, Yang C. Collective effects on thermophoresis of colloids: A microfluidic study within the framework of DLVO theory. Soft Matter. 2013;**9**(32): 7726-7734

[69] Yan Z, Huang X, Yang C. Rapid prototyping of single-layer microfluidic PDMS devices with abrupt depth variations under non-clean-room conditions by using laser ablation and UV-curable polymer. Microfluidics and Nanofluidics. 2017;**21**(6):108

[70] Vigolo D, Rusconi R, Piazza R, Stone HA. A portable device for temperature control along microchannels. Lab on a Chip. 2010;**10**(6):795-798

[71] Yan Z, Huang X, Yang C. Hydrodynamic effects on particle deposition in microchannel flows at elevated temperatures. Journal of Heat Transfer. 2017;**140**(1):012402-012410

[72] Yan Z, Huang X, Yang C. Particulate fouling and mitigation approach in microchannel heat exchanger. In: ASME 2016 5th International Conference on Micro/Nanoscale Heat and Mass Transfer. American Society of Mechanical Engineers; 2016

[73] Yan Z, Huang X, Yang C. Deposition of colloidal particles in a microchannel at elevated temperatures. Microfluidics and Nanofluidics. 2015;**18**(3):403-414

[74] Matsui T, Franzke J, Manz A, Janasek D. Temperature gradient focusing in a PDMS/glass hybrid microfluidic chip. Electrophoresis. 2007;**28**(24):4606-4611

[75] Jiao ZJ, Huang XY, Nguyen NT. Manipulation of a droplet in a planar channel by periodic thermocapillary actuation. Journal of Micromechanics and Microengineering. 2008;**18**(4):045027

[76] Yap Y-F, Tan S-H, Nguyen N-T, Sohel Murshed SM, Wong T-N, Yobas L. Thermally mediated control of liquid microdroplets at a bifurcation. Journal of Physics D: Applied Physics. 2009;**42**(6):065503

[77] Selva B, Cantat I, Jullien M-C. Temperature-induced migration of a bubble in a soft microcavity. Physics of Fluids. 2011;**23**(5):052002

[78] Miralles V, Huerre A, Williams H, Fournie B, Jullien M-C. A versatile technology for droplet-based microfluidics: Thermomechanical actuation. Lab on a Chip. 2015;**15**(9): 2133-2139

[79] Yan Z, Huang X, Yang C. Kinetics of Colloidal Particle Deposition in Microfluidic Systems under Temperature Gradients: Experiment and Modelling. Submitted

[80] Newson IH, Miller GA, Haynes JW, Bott TR, Williamson RD. Particulate fouling: Studies of deposition, removal and sticking mechanisms in a haematite/water system. In: UK national conference on heat transfer. Glasgow, United Kingdom; 1988

[81] Gu Y, Li D. Deposition of spherical particles onto cylindrical solid surfaces: II. Experimental studies. Journal of Colloid and Interface Science. 2002;**248**(2):329-339

[82] Gu Y, Li D. Deposition of spherical particles onto cylindrical solid surfaces: I. numerical simulations. Journal of Colloid and Interface Science. 2002;**248**(2):315-328

[83] Kar A, Guha R, Dani N, Velegol D, Kumar M. Particle deposition on microporous membranes can be enhanced or reduced by salt gradients. Langmuir. 2014;**30**(3):793-799

[84] Guha R, Shang X, Zydney AL, Velegol D, Kumar M. Diffusiophoresis contributes significantly to colloidal fouling in low salinity reverse osmosis systems. Journal of Membrane Science. 2015;**479**(0):67-76

[85] Salim M, Wright PC, McArthur SL. Studies of electroosmotic flow and the effects of protein adsorption in plasma-polymerized microchannel surfaces. Electrophoresis. 2009;**30**(11):1877-1887

[86] Mustin B, Stoeber B. Deposition of particles from polydisperse suspensions in microfluidic systems. Microfluidics and Nanofluidics. 2010;**9**(4–5):905-913

[87] Mustin B, Stoeber B. Single layer deposition of polystyrene particles onto planar polydimethylsiloxane substrates. Langmuir. 2016;**32**(1):88-101

[88] Yiantsios SG, Karabelas AJ. The effect of gravity on the deposition of micron-sized particles on smooth surfaces. International Journal of Multiphase Flow. 1998;**24**(2):283-293

[89] Yiantsios SG, Karabelas AJ. Deposition of micron-sized particles on flat surfaces: Effects of hydrodynamic and physicochemical conditions on particle attachment efficiency. Chemical Engineering Science. 2003;**58**(14):3105-3113

[90] Stamm MT, Gudipaty T, Rush C, Jiang L, Zohar Y. Particle aggregation rate in a microchannel due to a dilute suspension flow. Microfluidics and Nanofluidics. 2011;**11**(4):395-403

[91] Gudipaty T, Stamm MT, Cheung LSL, Jiang L, Zohar Y. Cluster formation and growth in microchannel flow of dilute particle suspensions. Microfluidics and Nanofluidics. 2011;**10**(3):661-669

[92] Unni HN, Yang C. Colloidal particle deposition from electrokinetic flow in a microfluidic channel. Electrophoresis. 2009;**30**(5):732-741

[93] Spielman LA, Friedlander SK. Role of the electrical double layer in particle deposition by convective diffusion. Journal of Colloid and Interface Science. 1974;**46**(1):22-31

[94] Adamczyk Z, Van De Ven TGM. Deposition of particles under external forces in laminar flow through parallel-plate and cylindrical channels. Journal of Colloid and Interface Science. 1981;**80**(2):340-356

[95] Song L, Elimelech M. Calculation of particle deposition rate under unfavourable particle-surface interactions. Journal of the Chemical Society, Faraday Transactions. 1993;**89**(18):3443-3452

[96] Song L, Elimelech M. Particle deposition onto a permeable surface in laminar flow. Journal of Colloid and Interface Science. 1995;**173**(1):165-180

[97] Nazemifard N, Masliyah JH, Bhattacharjee S. Particle deposition onto micropatterned charge heterogeneous substrates: Trajectory analysis. Journal of Colloid and Interface Science. 2006;**293**(1):1-15

[98] Nazemifard N, Masliyah JH, Bhattacharjee S. Particle deposition onto charge heterogeneous surfaces: Convection-diffusion-migration model. Langmuir. 2006;**22**(24):9879-9893

[99] Chatterjee R, Bhattacharjee S, Mitra SK. Particle transport in patterned cylindrical microchannels. Microfluidics and Nanofluidics. 2012;**12**(1–4):41-51

[100] Marshall JS. Particle aggregation and capture by walls in a particulate aerosol channel flow. Journal of Aerosol Science. 2007;**38**(3):333-351

[101] Jin C, Ren CL, Emelko MB. Concurrent modeling of hydrodynamics and interaction forces improves particle deposition predictions. Environmental Science & Technology. 2016;**50**(8):4401-4412

[102] Jin C, Glawdel T, Ren CL, Emelko MB. Non-linear, non-monotonic effect of nano-scale roughness on particle deposition in absence of an energy barrier: Experiments and modeling. Scientific Reports. 2015;**5**:17747

[103] Unni HN. Transport and deposition of colloidal particles in microchannel flow. In: School of Mechanical and Aerospace Engineering. Singapore: Nanyang Technological University; 2007

[104] Patel MV, Tovar AR, Lee AP. Lateral cavity acoustic transducer as an on-chip cell/particle microfluidic switch. Lab on a Chip. 2012;**12**(1):139-145

[105] Ashkin A, Dziedzic JM, Bjorkholm JE, Chu S. Observation of a single-beam gradient force optical trap for dielectric particles. Optics Letters. 1986;**11**(5):288-290

[106] Sheikholeslami M, Hayat T, Alsaedi A. Numerical simulation of nanofluid forced convection heat transfer improvement in existence of magnetic field using lattice Boltzmann method. International Journal of Heat and Mass Transfer. 2017;**108**:1870-1883

Chapter 6

Microdroplets Advancement in Newtonian and Non-Newtonian Microfluidic Multiphase System

Kai Seng Koh, Voon Loong Wong and Yong Ren

Additional information is available at the end of the chapter

http://dx.doi.org/10.5772/intechopen.75358

Abstract

With recent advancement in droplet microfluidics for both microdroplet encapsulation and fission, it is of paramount importance to understand the flow physics for both Newtonian and non-Newtonian fluids in microdroplet encapsulation and fission as the development of the field is approaching to its maturity. The chapter aims to review and discuss the fluid flow behavior of the multiphase system, mathematical models as well as the fundamental phenomena driving force of microdroplet encapsulation and fission multiphase system. Together, the recent advances in technologies that enable fabrication and application of droplets encapsulation and fission from both Newtonian and non-Newtonian microfluidic multiphase system will be reviewed as well.

Keywords: Newtonian fluids, non-Newtonian fluids, microfluidic, multiphase system, microdroplet encapsulation, microdroplet fission

1. Introduction

Emulsion is a fine dispersion of small droplet made from two immiscible liquids, where one liquid phase is dispersed into another continuous flow of liquid phase in which both liquids are not soluble with one another [1]. The fundamental liquids combination for emulsion to form is rather simple, i.e., where the multiphase system usually have large liquids density difference (at least >100 kg/m^3) using the example of water and cooking oil; the larger the density difference, the better the emulsion can be formed. In addition, the multiphase system of the liquids selection can be classified into Newtonian fluid systems and non-Newtonian fluid systems. In the case of Newtonian fluids such as water, oil, glycerol and salt solutions with low molecular weight, due to the independent of fluid viscosity from shear rate and fluid velocity flow, the shear stress in a steady, laminar flow condition is having a linear relationship with its shear rate. In contrast, non-Newtonian fluids, examples of which in daily life including chocolate, toothpaste, lubricating oils experience a non-linear relationship for its shear stress versus shear rate curve due to the fact that the fluid viscosity is a variable at any given flow condition, i.e., temperature and pressure. Such fluid properties lead to much differences in term of the emulsion productions thus far in all engineering approaches. Conventionally, emulsion is prepared using high-pressure homogenizers and colloid mill. These devices apply high mechanical shear force to break up the large emulsion into smaller ones that are subsequently stabilized by the use of emulsifier [2]. However, such method contributed to large size distribution of emulsions formed [3], leading to material loss as well as emulsion function efficiency issue. This dispersity issued was then resolved by integration of microfluidic technology that was firstly developed in 1950s [4]. Microfluidic technology is defined as a branch of fluid mechanics that focuses on the understanding, designing, fabrications and operations of system that convey liquids inside channels with two of the three geometry length scales in the order of microns [5]. Furthermore, with the length scale associated within the microchannel, the flow regime formed in a microfluidic channel will not develop into turbulent flow, enables fluid to be manipulated that will form emulsion with high monodispersity. The wide range of technology options from decades of microfluidic multiphase system developments has allowed emulsion to be generated and manipulated.

In this chapter, we aim to summarize the main technologies for emulsion formation non-Newtonian microfluidic multiphase system using Newtonian fluid as a comparison. The chapter will start with the review of fundamental two-phase flow in microfluidic followed by discussion of the fundamental flow physics of microdroplets translocation and breakup phenomena in microfluidics. The detailed differences between Newtonian and non-Newtonian flow system will be illustrated and compared. Emphasis will be placed on the advancement of emulsions formed, i.e. encapsulation and fission from single emulsion. Finally, we conclude with an outlook to the future of the field. This chapter is meant to familiarize readers who may be new to the field of microdroplets formation in Newtonian and non-Newtonian fluid systems, as well as those readers who are new to the field of microdroplets formation via encapsulation and fission approaches, and eventually bridge the knowledge gap between the two correlations, disciplinary fields.

2. Two-phase flow in microchannels

In two-phase flow microfluidics, dispersed and continuous phase fluids generally were dispensed separately into the microfluidic device. The continuous and dispersed phase channels typically meet at a junction, depending on the specific geometry of a microfluidic device. Each shape of the junction helps to define the local flow fields that deform the interface between the two immiscible fluids [6, 7]. These configurations are also shown schematically in **Table 1**. When the instabilities of free surface between phases are sufficiently large, drops in microscale emerge and eventually pinch-off from the dispersed phase. There are many controlling parameters that will affect the microdroplet generation regime such as, interfacial tension [8, 9], surface

Approaches	**Breakup regimes**	**Flow regime description**	**Diagrams**
Co-flowing streams	Dripping	Droplet forms at the exit of the capillary tube tip and propagates downstream with the flow.	
	Jetting	Droplets pinch off from an extended thread downstream of the capillary tube tip.	
Cross-flowing streams	Squeezing	Dispersed threads block the outlet channel leads to dramatic increase in the dynamic pressure upstream of the droplet, thus forcing the interface to neck and pinch off into a droplet.	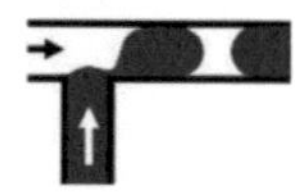
	Dripping	Dispersed thread does not touch the wall of the channel in the entire droplet formation process.	
	Jetting	Droplet breakup point moves progressively downstream of the outlet channel.	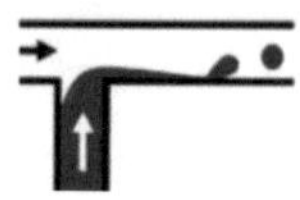
Elongational strained flows	Squeezing	The tip of the droplet phase effectively blocks the cross-section of the orifice.	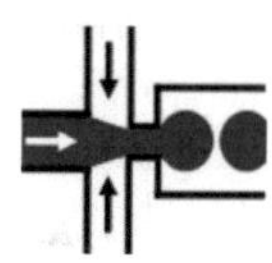
	Dripping	Droplets are periodically formed in the orifice.	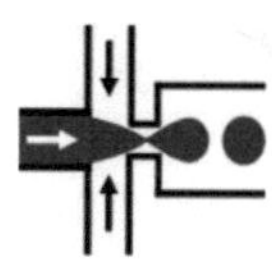
	Jetting	The tip of the droplet phase extends downstream of the orifice and small droplets are formed at its terminus.	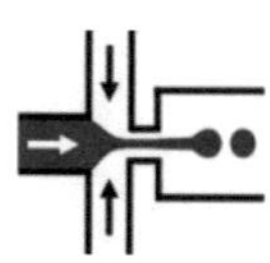
	Threading	Each droplet is formed a long and thin thread of the inner liquid is dragged behind the droplet. This thread subsequently breaks up into a group of tiny secondary droplets	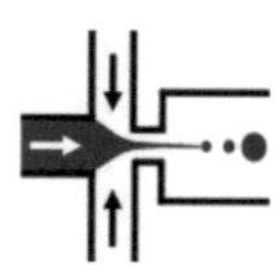

Reproduced from Ref. [7, 15–17] with permission from the Physics of Fluids, Physical Review Letter and Lab on Chip.

Table 1. Main approaches for droplet breakup in microchannel.

wettability or wall adhesion [10], the volumetric flow rate [11, 12], viscosities of both immiscible fluids [9, 11, 13] and channel geometry [14].

For microdroplet formation to be made possible in microfluidics, a defined flow conditions must be fulfilled in order to achieve the dripping regime [16, 18–20]. In the case of Newtonian fluids, the dripping regime occurs when the Weber number (We) of the dispersed phase and the Capillary number (Ca) of the continuous phase are less than one. Under these conditions, droplets formed are highly uniform and their uniformity is unaffected over a wide range of flow rates. However, non-Newtonian fluids rarely fulfill these conditions and pose a challenge in achieving monodispersed microdroplets due to their complex rheological characteristics. As general thumb rule, non-Newtonian fluids can be further classified intro three groups, i.e. purely viscous fluids, time-dependent fluids, and viscoelastic fluids. Each fluid group possess distinct characteristic respectively; however, there is no single constitutive equation that has been established to describe the rheogram for these fluids. For instance, the extensional viscosity of the fluid can resist the pinching at the tip of the capillary that is required for a microdroplet to form. This results in a long cylinder of fluid forming from the capillary tip to a distance downstream from the tip before breaking into non-uniform sized or polydispersed microdroplets formation due to the Rayleigh-Plateau instability [21]. Non-Newtonian fluids also exhibit a variety of behaviors that are unique to their chemical compositions, mixture combinations and many physical conditions that include flow rate, temperature, etc. Such complexities in the characteristics of non-Newtonian fluids prevent a thorough understanding of the dispersion stability and break-up of individual emulsions in microfluidics. Moreover, it is also known that an emulsifying non-Newtonian solutions in microfluidics in a controllable manner is a persistent problem [21] that prevents these techniques from being suitable for industrial applications. Understanding the dynamical mechanisms of microdroplets formation of non-Newtonian fluids in microfluidic channels is essential to ensure microdroplets can be created based on the droplet size, patterns, and productivity.

In microfluidic systems, the length scales demands that all flow to be laminar. Based on the Newtonian and non-Newtonian flow through the microchannel with rectangular cross-section, the axial velocity in the fully developed region is a function of two independent variables and the study of the hydrodynamic behavior in a rectangular microchannel requires a two-dimensional or three-dimensional analyses. The non-linear partial differential momentum Navier-Stokes equation with the associated boundary condition of zero velocity at the wall for a pressure-driven, steady-state, incompressible, constant viscosity, known as Newtonian fluid and the flow without body forces in microchannel is being presented as follows:

$$0 = -\frac{\partial P}{\partial x} + \eta\left(\frac{\partial^2 u}{\partial y^2} + \frac{\partial^2 u}{\partial z^2}\right) \text{ for } -\frac{w}{2} < y < \frac{w}{2} \text{ and } 0 < z < h \tag{1}$$

where u is average velocity characteristic of the flow, dynamic viscosity is η, and P is the pressure, x is the coordinate axes along the channel length, y is the coordinate axes along the channel width, z is the coordinate axes along the channel height, w is the channel width, and h is the channel height, respectively. With no-slip boundary condition, the geometry of the cross-section of the rectangular is fixed and an analytical solution solutions is possible [22, 23].

In applying the Navier-Stokes equation, the analytical solutions of the Poiseuille flow for rectangular cross-section are shown as follows:

$$u_x(y,z) = \frac{4h^2 \Delta P}{\eta \pi^3 l} \sum_{k=1,3,5\ldots}^{\infty} \frac{1}{k^3} \left[1 - \frac{\cosh\left(k\pi \frac{y}{h}\right)}{\cosh\left(k\pi \frac{w}{2h}\right)}\right] sin\left(k\pi \frac{z}{h}\right) \tag{2}$$

where l, w and h are the length, width and height of the channel, respectively, ΔP is the driven pressure drop, η is the viscosity of liquid. Most real fluids exhibit non-Newtonian behavior, which means that the flow curve presents a nonlinear relationship between shear stress and shear rate or does not pass through the origin. The laminar velocity profile of a power-law fluid flowing through a rectangular duct to the governing equation is shown as below [24]:

$$\frac{u(y,z)}{U} = \sum_{i=1}^{6} A_i sin\left[\alpha_i \pi \frac{(z/h)}{2} + 1\right] sin\left[\beta_i \frac{(y/w + 1)}{2}\right] \tag{3}$$

where constant α_i and β_i and A_i is the constants selected to minimize an integral. Six constants which are to be computed by the Ritz-Galerkin method are identical to the corresponding Fourier coefficient and tabulated as a function of aspect ratio and fluid-behavior index (**Figure 1**).

2.1. Microdroplets translocation and breakup phenomena in microfluidics

An emulsion contains a mixture of two immiscible liquids as one phase being dispersed throughout the other phase in small droplets. Most common emulsions include oil-in-water, or direct emulsions, and water-in-oil, or inverted emulsions [19]. The characteristics of emulsion products, foremost the droplet size distribution is the most important parameters that affect the stability, rheology, chemical reactivity, and physiological efficiency of any emulsion

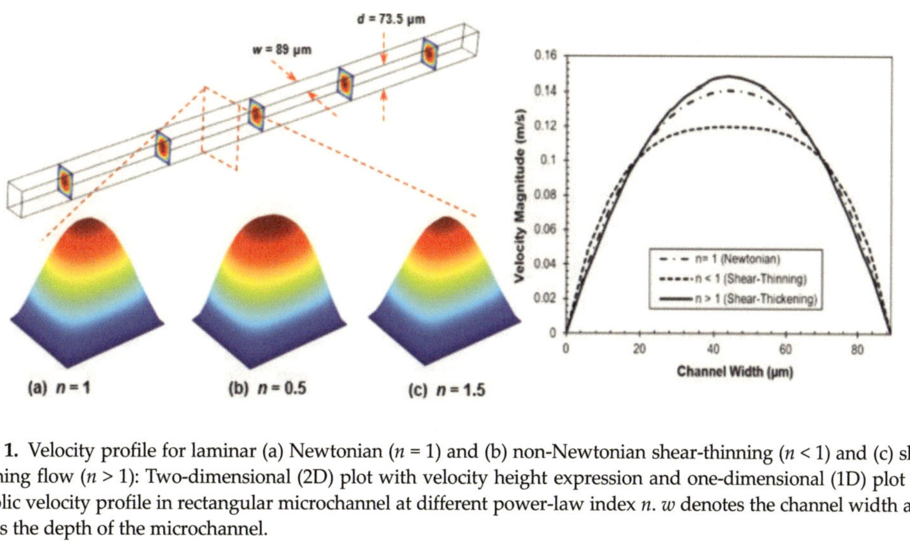

Figure 1. Velocity profile for laminar (a) Newtonian ($n = 1$) and (b) non-Newtonian shear-thinning ($n < 1$) and (c) shear-thickening flow ($n > 1$): Two-dimensional (2D) plot with velocity height expression and one-dimensional (1D) plot with parabolic velocity profile in rectangular microchannel at different power-law index n. w denotes the channel width and d denotes the depth of the microchannel.

[25–27]. For most microfluidic applications, the Reynolds number is much smaller than 1, indicating the effects of volume-based inertia and gravity are not as significant as that in macroscale. The surface-based interfacial tension, flow rates, surface chemistry, and the viscosity become more significant and play in controlling flow behavior of multiphase flow in microscale. Moreover, the degree of confinement, channel aspect ratio, and geometrical structure also significantly induce the impact on capillary pressure, which promote the destabilization of the interface and the drop formations.

2.1.1. Newtonian flow systems

Under this system, interfacial tension between two liquid phases becomes dominant above all physical forces such as gravitational forces, viscous forces and inertia forces as the interest of dimension gets smaller due to the existence of a high surface-to-volume ratio within a microscale device. With that, an approximation of the magnitude of surface tension force, which has the stabilizing effect on the emerging tip [13], arising from

$$\Delta P_L = \sigma\left(\frac{1}{r_\alpha} + \frac{1}{r_\gamma}\right) \tag{4}$$

where r_a and r_r are the radii of axial and radial curvature across the interface in a squeezing regime.

Much studies were conducted in particular on the predominance of interfacial tension on characters of microfluidic Newtonian droplet formations [8, 9, 28, 29] and its changes in presence of surfactants in microfluidic systems [30, 31]. Surface flows of surfactant can induce variations in surface tension, i.e. Marangoni effect, which can substantially alter the interfacial morphology and resulting droplets size [32]. Apart from the presence of surfactant, depending on the fluid viscosity or concentration [12, 33], temperature [34] and the presence of micro- or nanoparticles in the fluid [34, 35] can modify considerably the value of surface tension of the fluid in nature and thus tailor the two-phase flow behavior within the microfluidic system. The effect of aforementioned is summarized in **Figure 2**.

The interaction between the solid surface and fluid in a microchannel has also been a major research focal point as the interaction impacts the dynamics of the droplet formation process. **Figure 2** (V) illustrates the flow patterns for different wetting conditions in a T-shaped microchannel. Most studies of droplet generation in microfluidic devices involve the numerical studies of the contact angle effect on the shape, size, the distance between two neighboring droplets, detachment point and the generation frequency of droplets in microfluidic system [12, 29], in which the channel wall surface plays significant role in generating larger droplets with smaller contact angle provide longer contact time with the surface, especially for small values of capillary number, in consistent with the numerical results.

The flows in a microscale or nanoscale device naturally emphasize phenomena associated with interfacial tension and wetting properties. In addition, inertial and viscosity effects are two important parameters in characterizing the role of the shear-stress exerted from the continuous phase acts to deform the interface during the microdroplet breakup process, i.e. **Figure 2** (I)

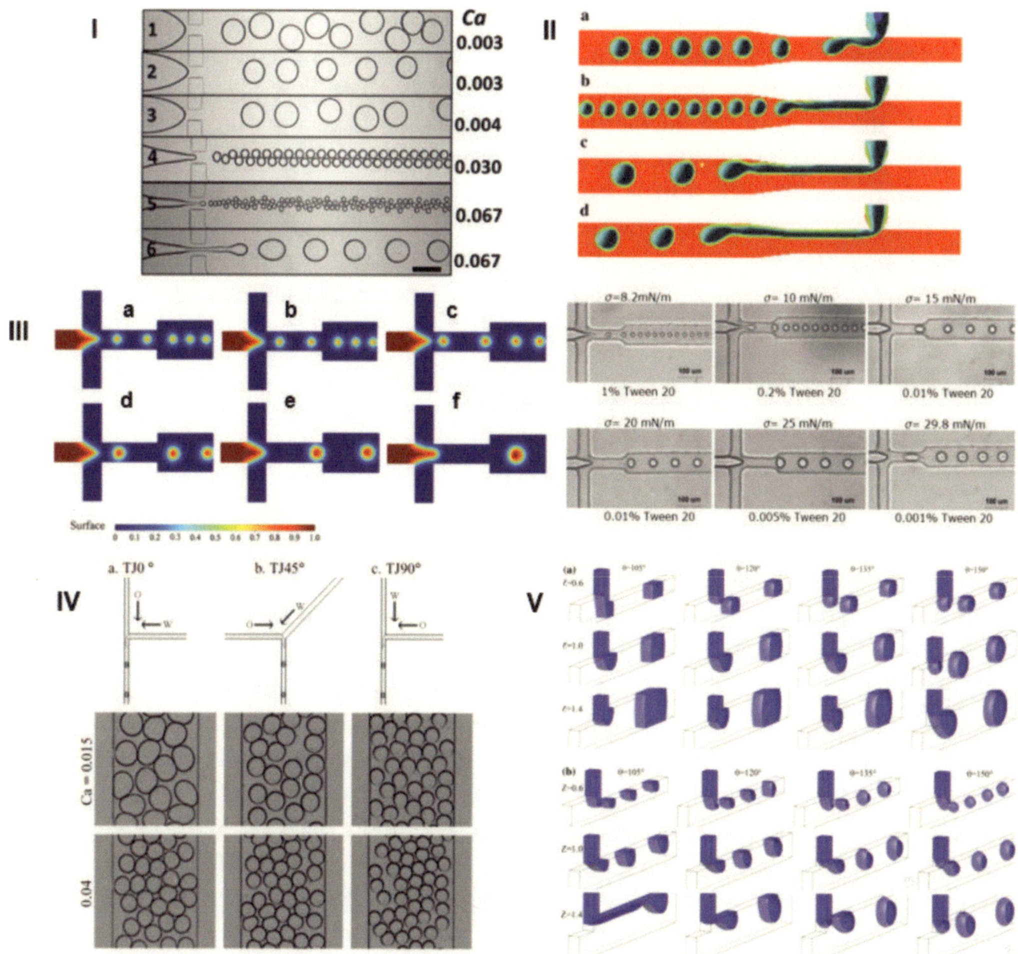

Figure 2. Passive Newtonian droplet generation with microfluidics. **(I)**. Oleic acid droplet production at aqueous glycerol solution with varying flow rate ratios: (1) 0.16, (2) 0.8, (3) 0.667, (4) 0.11, (5) 0.1 (6) 0.2. **(II)** Drop images at different values of continuous phase viscosity in Newtonian System. The images a–d are μ_{CP} = 0.00332, 0.008, 0.01 and 0.018 Pa.s, respectively. **(III)** The effect of interfacial tension on droplet breakup dynamics with experimental and numerical justification: (a) 8.2 mN/m, (b) 10 mN/m, (c) 15 mN/m, (d) 20 mN/m, (e) 25 mN/m and (f) 29.8 mN/m. **(IV)** The impact of different inlet channel geometries and angle on microfluidic drop formation. **(V)** Droplet formation mode under various contact angles. **I** is reproduced with permission from [36], Copyright 2015, Royal Society of Chemistry. **II** is reproduced with permission from [37], Copyright 2009, Springer. **III** is reproduced with permission from [31], Copyright 2011, Springer. **IV** is reproduced with permission from [21], Copyright 2009, American Physical Society. **V** is reproduced with permission from [38], Copyright 2015, Elsevier.

and (II). The influence of two phase flow rates and viscosities on the size of water-in-oil emulsion study at a T-shape microfluidics concluded that the relationship between shear stress and flow rate of continuous phase, thus quantifies the magnitude of shear stress ($\tau \propto \eta_c Q_{oil}$, where η_c is the viscosity of continuous phase) acting on the interface [13]. As Q_{water}, the volumetric flow rate of dispersed phase increases, the shear stress exerted on the droplet is increased due to the difference of the flow rate of the two fluids decreases with increasing

Q_{water}. These numerical and experimental results imply that the flow rate of the both phases considerably alter the microdroplet formation mechanisms under laminar flow in different microfluidic configurations. Apart from the effect of two phase flow rates or flow rate ratio, $Q = Q_d/Q_c$, the length of the microdroplet formed is also depending on the viscosity ratio ($\lambda = \eta_d/\eta_c$). Previous investigations reported that an increase of the viscosity of continuous phase gives rise to the viscous stress added on the dispersed phase, and thus decreasing the microdroplet size and alter the flow pattern [11, 20, 39]. This was further validated by a numerical study on the microdroplet detachment mechanism for Newtonian fluids that subjected to cross-flow drag, continuous phase inertia, interfacial tensions, and viscosity ratio, which are shown by the following equations

$$\text{Interfacial tension force, } F_\sigma = \frac{\pi \sigma w_d{}^2}{d} \tag{5}$$

$$\text{Cross} - \text{Flow drag force, } F_D = 3\pi\eta_c(v^* - v_d)df(\lambda_\eta) \tag{6}$$

where d is the droplet diameter, v^* is the continuous phase velocity at the height of the microdroplet center, v_d is the microdroplet velocity, $\lambda_\eta = \eta_d/\eta_c$ is the viscosity ratio and $f(\lambda_\eta) = \left(\frac{2}{3} + \lambda_\eta\right)/(1 + \lambda_\eta)$ captures the effect of the disperse phase viscosity on the microdroplet drag. When $\lambda_\eta \gg 1$, the disperse phase viscosity is sufficiently high that the microdroplet is solid-like and the interfacial tension force reduces to the drag from Stokes flow around a solid sphere. On the contrary, internal flow within the microdroplet becomes possible and this acts to reduce the drag on the microdroplet from the continuous phase for smaller values of λ_η [40]. Meanwhile, a detailed studies of droplet microfluidics performance as function of flow conditions as it is crucial when designing the geometry of the microchannel, particularly is one of the major parameter which can be very sensitive to flow conditions and drop properties [41]. Specifically, the cross-sectional area of the main channel can influence the volume of microdroplets detached significantly in both the squeezing regime as well as in the dripping regime as presented here.

2.1.2. Non-Newtonian flow systems

Although there are many methods of producing monodispersed microdroplets, controlling the size of the droplets within the range of 1–100 μm remains to be a challenge that has yet to be fully resolved. Non-Newtonian fluids have far more complex rheological properties that are normally physical and chemical conditions dependent. It also exhibits more complex behaviors in terms of its dynamics in comparison to its Newtonian counterpart, particularly in the field of microfluidics [17]. This is due to the non-linear relationship between the shear rate of the fluid and its viscosity as compared to Newtonian fluids, whose viscosities remain constant regardless of the shear rate. Generally, non-Newtonian fluid can be classified into three general groups in shear flow, which are shown in **Table 2**.

In an attempt to develop an efficient and reliable drug delivery system, researches have been carried out using various types of non-Newtonian fluids in order to determine which materials would yield the most desirable results. Of all, shear thinning fluids are the most commonly encountered fluids in our everyday lives, many existing as ordinary fluids such

Fluids	Description	Fluid type	Model	Constitutive equations
Purely viscous/Time-Independent	The rate of shear at any point is determined only by the value of the shear stress at that point at that instant; can be known as time-independent fluids/ generalized Newtonian fluids (GNF).	Dilatant (*Shear-Thickening*)	Ostwald Waele	$\mu = m\,(\dot{\gamma})^{n-1}$
			Carreau-Yasuda	$\frac{\mu-\mu_\infty}{\mu_o-\mu_\infty} = \left\{\left(1+(\lambda\dot{\gamma})^2\right\}^{(n-1)/2}\right.$
		Pseudoplastics (*Shear-Thinning*)	Cross	$\frac{\mu-\mu_\infty}{\mu_o-\mu_\infty} = \frac{1}{1+k(\dot{\gamma})^n}$
			Ellis	$\mu = \frac{\mu_o}{1+\left(\tau/\tau_{1/2}\right)^{\alpha-1}}$
		Viscoplastics	Bingham Plastics	$\tau = \tau^B + \mu(\dot{\gamma})$ for $\lvert\tau\rvert > \lvert\tau^B\rvert$ $\dot{\gamma} = 0$ for $\lvert\tau\rvert < \lvert\tau^B\rvert$
			Herschel-Bulkley	$\tau = \tau^H + m(\dot{\gamma})^n$ for $\lvert\tau\rvert > \lvert\tau^H\rvert$ $\dot{\gamma} = 0$ for $\lvert\tau\rvert < \lvert\tau^H\rvert$
			Casson	$(\lvert\tau\rvert)^{1/2} = (\lvert\tau_o{}^c\rvert)^{1/2} + (\mu\lvert\dot{\gamma}\rvert)^{1/2}$ for $\lvert\tau\rvert > \lvert\tau^c\rvert$ $\dot{\gamma} = 0$ for $\lvert\tau\rvert < \lvert\tau^c\rvert$
Time-dependent	The relation between shear stress and shear rate further dependence on the duration of shearing and their kinematic history.	Thixotropy Rheopexy	Generalized Herschel-Bulkley	$\tau = (\tau_o + \tau_1) + (m_o + \xi m_1)(\dot{\gamma})^n$
Viscoelastic*	Substances exhibit characteristics of both viscous fluids and elastic solids and showing partial elastic recovery after deformation.	Viscoelastic	Oldroyd-B	$\tau_1 + \lambda_1\overset{\nabla}{\tau}_1 - 2\mu_1\left(\boldsymbol{D} - \lambda_2\overset{\nabla}{\boldsymbol{D}}\right) = 0$
			Giesekus-Leonov	$\boldsymbol{\tau}_p + \lambda_1\boldsymbol{\tau}_p - \frac{\alpha\lambda_1}{\mu}\boldsymbol{\tau}_p{}^2 + \mu\dot{\gamma} = 0$
			Phan-Thien-Tanner	$\overset{\nabla}{\tau} + \xi\left(\boldsymbol{D}\cdot\tau - \tau\cdot\boldsymbol{D}^T\right) + \frac{Y}{\lambda}\cdot\tau = 2\mathrm{GD}$ $\boldsymbol{\tau}=\boldsymbol{\tau}_1 + \boldsymbol{\tau}_2$ $\boldsymbol{\tau}_2 = 2\mu_2\boldsymbol{D}$
			White-Metzner	$\tau_1 + \lambda(\gamma)\overset{\nabla}{\tau} = 2\mu(\gamma)\boldsymbol{D}$
			K-BKZ	$\tau_f = \mu(\gamma)\frac{nL^3}{24\ln\left(\frac{2h}{d}\right)}u_{1,2}S_{12}S_{ij}$ $\tau_p = \int_{-\infty}^{t}\sum_{k=1}^{N}\frac{G_k}{\lambda_k}\exp\left(-\frac{t-t'}{\lambda_k}\right)H(I_{c-1}, II_{c-1})C_t^{-1}(t')dt'$
			Upper-Convected Maxwell	$\tau + \lambda_1\overset{\nabla}{\tau} = 2\mu\boldsymbol{D}$ $\boldsymbol{D} = \frac{1}{2}\left[(\nabla u) + (\nabla u)^T\right]$

*Note: Most common models for viscoelastic fluids.μ denotes the shear viscosity; γ denotes the shear rate; f denotes the fiber volume fraction; d and L are the diameter and fiber length; n is the number of the suspension; h is the average distance from a given fiber to its nearest neighbor; $\boldsymbol{\tau}_p$ denotes the shear stress for the polymer; λ denotes relaxation times and G_k denotes relaxation moduli; N denotes the number of relation modes; C_t^{-1} denotes finger strain tensor; $I_{c-1}$and II_{c-2}denotes first and second invariants; H is the strain memory function; $\overset{\nabla}{\tau}$ denotes upper convected time derivative of the stress tensor; $\boldsymbol{D}$ denotes the strain rate tensor; ξ denotes the adjustable parameters of the model.

Table 2. Comparison of non-Newtonian power-law model fluid behavior [18, 24, 42, 43].

as paints and blood. Due to the rheological property of the latter, the common rule of thumb in designing and developing drug delivery system usually consider shear thinning approach as working mechanism. A mathematical model to simulate the deformation of droplets through an axisymmetric contraction using shear thinning fluids. When the dispersed phase was a shear thinning fluid and the continuous phase was a Newtonian fluid, the local viscosity of the microdroplet decreased upon entering the contraction, remained at a low viscosity within the contraction and increased upon exiting, resulting in a compact bullet-shaped microdroplet at the exit. In contrast, when the continuous phase was shear thinning and the dispersed phase was Newtonian, the microdroplet exited the contraction with a very irregular shape [44, 45].

The influence of the dispersed phase viscosity between Newtonian and shear-thinning fluid on generated droplet size in microfluidic T-junctions are illustrated in **Figure 3**. The experimental results indicate that the larger viscosity of dispersed phase fluid brings the significant effect on reducing the size of generated microdroplet at each constant value of Q_d. Also, when the concentration of dispersed fluid is increasing from 0 to 60 wt% at constant value of Q_d of 0.6 mL/h; there will be a reduction in microdroplet size formed due to the change in the viscosity from 0.928 to 8.406 cP. Nevertheless, the similar phenomenon is not found when Q_d exceeds 1.2 mL/h in the condition of 60 wt%, viscosity of 8.406 cP for dispersed fluid. At Q_d of 1.2 mL/h, for the lower dispersed fluid viscosity ranging from 0.928 to 3.191 cP, the microdroplet volume is decreased. However, it is increased for the higher viscosity dispersed fluid (8.406 cP) due to a convectively unstable jetting regime that took place in which microdroplet formation lacks both periodicity and size uniformity therefore highly polydispersed in size were produced. When Q_d is further increased to a high relatively flow rate, a transition between convectively unstable flow and droplet breakup begins to prevail. This can be explained by the instabilities of microdroplets due to the inertia effects at the T-junction, begin to dominate in the dispersed phase and then evolve in the microchannel which is characterized by long instability wavelengths [46]. The behavior of the perturbations propagates in the direction of the flow and the dispersed phase does not break into microdroplet which is also defined as convective instability or unstable stratified flow. In contrast, reduction in carboxymethylcellulose (CMC) microdroplet size relative to drop volume was observed in larger power-law index; however, the size of droplet shrinks

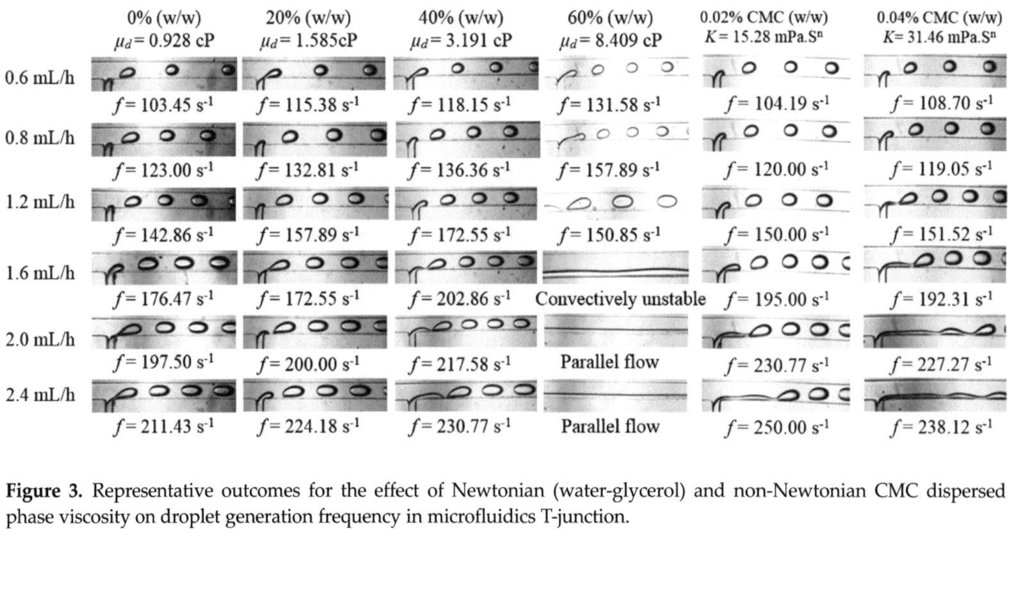

Figure 3. Representative outcomes for the effect of Newtonian (water-glycerol) and non-Newtonian CMC dispersed phase viscosity on droplet generation frequency in microfluidics T-junction.

when the viscosity of the power law fluid is increased with lower behavior index. In the case of CMC, the frequency of microdroplet formation increased when viscosity of dispersed phase is increased (0.02 wt% (w/w)) at each constant rate of Q_d; however, the reduction in frequency of droplet formation was observed when the viscosity of CMC polymer fluid is further increased to 0.04% (w/w) where power-law index (n) is smaller than unity contributes to significant effect of shear thinning behavior. This was explained by the other non-Newtonian fluid when the fluid elasticity also plays a key role in resisting the drop pinch off contributes an increase in droplet breakup time.

While shear thinning fluids are described as the most commonly encountered non-Newtonian fluid, viscoelastic fluids take the center stage in the synthesis of microdroplets. Such microdroplets are often produced using polymer solutions, such as alginate compounds [47] and polyethylene glycol [48], which have viscoelastic characteristics. This phenomenon was clearly observed in the multiple beads-on-a-string formation. It is determined that for polymer solutions that have high elasticity and extensibility, an extensional response was obtained, even at very low viscosities [49]. An observation on viscoelastic polymers that have larger molecular weight result in a higher elasticity number, El, which consequently increases the pinch-off times of the microdroplets. It was found that the size of the aperture affects the pinch-off times, whereby increasing the size of the apertures increases the pinch-off times [50]. This was validated by a demonstration on the production of monodisperse double emulsion microparticles from various non-Newtonian polymer solutions, one of which exhibited viscoelastic characteristics [21]. It was determined that when the viscoelastic solution was used as the inner fluid, the viscoelasticity of the fluid prevented microdroplets from pinching off from the orifice. This resulted in oscillations that are brought downstream by the continuous phase, which consequently brought about the beads-on-a-string phenomenon [49].

3. Microdroplets formation

3.1. Microdroplets encapsulation

The development of emulsification via microfluidic in the past decades is embodied in various fields. In the aspect of system design and fabrication, continuous reports with regards to new material development coupled with creative, novel fabrication techniques, enable evolutional microfluidic system from conventional two-dimensional (2D) straight microchannel to multifunctional three-dimensional (3D) systems [51]. In the aspect of theory and applications, in-depth understanding of the flow dynamic as mentioned in the previous section leads to much unique systems design for generation and manipulation of droplets with diverse behaviors and surface morphologies in various applications being reported [52–54]. Although the encapsulation process may seem straightforward, there are few considerations need to be noted. One common technique used is the addition of surface active agent to either continuous or dispersed phase. Such addition can create distinctive microdroplets formation, but not always an ideal solution to the development of more complex microfluidic systems with multi-inputs. Various innovative proposals have been attempted to resolve this issue, both for passive and active emulsion encapsulation system. An active encapsulation involves usage of external forces such as electricity to encapsulate emulsion, which is not favorable when it is deployed in biology-related application. While passive emulsion encapsulation method more often carries low throughput as its main shortcoming, the simplicity in encapsulation mechanism makes it a popular choice for microdroplet encapsulation.

There have been much microdroplet encapsulation attempts done in both Newtonian and non-Newtonian fluids system. In recent years, focus on microdroplets encapsulation had been given to non-Newtonian fluids especially in the medical field and in drug synthesis and released related work. In the case of non-Newtonian fluid systems, an integrated comprehensive droplet digital detection (IC 3D) was used to encapsulate bacteria such as E.coli in the blood content for detection purpose [55]. The device is able to encapsulate bacterial into microdroplets that will allow real time disease detection with high result accuracy. Such approach has shorten the conventional detection method that requires longer time. Much of other developments on microdroplet encapsulation can be found in various literatures.

3.2. Microdroplets fission

Generally, microdroplet splitting can be divided into two main approaches, i.e. active fission and passive fission, while the latter one is more widely accepted due to advantages including ease of implementation and low cost. Passive fission mainly depends on the fluid flow resistance and geometries in respective channels of the microfluidic system. A variety of geometries have been demonstrated in passive fission where microdroplet is split as it flows past T-junction [56], arbitrary angle [57], obstacle [58] or through a narrow constriction [59]. It was found that the microdroplet fission occurs at bifurcation if length of the droplet in the microchannel is greater than the circumference on the edge of the microdroplet. The microdroplet splits evenly if the resistances of the two daughter channels have the same fluidic resistance downstream of the bifurcation. Since fluidic resistance is proportional to microchannel length, changing the length of one of the two daughter channels allows microdroplets to split unevenly. In this way, the volume ratio of the daughter droplets produced by the fission can be changed. A large post can be employed near the middle of a microchannel to induce microdroplet fission, the ratio of sizes of daughter droplets can be changed by adjusting the position of the post in the microchannel [60]. A repeating bifurcation structure can be used to split a single parent droplet into 8 or 16 daughter droplets of nanoliter volumes. Droplet breakup occurs because of a high surface tension pressure relative to the pressure drop in the microchannel [61]. Multiple monodisperse droplets can be generated using a side-branch structure with varied resistance in the microchannel; the size of the daughter droplets was controlled by the size of the original liquid plugs, thus providing a wide tuning range of droplet size [62]. A hydrophobic valve was used to arrest the flow of fluid into each daughter channel, while placed a waste channel at downstream of the daughter channels to drain excess fluid. The series of daughter channels can split a single liquid sample plug into multiple smaller plugs effectively [63]. This technology can allow for simultaneous screening for multiple viruses [64].

Despite of these latest advances, the relevant techniques to manipulate droplet fission by constriction are commonly performed in a 2D single planar microchannel. For instance, Rosenfeld et al. investigated splitting and deformations of a large number of drops in a concentrated emulsion when it flows through a narrow constriction in 2D monolayer Poly (dimethylsiloxane) (PDMS) microchannel [59]. There is no design employing 3D bilayer microchannel consisting of constriction formed by bifurcated junction. Bifurcated junction work should be focused because the flow of emulsions through porous media is important in many industrial processes, while the porous media normally have small constrictions with bifurcated junction [65]. For example, natural porous media such as oil reservoir features with complex interior structure formed by a number of constrictions or pores with heterogeneity of physiochemical characteristics. Industrial processes such as mobility control in enhanced oil recovery requires prediction of the evolution of the microstructure of the injected fluids in microstructure and their bulk rheological properties. However, there is still a lack of complete understanding towards the stability and break-up of individual emulsions as they flow through simple constrictions or pores, especially when the interactions among the emulsions are important. Moreover, in some biomedical applications such as drug delivery, the encapsulated drugs must go through media with complex 3D constrictions, such as blood capillaries or the porous material of a tissue structure in the human body. These constrictions normally feature with bifurcated junction. In some cases, these constrictions are the locations for drug release, thus the emulsion droplets must burst to release its contents. In other cases, the emulsion droplets must be transported through the constrictions to reach the targeted release locations, thus breakup must not occur in the constrictions. Therefore, it demands a thorough understanding of the physics of emulsion flow through constrictions to predict and manipulate the release of active ingredients from such emulsions (**Figure 4**).

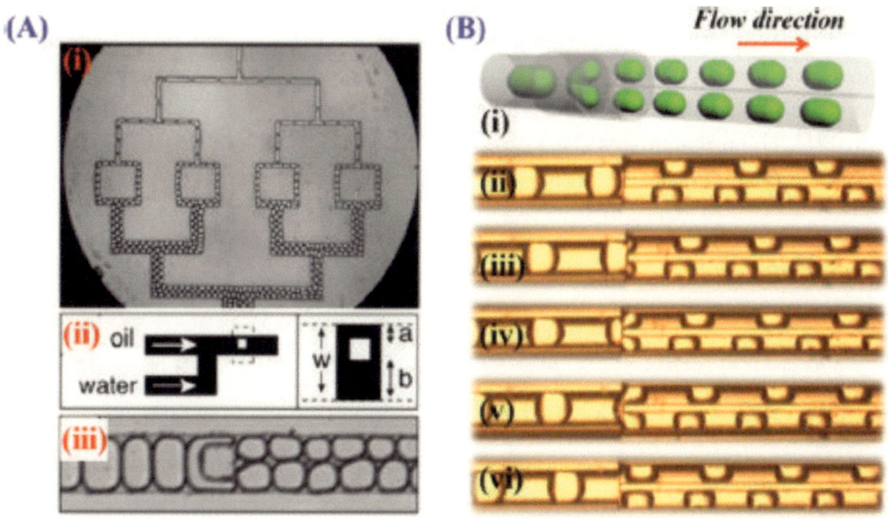

Figure 4. Droplet splitting in a microfluidic device. (A) 2D splitting: (i) droplets split repeatedly through Y-shaped junctions; (ii, iii) scheme illustration and microscopic image of droplet splitting mediated by a square obstruction. (B) 3D splitting: (i) scheme of the splitting process; (ii – vi) microscopic images showing droplets entering the junction and split into two daughter droplets. Reproduced with permission from [51]. Copyright 2017 American Chemistry Society.

4. Outlook

Emulsions, such as microcapsules and colloids, are crucial in daily life as well as playing vital role in different industrial applications such as pharmaceutical and food industry. Therefore, it is of important to understand the fundamental concept of Newtonian and non-Newtonian fluid behavior in the aspect of emulsification using microfluidic approach. Such understanding will bridge the practical application of the approach which aimed to control structure and monodispersed size in multiphase microsystem using emulsion as template. The two-phase flow understanding will stir the continuous development of emulsion to other applications such as microbeads for carbon capture and storage, wastewater treatment in environmental application as well as pharmaceutical drugs control. In this chapter, we have reviewed the fundamental two-phase multiphase model in emulsification process using microfluidic approach by outlining the main difference between Newtonian and non-Newtonian fluids and its corresponding pros and cons of each system. More effort should be given to non-Newtonian multiphase system research as non-Newtonian has more impact in daily life such as drug release kinetics in human blood, etc. Furthermore, more focus should also be given in the advancement of droplet size distribution control when both Newtonian and non-Newtonian fluids are used in a microfluidic multiphase system. This will eventually help to achieve applications such as advanced drug control and release as well as development of new food product that has better nutrient control.

Acknowledgements

We gratefully acknowledge the funding support from Zhejiang Provincial Natural Science Foundation of China (LQ15E090001), Young Scientist Program from National Natural Science Foundation of China (NSFC51506103/E0605) and Ningbo Natural Science Foundation (2015A610281). The research was also supported by Inspiration Grant from Faculty of Science and Engineering, University of Nottingham Ningbo China.

Author details

Kai Seng Koh[1]*, Voon Loong Wong[1] and Yong Ren[2]

*Address all correspondence to: k.koh@hw.ac.uk

1 School of Engineering and Physical Sciences, Heriot-Watt University Malaysia, Putrajaya, Malaysia

2 Department of Mechanical, Materials and Manufacturing Engineering, The University of Nottingham Ningbo China, China

References

[1] Schramm LL. Emulsions, Foams, and Suspensions: Fundamentals and Applications. Germany: Wiley-VCH; 2005

[2] Sami S, Olesya B, Ana RC, Karin S. Microfluidic EDGE emulsification: the importance of interface interactions on droplet formation and pressure stability. Scientific Reports. 2016; **6**:26407. DOI: 10.1038/srep26407

[3] Ajay M, Abhijit S, Achinta B, Keka O. Characterization of oil−water emulsion and its use in enhanced oil recovery. Industrial and Engineering Chemistry Research. 2010;**49**(24): 12756-12761. DOI: 10.1021/ie101589x

[4] Mark D, Haeberle S, Roth G, von Steffen F, Zengerle R. Microfluidic lab-on-chip platforms: requirements, characteristics, and applications. Chemical Society Reviews. 2010; **39**:1153-1182. DOI: 10.1039/B820557B

[5] Garstecki P. Formation of droplets and bubbles in microfluidic systems. Bulletin of the Polish Academy Sciences. 2005;**53**(4)

[6] Gu H, MHG D, Mugele F. Droplet formation and merging in two-phase flow microfluidic. International Journal of Molecular Sciences. 2011;**12**(4):2572-2597. DOI: 10.3390/ijms12042572

[7] Nunes JK, Tsai SSH, Wan J, Stone HA. Dripping and jetting in microfluidic multiphase flows applied to particle and fibre synthesis. Journal of Physics D: Applied Physics. 2013; **46**(11):23626378. DOI: 10.1088/0022-3727/46/11/114002

[8] Glawdel T, Elbuken C, Ren CL. Droplet formation in microfluidic T-junction generators operating in the transitional regime. II. Modeling. Physical Review E. 2012;**85**:016322. DOI: 10.1103/PhysRevE.85.016323

[9] Wehking JD, Gabany M, Chew L, Kumar R. Effects of viscosity, interfacial tension, and flow geometry on droplet formation in a microfluidic T-junction. Microfluidics and Nanofluidics. 2014;**16**(3):441. DOI: 10.1007/s10404-013-1239-0

[10] Raj R, Mathur N, Buwa VV. Numerical simulations of liquid-liquid flows in microchannels. Industrial & Engineering Chemistry Research. 2010;**49**(21):10606-10614. DOI: 10.1021/ie100626a

[11] DeMenech M, Garstecki P, Jousse F, Stone HA. Transition from squeezing to dripping in a microfluidic T-shaped junction. Journal of Fluid Mechanics. 208;**595**:141-161. DOI: 10.1017/S002211200700910X

[12] Liu H, Zhang Y. Droplet formation in a T-shaped microfluidic junction. Journal of Applied Physics. 2009;**106**:034906. DOI: 10.1063/1.3187831

[13] Garstecki P, Fuerstman MJ, Stone HA, Whitesides GM. Formation of droplets and bubbles in a microfluidic T-junction-scaling and mechanism of breakup. Lab on Chip. 2006; **6**(3):437-446. DOI: 10.1039/b510841a

[14] Gupta A, Kumar R. Flow regime transition at high capillary numbers in a microfluidic T-junction: Viscosity contrast and geometry effect. Physics of Fluids. 2010;**22**:1-11. DOI: 10.1063/1.3523483

[15] Anna SL, Mayer HC. Microscale tipstreaming in a microfluidic flow focusing device. Physics of Fluids. 2006;**18**:121512. DOI: 10.1063/1.2397023

[16] Utada AS, Fernandez-Nieves A, Stone HA, Weitz DA. Dripping to jetting transitions in coflowing liquid streams. Physical Review Letter. 2007;**99**:094502. DOI: 10.1103/PhysRevLett.99.094502

[17] Baroud CN, Gallaire F, Dangla R. Dynamics of microfluidic droplets. Lab on Chip. 2010; **10**(16):2032-2045. DOI: 10.1039/c001191f

[18] Chhabra RP, Richardson JF. Non-Newtonian Flow and Applied Rheology. 2nd ed. Elsevier: Butterworth-Heinemann; 2008. 536 p. DOI: 10.1016/B978-0-7506-8532-0.00013-5

[19] Umbanhowar PB, Prasad V, Weitz DA. Monodisperse emulsion generation via drop break off in a coflowing stream. Langmuir. 2000;**16**(2):347-351. DOI: 10.1021/la990101e

[20] Cramer C, Fischer P, Windhab EJ. Drop formation in a co-flowing ambient fluid. Chemical Engineering Science. 2004;**59**(15):3045-3058. DOI: 10.1016/j.ces.2004.04.006

[21] Abate RA, Weitz DA. Faster multiple emulsification with drop splitting. Lab on Chip. 2011;**11**:1911-1915. DOI: 10.1039/C0LC00706D

[22] Nguyen NT, Wereley ST. Fundamentals and Applications of Microfluidics. 2nd ed. Artech House: Boston; 2002. p. 497

[23] Tanyeri M, Ranka M, Sittipolhul N, Schroeder CM. A microfluidic-based hydrodynamic trap: Design and implementation. Lab on Chip. 2011;**11**(10):1786-1794. DOI: 10.1039/c0lc00709a

[24] Cho YL, Hartnett JP. Non-Newtonian fluids. In: Cho YL, Hartnett JP, editors. Handbook of Heat Transfer Applications. 3rd ed. New York: McGraw-Hill; 1998. pp. 732-784

[25] Xu JH, Luo GS, Li SW, Chen GG. Shear force induced monodisperse droplet formation in a microfluidic device by controlling wetting properties. Lab on Chip. 2006;**6**(1):131-136. DOI: 10.1039/B509939K

[26] Xu JH, Li SW, Tan J, Wang YJ, Luo GS. Preparation of highly monodisperse droplet in a T-junction microfluidic device. American Institute of Chemical Engineers Journal. 2008; **52**(9):3005-3010. DOI: 10.1002/aic.10924

[27] Qiu D, Silva L, Tonkovich AL, Arora R. Micro-droplet formation in non-Newtonian fluid in a microchannel. Microfluidics and Nanofluidics. 2010;**8**(4):531-548. DOI: 10.1007/s10404-009-0487-5

[28] Wang K, Lu YC, Luo GS. Determination of dynamic interfacial tension and its effect on droplet formation in the T-shaped microdispersion process. Langmuir. 2009;**25**(4):2153-2158. DOI: 10.1021/la803049s

[29] Bashir S, Rees JM, Zimmerman WB. Simulations of microfluidics droplet formation using the two-phase level set method. Chemical Engineering Science. 2011;**66**(20):4733-4741. DOI: 10.1016/j.ces.2011.06.034

[30] Shui LL, Berg AVD, Eijkel JCT. Interfacial tension controlled W/O and O/W 2-phase flows in microchannel. Lab on Chip. 2009;**9**:795-801. DOI: 10.1039/B813724B

[31] Peng L, Yang M, Guo SS, Liu W, Zhao XZ. The effect of interfacial tension on droplet formation in flow-focusing microfluidic device. Biomedical Microdevices. 2011;**13**(3):559-564. DOI: 10.1007/s10544-011-9526-6

[32] Cubaud T, Mason TG. Capillary threads and viscous droplets in square microchannels. Physics of Fluids. 2008;**20**(5):053302. DOI: 10.1063/1.2911716

[33] Gu ZP, Liow JL.Microdroplet formation in a T-junction with xanthan gum solutions. In: Chemeca 2011: Engineering a Better World; 18-21 September; Sydney Hilton Hotel, NSW. Australia: Engineers Australia; 2011. pp. 1442-1451

[34] Murshed SMS, Tan SH, Nguyen NT. Temperature dependence of interfacial properties and viscosity of nanofluids for droplet-based microfluidics. Journal of Physics D: Applied Physics. 2008;**41**(8):085502. DOI: 10.1088/0022-3727/41/8/085502

[35] Wang RJ. Nanoparticles influence droplet formation in a T-shaped microfluidic. Journal of Nanoparticle Research. 2013;**15**:2128. DOI: 10.1007/s11051-013-2128-x

[36] Fang C, Lee D, Stober B, Fuller GG, Shen AQ. Integrated microfluidic platform for instantaneous flow and localized temperature control. RSC Advances. 2015;**5**(104):85620-85629. DOI: 10.1039/C5RA19944A

[37] Sang L, Hong Y, Wang F. Investigation of viscosity effect on droplet formation in T-shaped microchannels by numerical and analytical methods. Microfluidics and Nanofluidics. 2009;**6**(5):621-635. DOI: 10.1007/s10404-008-0329-x

[38] Ba Y, Liu H, Sun J, Zheng R. Three dimensional simulations of droplet formation in symmetric and asymmetric T-junctions using the color-gradient lattice Boltzmann model. International Journal of Heat Transfer. 2015;**90**:931-947. DOI: 10.1016/j.ijheatmasstransfer.2015.07.023

[39] Husny J, Cooper-White JJ. The effect of elasticity on drop creation in T-shaped microchannels. Journals of Non-Newtonian Fluid Mechanics. 2006;**137**(1–3):121-136. DOI: 10.1016/j.jnnfm.2006.03.007

[40] Wong VL, Loizou K, Lau PL, Graham RS, Hewakandamby BN. Numerical studies of shear-thinning droplet formation in a microfluidic T-junction using two-phase level-set method. Chemical Engineering Science. 2017;**174**:157-173. DOI: 10.1016/j.ces.2017.08.027

[41] Abate AR, Poitzsch A, Hwang Y, Lee J, Czerwinska J, Weitz DA. Impact of inlet channel geometry on microfluidic drop formation. Physical Review E. 2009;**80**(2). DOI: 026310, 10.1103/PhysRevE.80.026310

[42] Cherizol R, Sain M, Tjong J. Review of non-Newtonian mathematical models for rheological characteristics of viscoelastic composites. Green and Sustainable Chemistry. 2015;**5**(1): 6-14. DOI: 10.4236/gsc.2015.51002

[43] Kostic M. Influence of viscosity function simplification on non-Newtonian velocity and shear rate profiles in rectangular ducts. International Communications in Heat and Mass Transfer. 1993;**20**(4):515-525. DOI: 10.1016/0735-1933(93)90063-2

[44] Harvie DJE, Davidson MR, Cooper-White JJ, Rudman M. A parametric study of droplet deformation through a microfluidic contraction: Low viscosity Newtonian droplets. Chemical Engineering Science. 2006;**61**(15):5149-5158. DOI: 10.1016/j.ces.2006.03.011

[45] Fu TT, Ma YG, Li HZ. Breakup dynamics of slender droplet formation in shear-thinning fluids in flow-focusing devices. Chemical Engineering Science. 2016;**144**:75-86. DOI: 10.1016/j.ces.2015.12.031

[46] Mbanjwa MB, Land KJ, Jewell L, Moss EA, Gledhill IMA. Experimental observation of capillary instabilities of two phase flow in a microfluidic T-junction. In: Seventh South African Conference on Computational and Applied Mechanics (SACAM10); 10-13 January; Pretoria. Republic of South Africa:2010

[47] KhosraviZanjani MA, Tarzi BG, Sharifan A, Mohammadi N. Microencapsulation of probiotics by calcium alginate-gelatinized starch with chitosan coating and evaluation of survival in simulated human gastro-intestinal condition. Iranian Journal of Pharmaceutical Research. 2014;**13**(3):843-852

[48] Villa C, Manzoli V, Abreu MM, Verheyen CA, Seskin M, Najjar M, Molano RD, Torrente Y, Ricordi C, Tomei AA. Effects of composition of alginate-polyethylene glycol microcapsules and transplant site on encapsulated islet graft outcomes in mice. Transplantation. 2017;**101**(5):1025-1035. DOI: 10.1097/TP.0000000000001454

[49] Oliveira MS, McKinley GH. Iterated stretching and multiple beads-on-a-string phenomena in dilute solutions of highly-extensible flexible polymers. Physics of fluids. 2005;**17**: 071705. DOI: 10.1063/1.1949197

[50] Steinhaus B, Shen AQ. Dynamics of viscoelastic fluid filaments in microfluidic devices. Physics of Fluids. 2007;**18**:073103. DOI: 10.1063/1.2747660

[51] Shang LR, Cheng Y, Zhao YJ. Emerging droplet microfluidics. Chemical Reviews. 2017; **117**(12):7964-8040. DOI: 10.1021/acs.chemrev.6b00848

[52] Zhu PG, Wang LQ. Passive and active droplet generation with microfluidics:A review. Lab on Chip. 2017;**17**:34-75. DOI: 10.1039/C6LC01018K

[53] Takeuchi S, Garstecki P, Weibel DB, Whitesides GM. An axisymmetric flow-focusing microfluidic device. Advanced Materials. 2005;**17**(8):1067-1072. DOI: 10.1002/adma.200401738

[54] Sun JS, Xianyu YL, Jiang XY. Point-of-care biochemical assays using gold nanoparticle-implemented microfluidics. Chemical Society Reviews. 2014;**43**:6239-6253. DOI: 10.1039/C4CS00125G

[55] Kang DK, Ali MM, Zhang KX, Huang SS, Peterson E, Digman AM, Gratton E, Zhao WA. Rapid detection of single bacteria in unprocessed blood using integrated comprehensive droplet digital detection. Nature Communications. 2014;**5**:5427. DOI: 10.1038/ncomms6427

[56] Samie M, Salari A, Shafii B. Breakup of microdroplets in asymmetric T junctions. Physical Review E. 2013;**87**:053003. DOI: 10.1103/PhysRevE.87.053003

[57] Menetrier-Deremble L, Tabeling P. Droplet breakup in microfluidic junctions of arbitrary angles. Physical Review E. 2006;**74**:035303. DOI: 10.1103/PhysRevE.74.035303

[58] Protiere S, Bazant MZ, Weitz DA, Stone HA. Droplet breakup in flow past an obstacle: A capillary instability due to permeability variations. EPL (Europhysics Letters). 2010;**92**(5): 54002. DOI: 10.1209/0295-5075/92/54002

[59] Rosenfeld L, Fan L, Swoboda R, Tang SKY. Break-up of droplets in a concentrated emulsion flowing through a narrow constriction. Soft Matter. 2014;**10**:421-430. DOI: 10.1039/C3SM 51843D

[60] Link DR, Anna SL, Weitz DA, Stone HA. Geometrically mediated breakup of drops in microfluidic devices. Physical Review Letters. 2004;**92**:054503. DOI: 10.1103/PhysRev Lett.92.054503

[61] Adamson DN, Mustafi D, Zhang JXJ, Zheng B, Ismagilov RF. Production of arrays of chemically distinct nanolitre plugs via repeated splitting in microfluidic. Lab on Chip. 2006;**6**:1178-1186. DOI: 10.1039/B604993A

[62] Xiong S, Chin LK, Yu YF, Yu JQ, Chen Y, Zhang GJ, Lo GQ, Kwong DL, Liu AQ. Multi-size droplets generation via side-branch microfluidic channels. In: Verpoorte S, editors. 14th International Conference on Miniaturized Systems for Chemistry; 3–7 October; Groningen, The Netherlands. 2010

[63] Mielnik M, Voitel J, Solli L, Furuberg L. Sample metering and parallel liquid plug actuation for multiple biochemical assays. In: 11th International Conference on Miniaturized Systems for Chemistry and Life; Paris, France. p. 1513-1515

[64] Furuberg L, Mielnik M, Gulliksen A, Solli L, Johansen IR, Voitel J, Baier T, Riegger L. RNA amplification chip with parallel microchannels and droplet positioning using capillary valves. Microsystem Technologies. 2008;**14**(4–5):673-681. DOI: 10.1007/s00542-007-0515-x

[65] vanDinther AMC, Schroën CGPH, Boom RM. Separation process for very concentrated emulsions. Innovative Food Science & Emerging Technologies. 2013;**18**:177-182. DOI: 10.1016/j.ifset.2012.12.007

Chapter 7

Micro/Nanofluids in Sustainable Machining

Tran The Long and Tran Minh Duc

Additional information is available at the end of the chapter

http://dx.doi.org/10.5772/intechopen.75091

Abstract

Micro/nanofluids are the recent alternative solutions for cooling lubrication that can be defined as the fluids containing microparticles or nanoparticles, which own superior lubrication and cooling characteristics. For these reasons, they have gained significant attention in industrial applications, such as automotive, machining, and biomedical industries. In this chapter, the authors mainly present the recent progress and applications of nanofluids in machining processes, as well as some initial researches about microfluids. Nanofluids provide an excellent media in cutting zone for enhancing the thermal conductivity and tribological characteristics. Therefore, they help to enhance the cutting performance by reducing the coefficient of friction, cutting temperature tool wear, and improving the surface quality. Moreover, the application of nanoparticles in vegetable oils, which are inherently nontoxic as well as biodegradable, gives them superior lubrication properties suitable for MQL application, especially for difficult-to-cut materials. The novel green technology definitely brings out many new solutions in machining practice.

Keywords: micro/nanofluids, nanofluid, sustainable machining, MQL, hard machining, vegetable oil, cutting

1. Introduction

Nanofluids are suspensions of nanoparticles in fluids that show significant enhancement of their thermophysical properties with proper volumetric fraction of nanoparticles. Much of the research on nanofluids are about understanding their behavior, so that they can be utilized effectively as an alternative solutions in many industrial applications, nuclear reactors, transportation, electronics, machining, as well as biomedicine and food [1].

Environmental friendliness has become one of the biggest issues in modern industry worldwide, especially in machining industry. In addition, recent regulations on environmental problems, such as ISO 14000 and Green Round, have become much stricter for promoting green manufacturing approaches. The necessity of reducing environmental loads should be increasingly considered, and many green manufacturing processes have been developed and studied.

According to the statistics of cost distribution in manufacturing shown in **Figure 1**, the coolant expense for usage and disposal represents about 15% of total production costs, depending on the workpart and the types of cooling system, as well as machining location [3]. In contrast, tooling cost contributes only a small value of 4%. On the other hand, health and environmental issues associated with the airborne cutting fluid particles on factory shop floors motivate manufacturing enterprises to drastically reduce coolant consumption and, if possible, eliminate it altogether.

As a result, the conception of dry cutting has been first considered to achieve environmental friendliness. Eliminating the cutting fluids in machining processes means that there is no cooling lubricating media, which has three essential functions (i.e., reduction of friction, absorption of the generated heat, and chip evacuation). Hence, these following problems must be considered:

- Workpart: deteriorate surface texture and need additional works (cleaning or deburring)
- Cutting tool: difficulty in chip formation, reduction of tool life or change to expensive ones
- Machine tool: high rigidity, equipment specialized in pushing chips away from the cutting zone as well as controlling temperature

Especially in cases of machining difficult-to-machine materials like high-strength and high-hardness steels, solving the mentioned problems has strong influence on leading industrial branches as automotive, roller bearing, hydraulic, and die and mold sectors. The term "hard machining" is a recent technology that can be defined as the machining operation of a workpiece that has a hardness value typically in the 45–70 HRC range, using directly tools with geometrically defined cutting edges [5]. Hard-cutting operations are capable of replacing, in some cases, grinding operations and produce comparable surface finish. Various machining operations in hard machining include milling, boring, broaching, hobbling, and others. Together

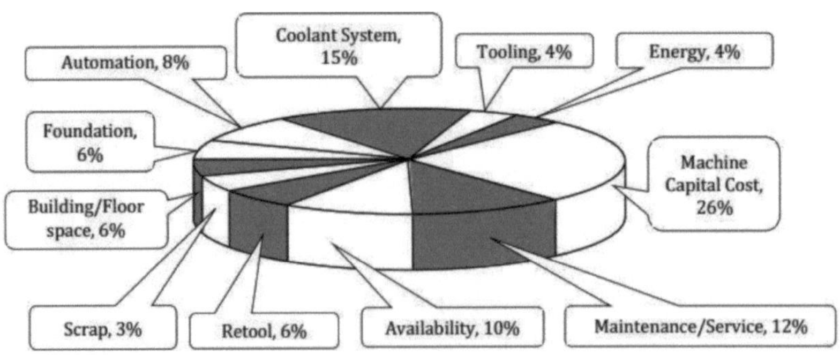

Figure 1. Distribution of manufacturing costs for wet machining [2].

with the developments of suitable rigid machine tools, superhard cutting-tool materials and special tool (toolholders) designs, and complete set-ups, the machining of hardened parts has become more easily accessible and widely applied for the modern machining industry. However, the challenge of selecting a cutting-tool insert to ensure tool life and high-precision machining of the component is the main problem, which slows down the development and application of hard machining. Since its broader introduction in the mid-1980s in the form of hard turning, people have seen that the cutting inserts, such as coated carbides (PVD (Ti, Al) N–TiN and CVD Ti(C, N)–Al_2O_3-coated tools, etc.), ceramics, and (P)CBN, are widely utilized in various dry hard cutting processes.

Without cooling lubricating media, the enormous amount of heat generated from cutting zone remains a big question, which limits the cutting condition, reduces the tool life, and deteriorates the surface finish (i.e., the so-called "white layer" formation in hard machining). This problem has promoted the development of minimum quantity lubrication (MQL) using a special nozzle to form oil mist directly supplied to the machining interface with a tiny amount of fluid consumption (5–500 ml/hr). Because the cutting fluid mostly vaporizes and leaves dry chips, it brings out cost effective and green machining [4]. The very small amount of cutting fluids is utilized and delivered effectively to cutting zone, and the formation of oil films in contact faces plays an important role in lubrication [7]. Numerous publications have been reported on the effectiveness of the MQL technique for enhancing cutting performance [2, 4, 6, 7]. However, the main drawback of MQL technology is low cooling effect, and so it does not work so well in cutting difficult-to-machine materials with high strengths and hardness. To improve the MQL technology, nanofluids containing nanoparticles (Al_2O_3, MoS_2, SiO_2, CuO, diamond, and so forth) with at least one of their principal dimensions smaller than 100 nm [3] used in MQL technique recently reach a significant attention of worldwide researchers and are up-to-date topics to increase the cutting performance and productivity.

2. The effects of nanofluids on machining processes

The applications of nanofluids for MQL machining have been proven to improve the interaction of friction in cutting zone due to the occurrence of nanoparticles. However, the direct evaluation of cutting performance faces many difficulties, and so numerous publications are focused on the indirect evaluation through machining outputs such as cutting forces, cutting temperature, tool wear, tool life, and surface integrity. In this section, the recent studies related to the effects of nanofluids on machining performance will be discussed.

2.1. Thermal properties of nanofluids

From previous investigations, nanofluids have been found to possess enhanced thermophysical properties such as thermal conductivity, thermal diffusivity, viscosity, and convective heat transfer coefficients. The thermal conductivity of more than 50 various nanofluids based on water, ethylene glycol, and engine oil containing particles of SiO_2, Al_2O_3, TiO_2, ZrO_2, CuO, and diamond was experimentally measured [8]. The obtained results had shown that the thermal conductivity coefficient of nanofluids enhances with increasing particle sizes. It has been confirmed that the lower the thermal conductivity of the base fluid, the higher the relative thermal conductivity coefficient of the nanofluids. The different researches have made to investigate convective heat transfer of nanofluids [10]. Based on the results, augmenting nanofluid volume fraction, Rayleigh and Magnetic numbers lead to improve of the temperature gradient, while it reduces with augment of Lorentz forces. Heat transfer improvement augments with increase in Kelvin forces, while it reduces with augment of Lorentz forces at high Rayleigh number, but different manners are detected for low Rayleigh number. The active method for nanofluid heat transfer enhancement by means of EHD was also studied [10]. The obtained results suggest that influence of electric field on forced convection improvement is more sensible for lower *Re* number. Temperature gradient enhances with rise of voltage supply. Moreover, throughout the experimental results, the convective heat transfer increases with the presence of nanoparticles in the base fluids [11, 12, 21, 27]. Based on the newest publications, the deeper understanding about thermal properties of NFs is studied. The shapes of NPs are proven to influence on the rate of heat transfer, and the effect of thermal radiation on CuO nanofluid behavior is successfully modeled via Control Volume–based Finite Element Method (CVFEM). Platelet shape nanoparticles reveal the highest heat transfer rate [22, 25, 28, 29]. Nanofluid motion, as well as flow circulation and thermal energy transport, enhances by increasing the volume fraction of NPs [23, 24, 26]. A novel research of melting temperature of CuO-water NF heat transfer enhancement is simulated by CVFEM. The highlighted results include (1) flow velocity of NF increases due to the presence of CuO nanoparticles; (2) heat transfer enhancement of NF improves at higher nanoconcentrations; (3) melting temperature rises with the increment of nanovolume fraction [30, 31, 33]. The same observations were obtained from the study of Fe_3O_4-water nanofluid [32, 34]. They contribute a very good understanding of nanofluid behavior as cutting fluid in various cutting processes.

2.2. The effects on cutting temperature

The cutting fluid can be useless if not delivered efficiently to contact zone, and so the methods of supplying the coolant in machining are the critical parameter. However, the effectiveness of supplying cutting fluids in wet cutting can help to dissipate relatively small amount of the generated heat. It is well known that only very small amount of cutting fluid can penetrate to contact zone although large amount is delivered. On the other hand, costs, as well as health and environmental issues, motivate manufacturing enterprises to drastically reduce consumption of cooling fluids.

Dry machining processes face the serious difficulties in heat dissipation and chip transportation though eliminating the use of cutting fluids. From **Figure 2**, nongeometrically defined machining processes, such as grinding, honing, etc., are considered cooling function the most important factor. When some of these processes can be replaced by geometrically defined hard machining methods (for instance hard turning, hard milling), successful machining with minimization or without fluids can be achieved [13].

Stainless steel, for instance, belongs to the difficult machining material, which is easy to stick tool leading to increasing the cutting temperature and intensifying the abrasion of the tool

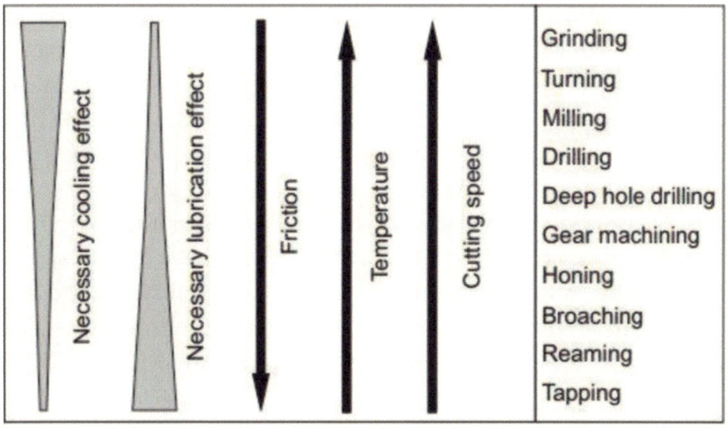

Figure 2. Machining operations and their needs for cooling and lubricating functions [3].

nose. Dry turning of AISI 304 stainless steel (85 HRB) at v = 75÷265 m/min; f = 0.1÷0.3 mm/rev; a_p = 0.8÷1.6 mm with two different groups of inserts: uncoated and TiN coated carbide inserts (rake angle $\gamma_o = 15°$, relief angle $\alpha_o = 8°$, inclination angle $\lambda_s = -4°$, and side cutting-edge angle $k_r = 75°$) [14].

From **Figure 3**, the cutting temperatures of two kinds of cutting tools increase with increasing cutting speed. The reason was that frictional heat generated from the contact zone of the bottom of chip and tool rake face was too late to transfer and was accumulated at the bottom of chip. Therefore, the cutting temperature increased. The comparison of cutting temperature is made among dry, wet, and MQL turnings of AISI 4140 steel (340 HV) at v = 50.2÷141.4 m/min; f = 0.09÷0.22 mm/rev; and a_p = 0.5÷1.5 mm with HSS tools [15]. The tool-chip interface temperature in which MQL fluid is supplied from both nozzles to the rake and flank faces is approximately

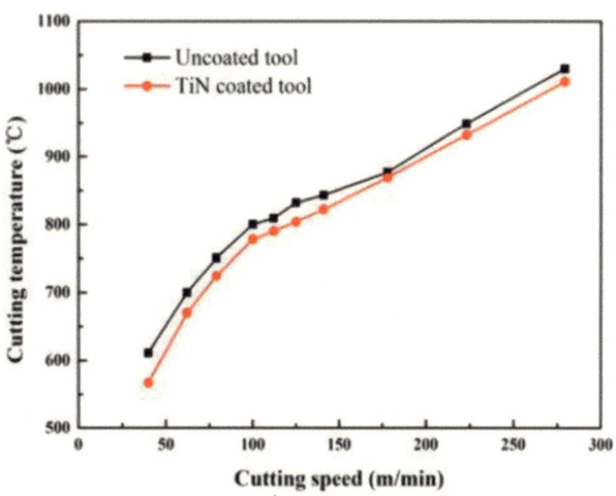

Figure 3. Evolution of cutting temperature with cutting speed in dry turning of AISI 304 stainless steel [14].

350°C lower than that in dry turning, and if supplied only to rake face, the tool temperature is about 200°C lower than that in dry turning. Additionally, the tool-chip interface temperature in wet turning is about 300°C lower than that in dry turning. The difference in cutting temperatures under dry, wet, and MQL conditions is closely related to the difference in cutting forces. The greater the cutting forces, the more heat and higher cutting temperatures are generated. Accordingly, the application of dry cutting processes is limited. It can also be observed that MQL techniques effectively provide oil mist directly to cutting zone to improve lubricant characteristics, but the main drawback of this technique is cooling character. It has a significant meaning for machining hard materials with the hardness range of 45 ÷ 70 HRC. Therefore, the application of nanofluids in MQL machining, an up to date research topic, brings out a novel substitution for dry and wet cutting, as well as the development of semi-dry machining (MQL technique).

The thermal conductivity of nanofluids has been found to be higher than that of the base fluid by using KD2 Pro Thermal analyzer to measure at room temperature (25°C) to note down the increased conductivity value (seen in **Figure 4**) [16]. It is also observed that the thermal conductivity of nanofluids enhances when increasing the nanoparticle concentration.

The comparison of six types of nanoparticles, namely molybdenum disulfide (MoS_2), zirconium dioxide (ZrO_2), carbon nanotube (CNT), polycrystalline diamond, aluminum oxide (Al_2O_3), and silica dioxide (SiO_2), mixed with palm oil to formulate nanofluids is made and used for MQL grinding of Ni-based alloys [17]. The grinding temperatures of six nanofluids are shown in **Figure 5**.

It can be clearly observed from **Figure 5** that the grinding temperatures sharply increase at the initiation of grinding process but decrease gradually to reach a stable temperature when six different nanofluids are supplied to contact zone.

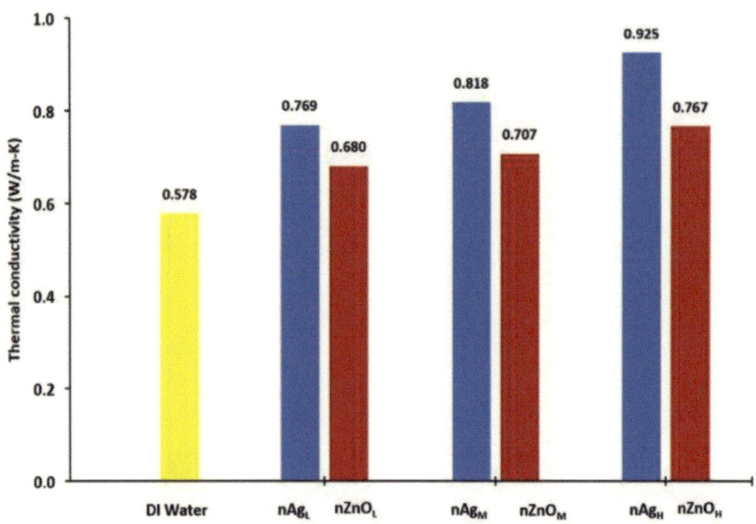

Figure 4. Thermal conductivity variation for silver and zinc oxide nanofluids with different volume fractions [16].

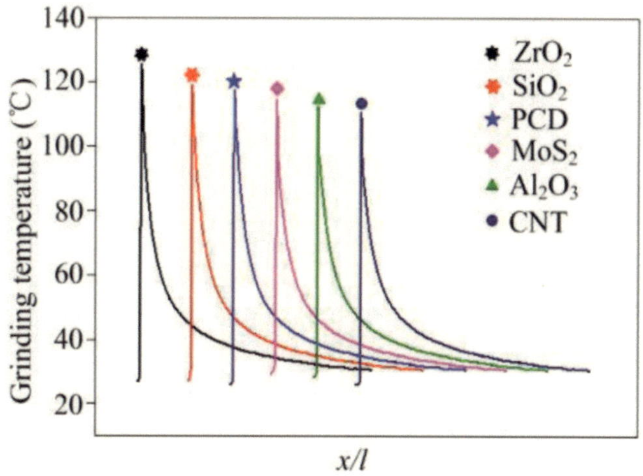

Figure 5. Grinding temperature of six nanofluids with respect to a dimensionless grinding distance *x/l* [17].

From **Figure 6**, it can be clearly seen that six different nanofluids help to effectively reduce MQL grinding temperature compared to the base fluid due to the presence of nanoparticles with hard property, as well as good heat transfer. The use of nanoparticles has a significant meaning in improving cooling and lubricating characteristics. CNT nanofluid shows the best cooling performance, presumably it has good heat transfer properties. On the other hand, the viscosity of cutting fluids is an important influencing factor of lubrication performance. **Figure 7** shows the relationship between six different nanofluids' viscosity and temperature.

The viscosity of all nanofluids decreases with the rise of temperature, especially before 70°C. SiO_2, Al_2O_3, and CNT nanofluids have higher viscosity than other ones. High viscosity allows the cutting fluids to stay in the cutting area for a longer time. This phenomenon improves the cooling

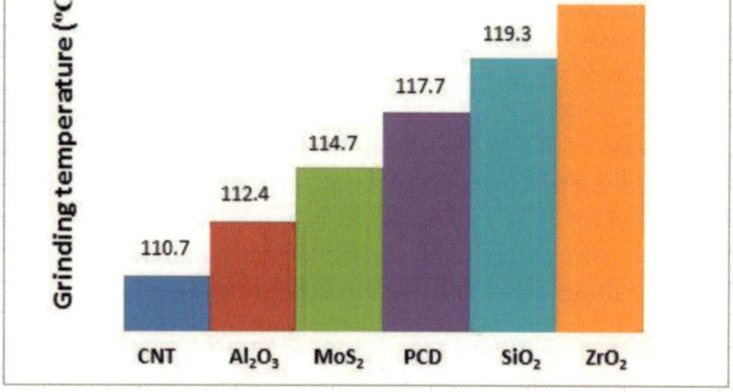

Figure 6. Grinding temperatures of six different nanofluids [17].

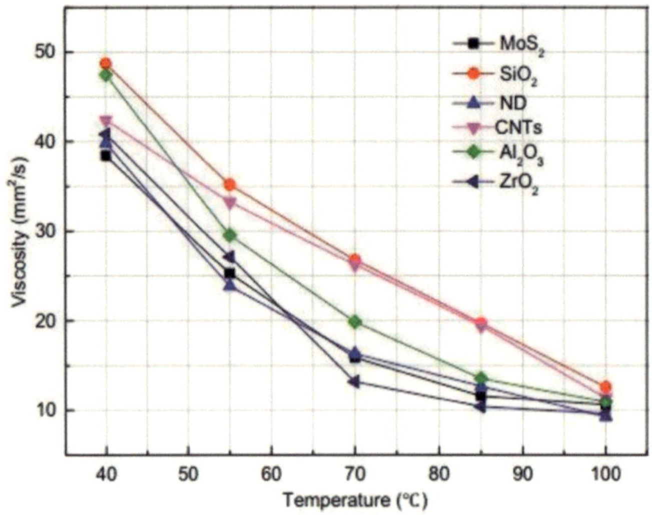

Figure 7. Relationship between nanofluids' viscosity and temperature [18].

lubrication of the contact area. In addition, the use of vegetable oils as the base nanofluids not only improves their cooling, lubricating, and viscous characteristics but also is the step toward sustainable manufacturing.

2.3. The effects on cutting forces

The tribological characteristic of cutting fluids has a significant meaning for investigating cutting forces. The nanofluids' tribological characteristics are improved by using Al_2O_3 and TiO_2 nanomaterials. The kinematic viscosity of Al_2O_3 and TiO_2 nanolubricants decreased slightly due to the presence of nanoparticles between the lubricant layers leading to an ease of relative movement with the nanoparticles acting as "rollers." On the other hand, the viscosity index increased with the use of nanolubricants [9]. Accordingly, nanoparticles in MQL fluid play an important role in converting sliding into rolling contact. That is the reason why the friction coefficient in cutting zone is much reduced, and the cutting temperature, cutting forces, and tool wear decrease. MQL hard milling of $60Si_2Mn$ steel (50÷52HRC) was done by using Al_2O_3 nanofluid (0.5 wt%) with carbide inserts at v = 110 m/min; f = 0.12 mm/tooth; a_p = 0.2 mm [19]. The cutting forces were directly measured during cutting process by Kistler quartz, three-component dynamometer (9257BA). **Figures 8–10** show the cutting force components F_x, F_y, and F_z of MQL hard milling process with/without Al_2O_3 nanoparticles. It is clearly observed that, compared to the case of MQL fluids without nanoparticles, all the cutting force components are much reduced when cutting with nanofluids. Interestingly, it is revealed that during the first 20 min, all the cutting forces F_x, F_y and F_z in both cases are low; therefore, in this time, the performance of Al_2O_3 nanoparticles in MQL hard milling is not really clear. After the first period, the rapid tool wear occurs, and wear land reaches some extent, which allows nanoparticles to penetrate to cutting zone. The formulation of oil film with nanoparticles in contact zone plays an important role in creating "roller effect." Rolling friction instead of sliding one occurs between flank face and machined surface, rake face and chip surface, and so forth. Hence, the cutting forces significantly reduce and the tool life extends much.

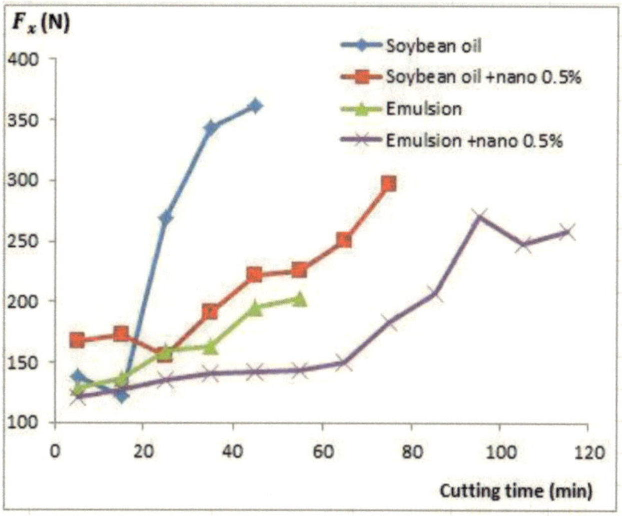

Figure 8. Cutting force component F_x under MQL conditions with/without nanofluids [19].

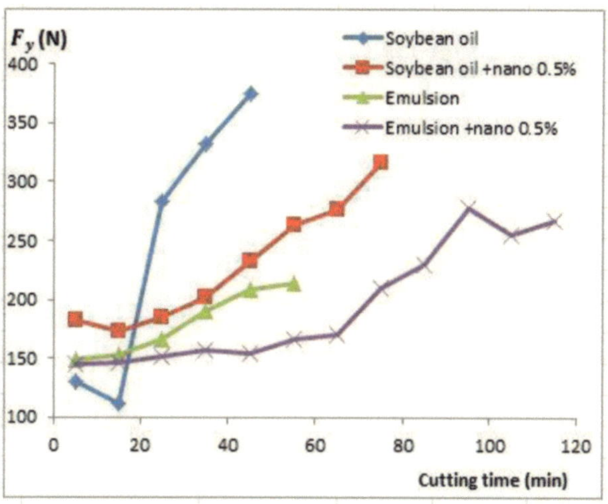

Figure 9. Cutting force component F_y under MQL conditions with/without nanofluids [19].

The use of soybean oil–based nanofluids in MQL hard milling was less effective than that of emulsion. However, both of them allow the normal APMT 1604 PDTR LT30 carbide inserts to use effectively for hard milling, and the economic and technological characteristics of cutting performance are achieved. Another promising research investigated lubrication properties of the wheel/workpiece interface in MQL nanofluids grinding compared with flood and MQL grinding without nanoparticles. The experiments were conducted at wheel speed v_s = 30 m/s, feed speed v_w = 3000 mm/min, and cutting depth a_p = 10 μm for machining the high-temperature nickel base alloy GH4169 [18]. **Figures 11** and **12** show the grinding forces obtained.

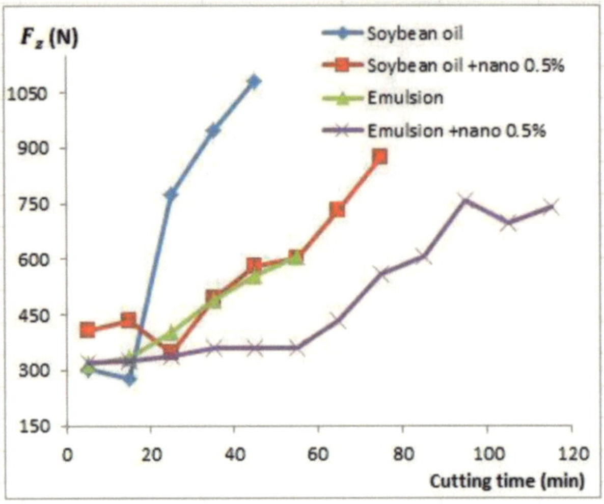

Figure 10. Cutting force component F_z under MQL conditions with/without nanofluids [19].

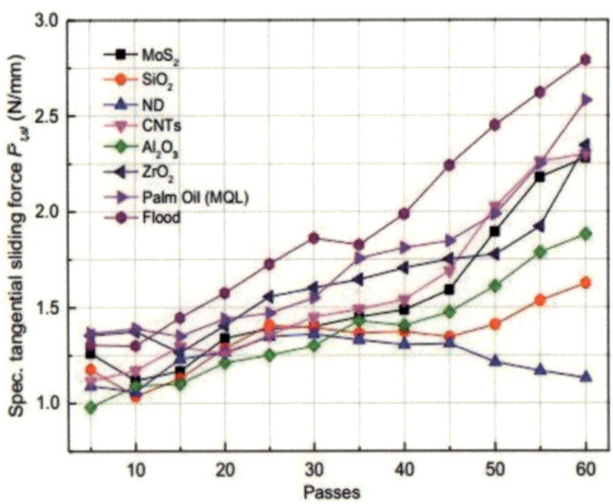

Figure 11. Specific tangential sliding grinding forces in the cases of flood, pure MQL, and nanofluids.

Both normal and tangential grinding forces increased with an increased number of passes. The flood grinding process shows the largest sliding grinding force. MQL grinding with pure palm oil achieves smaller sliding force because it effectively increases the lubrication effect of the grinding area due to oil-mist formation. Nanofluids are superior to pure palm oil in lubrication improvement. When making the comparison among six different nanofluids, the sliding grinding forces under MQL nanofluid (Al_2O_3, MoS_2, SiO_2, and ND) are lowest due to hard characteristic and small sliding friction coefficient. The effectiveness of nanofluids on reduction of cutting forces becomes a novel observation and has an important influence on tool wear and tool life, which directly affect the surface quality and manufacturing cost.

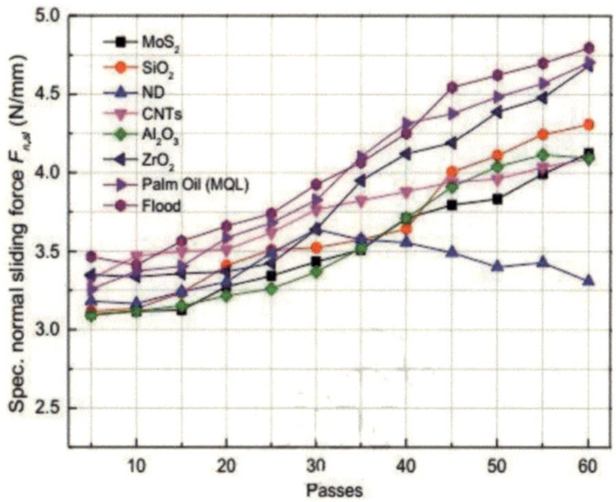

Figure 12. Specific normal sliding grinding forces in the cases of flood, pure MQL, and nanofluids.

2.4. The effects on tool wear and tool life

With the presence of nanoparticles between rake face and fresh chip, as well as flank face and machined surface, the mechanism of the tribological effect takes many forms, such as "roller effect," third body effect, chemical mechanical protective film effect, mending effect, and polishing effect [2, 3]. For instance, during gear hobbing process of AISI 4118 steel (spindle speed = 200 rev/min, depth of cut =4.375 mm, feed rate = 1.27 mm/rev), using nanofluid (Al_2O_3 with the size 80 nm suspended in ISO VG46 lubricant oil with volume fraction of 0.1÷ 0.2%) shows many promising results. Nanoparticles in the base oil effectively improve the heat transfer capability and reduce the friction by "roller effect" in cutting zone, leading to the reduction of tool wear, the much extension of tool life, and the enhancement of gear profile accuracy and gear surface roughness [20]. **Figures 13** and **14** show the flank wear of hob tools at different time of machining the 50th, 300th gears. It is clearly seen that during the period of machining first 50 gears, gear hobbing under nanolubrication exhibits the reduction of tool wear, but at the period of machining the 300th gear, the significant reduction of hob wear is observed. Moreover, the wear land at the time of machining the 300th gear with nanolubrication (39.93 μm) is nearly equivalent to that of the tool at the period of machining the 50th gear under flood lubrication (36.29 μm). Accordingly, it is clarified that there is a significant increment in tool life when machining under nanolubrication.

On the other hand, the tool wear is much reduced under nanolubrication, which leads to achieve higher gear profile accuracy (shown in **Figure 15**). It could be said that nanolubrication is the main factor contributed to preserve the tool profile accuracy. In addition, the spherical morphology of Al_2O_3 nanoparticles takes part in the decrease of friction force and cutting temperature.

Hard milling process of $60Si_2Mn$ steel (50÷52HRC) was done by using Al_2O_3 nanofluid (0.5 wt%) with carbide inserts at v = 110 m/min; f_t = 0.12 mm/tooth; and a_p = 0.2 mm. **Figures 16** and **17** illustrate the difference of tool wear between MQL hard milling with nanofluids and pure MQL. In **Figure 16**, the wear on cutting edge including rake and flank faces is dominant.

(a) $w_s = 36.29\ \mu m$

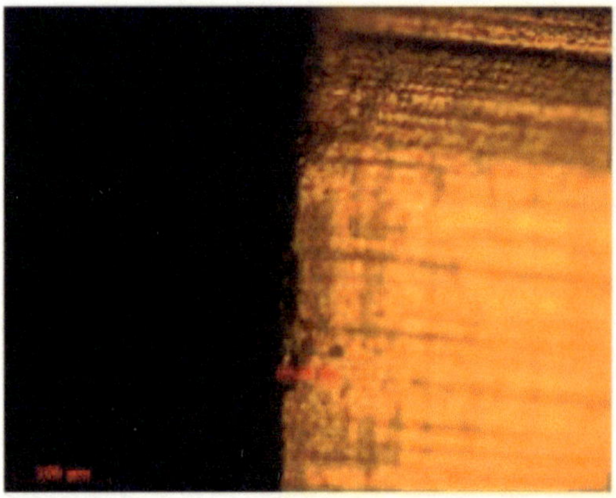

(b) $w_s = 24.23\ \mu m$

Figure 13. Flank wear after machining the 50th gear using: (a) flood lubrication; (b) nanolubrication [20].

In **Figure 17**, the friction between rake face and chip reduces due to "roller effect" of Al_2O_3 nanoparticles. Wear abrasion is not concentrated in cutting edge, and the abrasive area is formed on rake face (marked area shown in **Figure 17a**). On flank face, the formation of small wear land helps to form oil mist and contains nanoparticles to create "roller effect". The pressure on cutting edge decreases due to the reduction of friction. Therefore, the uniform wear occurs on cutting edge, which is different from the case without nanoparticles (**Figure 17b**). Moreover, the tool wear is reduced (about 26.4–33%) with the use of Al_2O_3 nanofluids.

From **Figure 18**, in case of soybean oil with Al_2O_3 nanoparticles, tool life is about 80 minutes (increase almost 177% compared to pure soybean oil). In case of emulsion 5% coolant with Al_2O_3 nanoparticles, tool life is about 115 minutes (increase almost 230% compared to pure emulsion 5% coolant). The promising results are supported to prove the explanation of "roller effect" of nanofluids.

http://dx.doi.org/10.5772/intechopen.75091

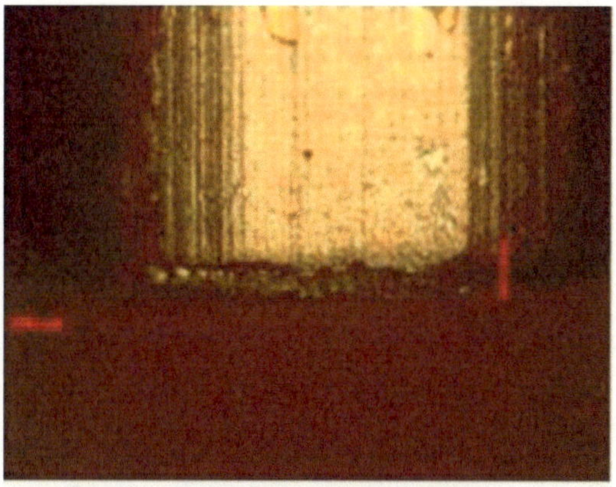

(a) $w_s=89.3\ \mu m$

(b) $w_s=39.93\ \mu m$

Figure 14. Flank wear after machining the 300th gear using: (a) flood lubrication; (b) nanolubrication [20].

The nanoparticles suspended in cutting fluids bring out the new trend in machining industries, which not only suggests many alternative solutions for conventional problems in metal cutting but also suits with green manufacturing industries, especially used as the base fluids of MQL techniques. Many publications have shown that the vegetable oils as the base fluids with MQL method, inherently nontoxic as well as biodegradable, can be effectively applied for machining processes, but their cooling characteristics is the main problem when cutting hard materials. During hard machining, the enormous amount of heat generated from cutting zone and strong adhesive wear between the tool and the work material will cause the reduction of hardness of cutting tool, increase the wear rate, and decrease the tool life. The occurrence of nanomaterials in MQL fluids has a strong meaning to overcome this problem. The difficulty of heat dissipation from cutting zone has been solved by the reduction of friction coefficient caused by "roller effect" of nanoparticles. Besides, MQL nanofluids also broaden the applicability of carbide tools in hard cutting with economic characteristic.

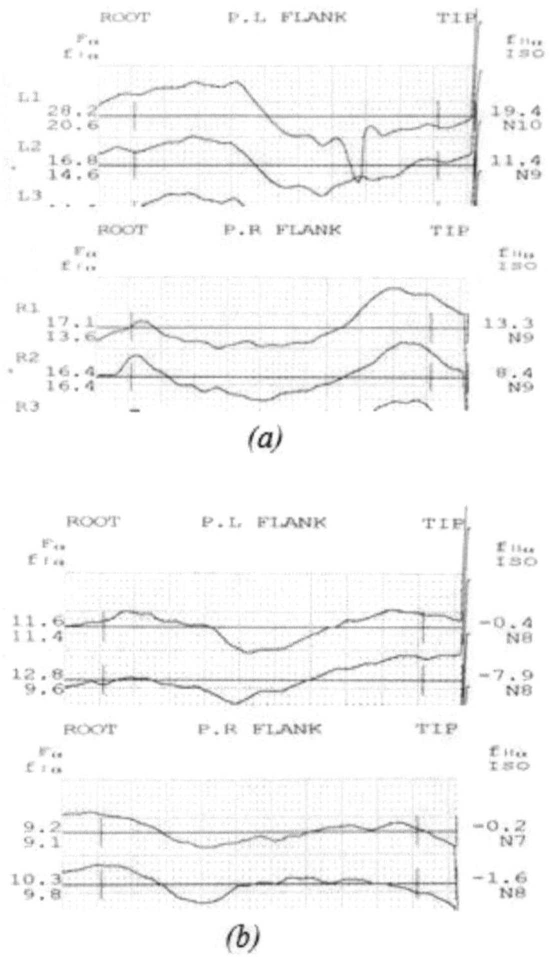

Figure 15. Measuring the gear profile error of the 300th machined gear by OSAKA SEIKI KIKAI gear measuring machine: (a) flood lubrication; (b) nanolubrication [20].

2.5. The effects on surface integrity

The surface integrity has become unquestionably a crucial parameter of any product in the past, present, and future. Quality characteristics must be tested during and after the manufacturing processes. With regard to components, the distinction is often made between macrogeometrical parameters and the surface quality. Macro-geometrical parameters refer to deviations of dimension, form and position. The surface quality is defined by roughness parameters. **Figures 19** and **20** show the surface roughness values of hard milling of $60Si_2Mn$ steel (50÷52HRC) under different MQL conditions. The values of surface roughness obtained from MQL nanofluids are better than those of MQL pure fluids. Furthermore, the good surface quality of hard milling under MQL nanofluid condition achieves and remains stable during longer cutting time. The best performance of nanoparticles is obtained when the flank wear land reaches to some extent called "appropriate wear land." This can be explained that the profile of machined surface of hardened steel reflects that of flank face of cutting tool with

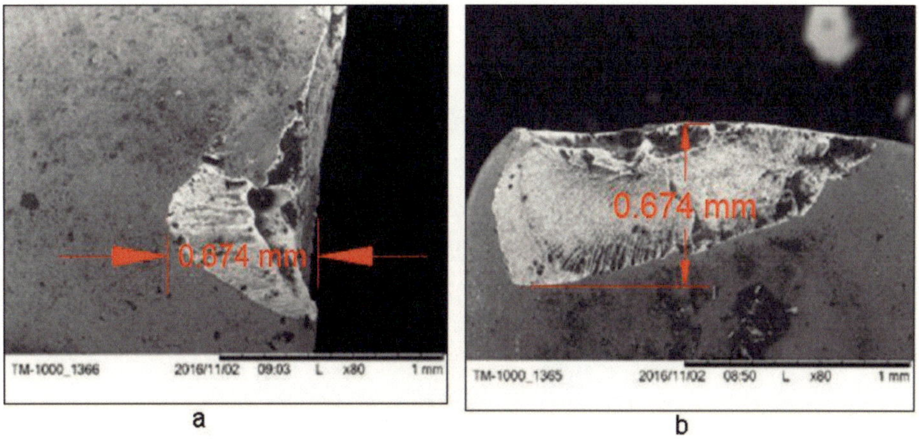

Figure 16. Tool wear under pure MQL cutting fluid with soybean oil (cutting time at 45 minutes): (a) rake face wear; (b) flank face wear [19].

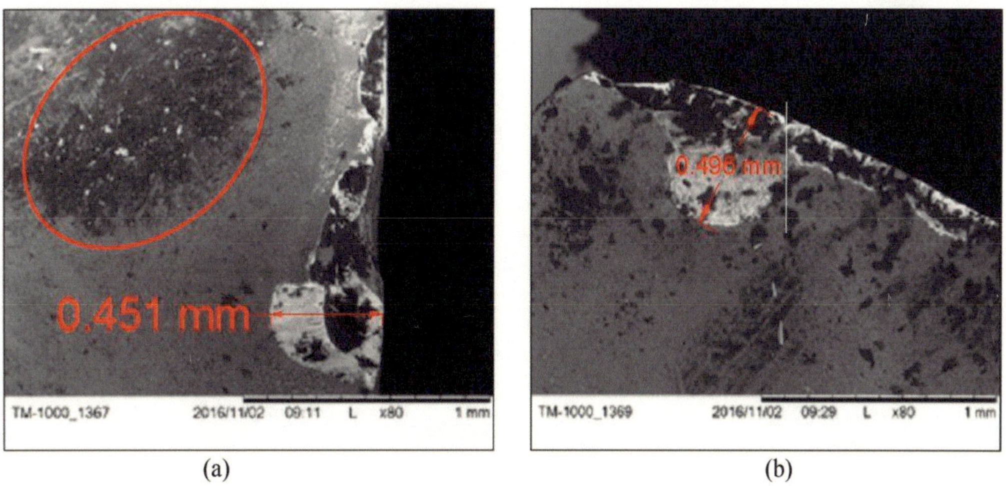

Figure 17. Tool wear under MQL nanofluid with soybean oil (cutting time at 80 minutes): (a) rake face wear; (b) flank face wear [19].

reasonable accuracy. As long as the flank wear profile is remained smooth, the flank wear to some extent not only deteriorates the surface finish but also somehow keeps or increases the surface quality [19]. This feature makes MQL hard machining utilizing nanofluids very different from other types of machining processes.

The surface roughness of grinding the high-temperature nickel base alloy GH4169 is shown in **Figure 21**. The comparison of different lubricating conditions reveals that the amount of surface quality improvement in the nanofluid MQL grinding is much higher. It is attributed to the more effective lubrication of nanofluids. The Al_2O_3 nanofluid MQL grinding achieved the best surface roughness.

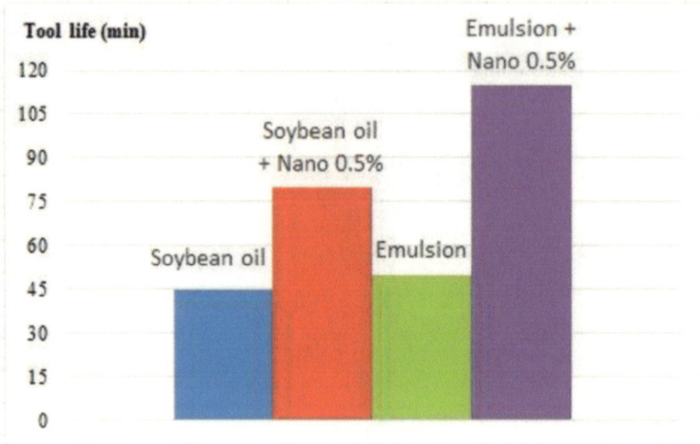

Figure 18. Tool life under MQL conditions with or without nanofluids [19].

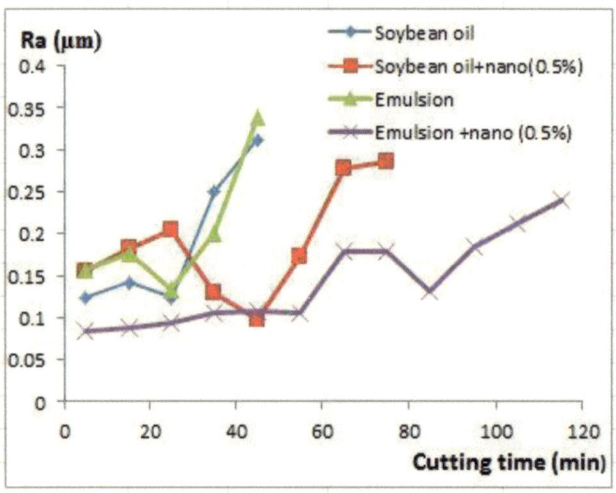

Figure 19. Surface roughness R_a under MQL conditions with or without nanofluids [19].

The nanofluid MQL grinding leads to a smoother surface and is better than either pure palm oil MQL or flood lubrication (seen in **Figure 22**). Interestingly, MQL grinding with Al_2O_3 nanofluid offers significant reduction in the sliding friction coefficient, specific sliding grinding energy, and best surface quality [18]. In addition, the SiO_2 and diamond nanofluids show relatively good lubrication effect.

The application of nanolubrication led to the formation of a tribo-film (seen in **Figure 23**) as a solid lubricant [9]. This observation can be made in machining field due to extremely high contact pressure and temperature in cutting zone, and so many nanoparticles are deformed and remained in the machined surface. The occurrence of tribo-film on the machined surface may lead to many new research topics needed to study. The deposition of a tribo-film on the surfaces could help to improve the operating function of the machined part. The very thin

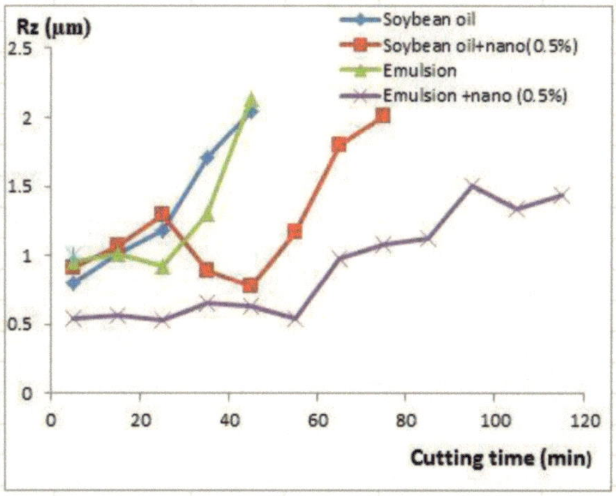

Figure 20. Surface roughness R_z under MQL conditions with or without nanofluids [19].

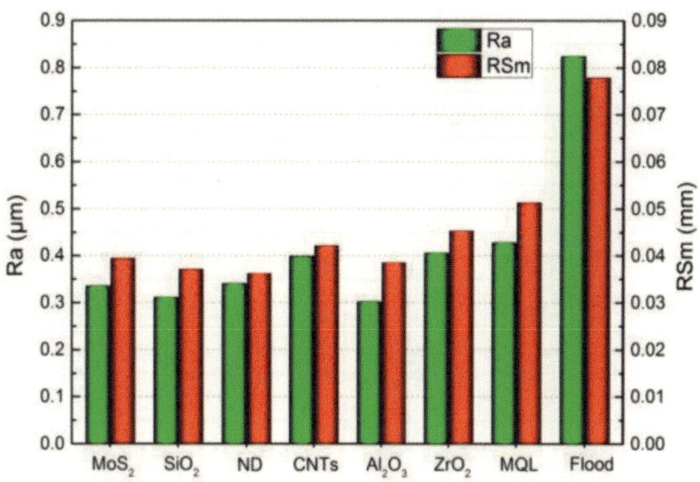

Figure 21. Surface roughness of grinding process at different lubrication conditions [18].

layer formed by nanoparticles has the same characteristics as nanopowder, and so through cutting processes by nanofluids, we can make further improvement for tribological effect on part surfaces by using proper nanofluids and cutting condition. These topics will be discussed and confirmed in many further researches.

2.6. Conclusion

An inclusive review on the application of nanofluids in various machining processes has been made. The nanofluid has achieved significant attention due to its capability to enhance the heat transfer and lubrication performance in cutting zone. The effects of nanofluids were proven to reduce the coefficient of friction and wear effect to enhance the cutting performance,

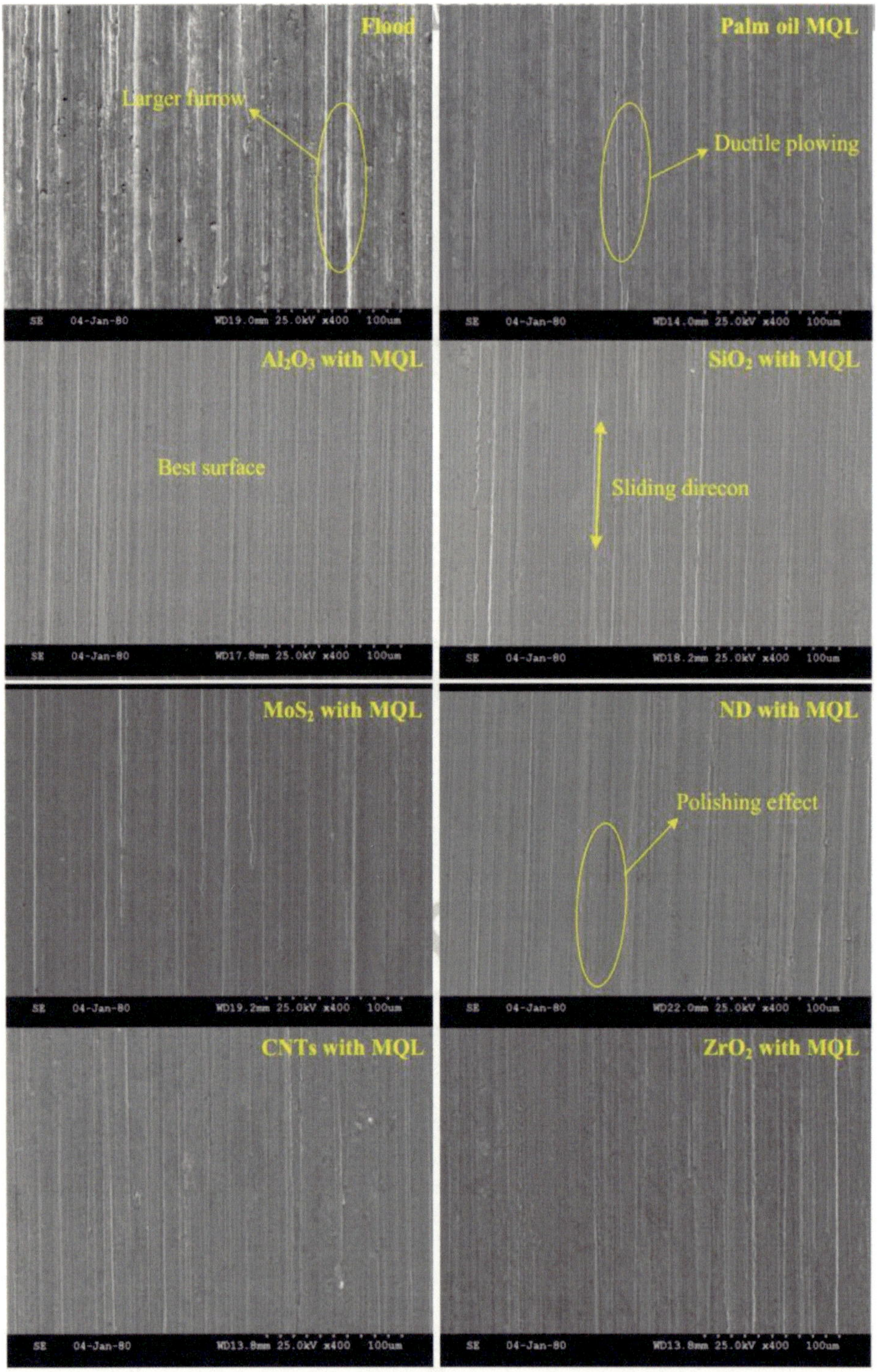

Figure 22. SEM micrographs of workpiece surface of nickel base alloy GH4169 at different lubricating conditions [18].

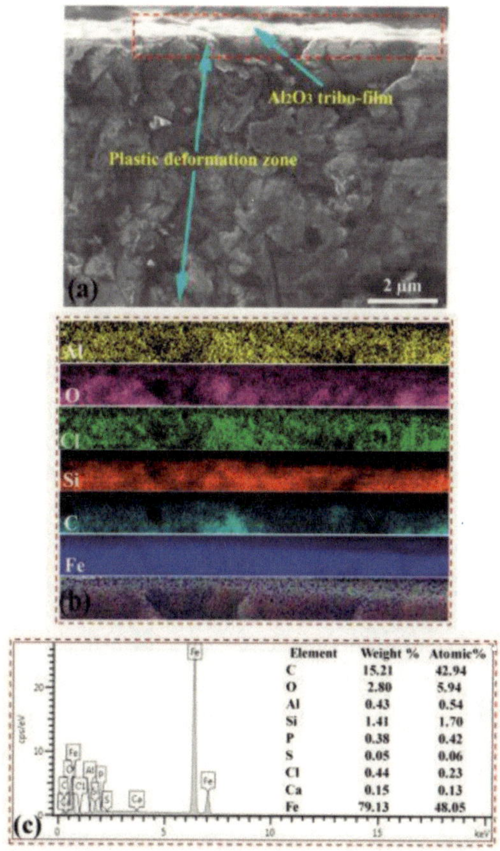

Figure 23. Formation Al_2O_3 tribo-boundary film on worn surface of the piston ring surface. (a) FE-SEM imaging on the cross-section, (b) EDS element mapping on tribo-boundary film, (c) EDS spectrum on tribo-boundary film [9].

tool life, and surface quality. Moreover, MQL technique with NFs makes a big improvement for some hard machining processes like hard turning, hard milling in term of surface quality, which is equivalent to that of finish grinding. Together with MQL using nontoxic fluids like water and vegetable oils, nanofluids have opened the new trend in machining and exhibited a wide range of application in different cutting processes. The promising results obtained definitely ensure the success of MQL machining with nanofluids. However, the performance and behavior of nanofluids may be affected by many parameters, such as the base fluid, nanoparticle type, nanoparticle size, nanoconcentration, and so on. Further research is necessarily made to optimize these parameters.

3. The effects of parameters of nanofluids

Each type of nanoparticles has with different structures, shapes, and sizes, which will vary in physical and morphological features, demonstrating diverse tribological performances [35].

From **Table 1**, it can be seen that the technical properties of each type of nanoparticles are different, and so the effectiveness of various nanofluids on MQL cutting performance will be an important investigated factor.

In this section, the authors will present the effects of six different types of nanoparticles on nanofluids mainly used in MQL machining in term of MQL base fluid, types of nanoparticles, size and morphology of nanoparticles, and nanoparticle concentration.

3.1. The base fluid of nanofluid

Nanofluids are formulated by suspending nanoparticles in various types of fluid, which can be the water, vegetable oil, industrial oil and so on. The properties of the medium are considered an important parameter, which directly influences on the activity of nanoparticles. Along with the trend of sustainable machining all over the world, the cutting fluids used in nanofluids should not be contained different toxic ingredients, and therefore water and vegetable oils are motivated to use as the alternative solution. There are some types of vegetable oils, which are commonly utilized in MQL machining: soybean, peanut (groundnut), maize, rapeseed, palm, castor, and sunflower oils. The ingredients, molecular structure, viscosity, and friction coefficient of the base fluids are the key parameters for vegetable oils [37]. **Figure 24** shows the relationship between vegetable oils and their coefficient of friction, which strongly influences in the contact area during machining. Vegetable oil is mainly composed of fatty acid and triglyceride -COOH in the fatty acid molecules and -COOR in triglyceride both belong to polar groups, which gives them excellent lubrication property [38]. **Figure 25** illustrates the typical polar molecule of vegetable oil.

Table 2 lists the basic ingredients of fatty acids of seven vegetable oils. The lubrication properties of saturated fatty acids are better than those of unsaturated fatty acids [41]. Saturated fatty acids have a strong effect on decreasing friction and wear, especially stearic acid, which

Types of nanoparticles technical properties	Al_2O_3	SiO_2	MoS_2	ZrO_2	CNT	ND
Morphology	Nearly spherical	Porous & nearly spherical	Ellipsoidal	Mainly spherical	Coaxial circular tubes	Spherical & flaky
Purity (%)	> 99	~99.5	> 99	99.9	> 95	93-95
Color	White	White	Black	White	Black	Gray
True density (g/cm^3)	3.97	2.4	4.8	5.89	~2.1	3.05–3.30
Thermal conductivity (W/m.K)	40	7.6	138	< 2	3000	2000
Friction coefficient	—	—	0.03 ~0.05	—	—	—
Melting temperature (°C)	2200	1600	1185	2715	3127	3727

Table 1. Technical properties of different types of nanoparticles.

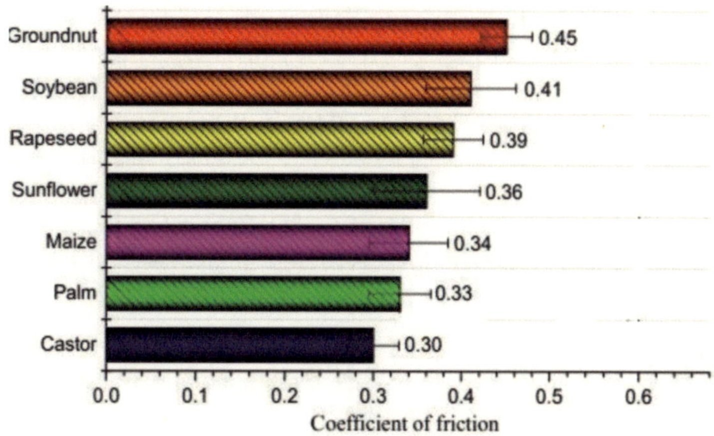

Figure 24. Friction coefficient of different vegetable oils [36].

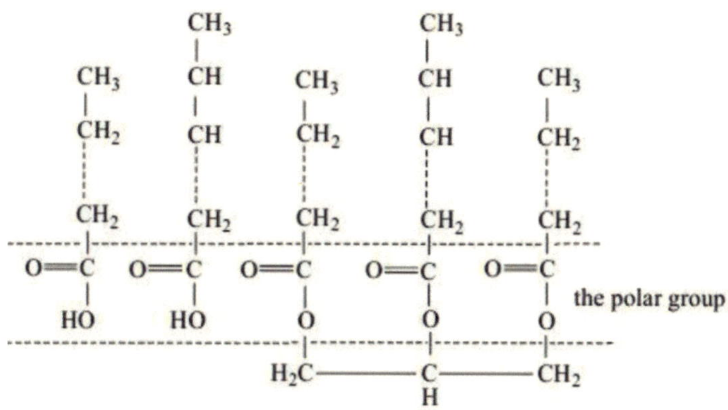

Figure 25. Polar molecule of the typical vegetable oil [36].

Oil types	Ingredient				
	Palmic acid	Stearic acid	Oleic acid	Linoleic acid	Linolenic acid
Peanut oil	6–9	3–6	55–71	13–25	0.5
Soybean oil	7–10	3–5	22–31	49–55	6–11
Maize	9–19	1–3	26–40	44–55	< 1
Castor oil	—	—	3–9	3–5	Trace
Palm	35–48	4–6	38–41	8–12	Trace
Rapeseed oil	2–4	1–2	40–60	19–20	7–8
Sunflower oil	4–19	3–6	14–35	50–75	0.1

Table 2. Ingredient of various vegetable oil (%) [36].

provides a more stable oil film in contact zone. Besides, they tightly contaminate to the molecular film and remain on metal surfaces, such as tiny magnets, to form lubricating film for anti-friction and anti-wear [39]. Hence, this characteristic gives vegetable oils good lubrication property and shows the excellent lubrication effects in the application of MQL fluids. The double bonds in unsaturated fatty acids are relatively unstable and easily generate chemical reactions, such as oxidization. Moreover, they can also weaken the acting force between molecules, which leads to poor lubrication properties [36]. Lubrication property of unsaturated fatty acids with a higher carbon atom number is stronger than those with a lower carbon atom number, and so the friction coefficient decreases as the chain length increases [42].

On the other hand, the selection of proper vegetable oil depends on the climate and soil texture of each country, but it contributes very little to total manufacturing cost because of very small amount of fluids used in MQL techniques. The small quantity lubrication (SQL) grinding process of Inconel 718 using silver and zinc oxide NPs mixed with DI water at cutting velocity (V) =18 m/s, table speed (V_w) = 6 m/min, depth of cut (a_p) = 10 mm. The flow rates of MQL nanofluids are 50, 100, 150, 200, 250 ml/h [40]. Minimum tangential forces have been obtained in the case of SQL grinding with nanofluids (seen in **Figure 26**), which is essentially due to better cooling, and lubrication that helps in preserving the cutting ability of the grits over a longer period. Additionally, hard NPs under grinding pressure might convert sliding to rolling friction.

Vegetable oil evidently has better lubrication property than water-soluble fluid, but the lubrication properties of seven typical vegetable oils also differ. Among these vegetable nanofluids, castor oil has the best lubrication property, followed by palm oil. In addition, peanut, sunflower, soybean, and rapeseed oils also exhibit excellent lubrication properties.

The viscosity of vegetable oils also has a strong influence on the machining performance and strongly affects its cooling and lubricating properties. They have a high natural viscosity as the machining temperature increases and drops more slowly than that of mineral oils [36]. These statements have significant meanings in machining difficult-to-cut materials such as hardened steel, tool steel, and so on. The formulation of oil film containing nanoparticles in cutting zone plays a key role in reducing the friction, which leads to decrease cutting temperature and tool wear. This film forms and loses continually, and so the higher the viscosity of cutting fluids, the more stable the film on contact faces.

3.2. The types and morphology of nanoparticles

Recently, nanomaterials have attained much attention because of their unique properties and tremendous application potentials in a wide range of industries. Currently, more and more researchers have been devoted to enhance the lubricant properties by using nanoparticles as lubricant additives (also called as nanolubrication or nanofluids). There are numerous types of nanoparticles in markets, and they are used in a wide range of industries. For machining processes, some types of nanoparticles are mainly used as Al_2O_3, MoS_2, ZrO_2, SiO_2, CNT, and ND. From **Table 1**, it is clearly seen that the morphology of six different NPs is different, and so it causes the various effects on cutting performance. Owing to the high cost of nanoparticles, the appropriate selection of nanoparticle type to suspend in MQL fluid is so important.

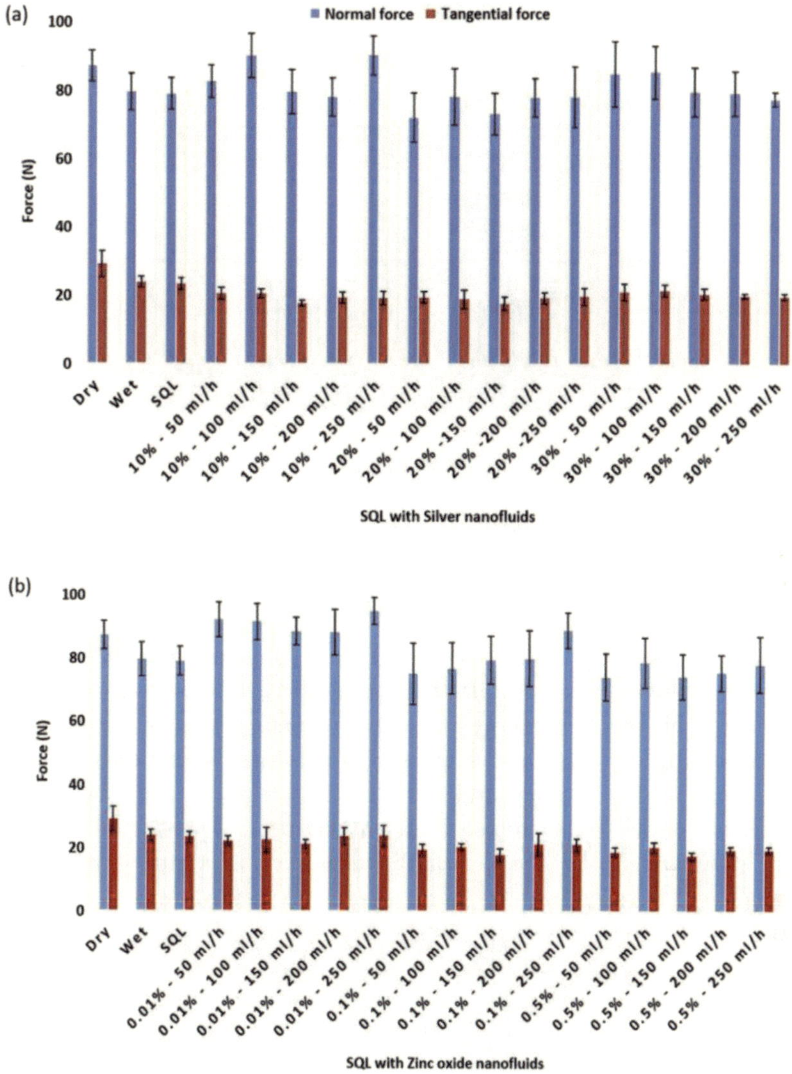

Figure 26. Grinding forces under different SQL nanofluids [36].

Al_2O_3 nanoparticles, one of hexagonal close-packed crystal materials, exhibit the good lubrication performance, which is related to its structures and characteristics. Al_2O_3 nanoparticles are spherical (shown in **Figure 27**) with characteristics of high strength, hardness, and heat resistance. The Al_2O_3 NPs are hard phase (HR = 2700-3000), showing good abrasive resistance during the friction process, and can carry some support to friction surface load between the area [43]. When Al_2O_3 nanoparticles are added into cutting fluid, they can easily move into the worn area under the compressive stress of nanocutting fluid, and then a self-laminating film can be formed, which results in micropolish and can self-mend the friction surface [44, 45]. Furthermore, the Al_2O_3 nanoparticles demonstrate good resistance to high temperature. The melting point of oil film can reach 2200°C. The morphology of Al_2O_3 nanoparticles are mostly

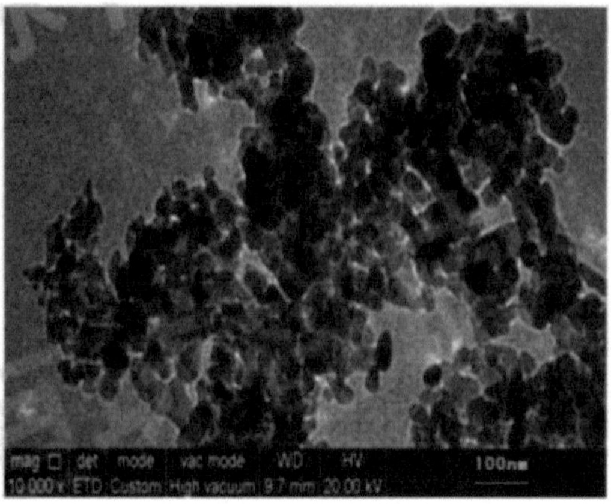

Figure 27. The SEM image of Al_2O_3 nanoparticles (30 nm) [57].

spherical; therefore, they can play the role of ball bearings that prevent the direct contact of friction pairs, and the sliding friction is changed to rolling friction in contact zone, which improves the cutting performance and the carrying capacity of lubricant [46].

The SiO_2 nanoparticles are spherical (shown in **Figure 28**), and the surface molecules exhibit a 3D network structure. Given the abundant unsaturated vacant bonds on the surface, SiO_2 nanoparticles exert high surface energy and activity, making them easy to sediment onto the workpiece friction surface [47]. For such machining processes having the intensive friction as grinding, hard turning, hard milling, and so on, the melting point of SiO_2 nanoparticles in contact faces decreases under local extremely high temperature and pressure. Therefore, they may be melted, semi-melted, or sintered to form the lubrication film. Furthermore, a ceramic-like

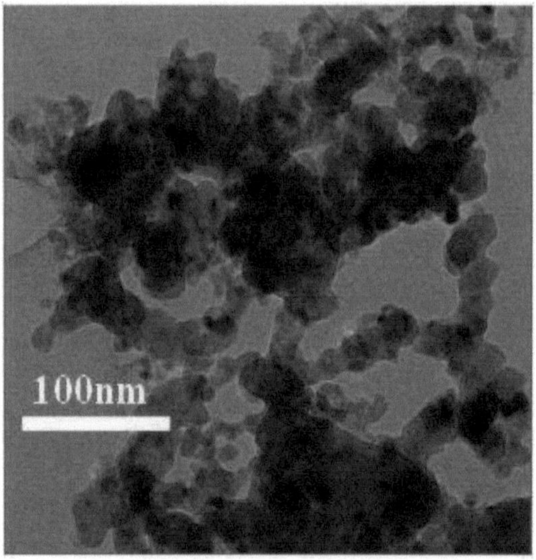

Figure 28. The SEM image of SiO_2 nanoparticles (8 nm) (www.us-nano.com).

nanofilm can be formed by diffusing and penetrating into machined surface or sub-surface of the elements in some SiO_2 NPs [48]. This phenomenon prevents direct contact between cutting tools and the workpiece, thus reducing the friction and achieving better surface quality.

Molybdenum disulfide or MoS_2 has been utilized in machining processes as solid lubricant for many years. MoS_2 has been considered to be the best solid lubricant material, because it can provide low coefficient of friction up to 0.03–0.05 or even lower. The MoS_2 nanoparticles are ellipsoidal (shown in **Figure 29**). The layer structure of MoS_2 is a hexagonal crystal system combining Mo and S through a covalent bond, and the bond between them is short, but the spacing between sulfur atoms is large. Accordingly, the bond between two adjacent sulfur atom layers is weak. That is the best explanation why a plane, so called "an easy-to-slide plane," will be generated from weak binding of sulfur atoms between molecular layers by shearing force caused by cutting processes. The numerous easy-to-slide planes make the contact faces sliding relatively to each other and they do not contact directly [43]. This unique characteristic makes MoS_2 good friction-reducing effect. Moreover, exposure sulfur atoms of the crystal surface on the metal surface have a very strong adhesion to form a very solid film, therefore lubrication is superior to other general-lubricating materials. When MoS_2 particle size becomes smaller, it is attached to the surface of the friction material and the coverage has increased significantly, and anti-wear friction properties have been significantly improved.

Carbon nanotubes or CNTs are coaxial circular tubes (shown in **Figure 30**) composed by layers and dozens of layers of carbon atoms in hexagonal arrangement. CNTs present a high modulus and strength because carbon atoms in CNTs adopt the sp^2 hybridization, which follows a higher proportion of S pathways than sp^3 hybridization [49]. During machining processes, CNTs will not be ground into hard film under high loads and pressure because of its high strength and hardness. As such, CNTs can reduce the friction force of the cutting area and improve the lubrication effect of nanofluids. Tool-workpiece interface thus provides efficient lubrication. Two types of CNTs include the single-walled carbon nanotubes (SWCNT) and multi-walled carbon nanotubes (MWCNT). The performance between CNTs found by them revealed better performances of MWCNT over the SWCNT in terms of cutting temperature

Figure 29. The SEM image of MoS_2 nanoparticles (135 nm) (www.us-nano.com).

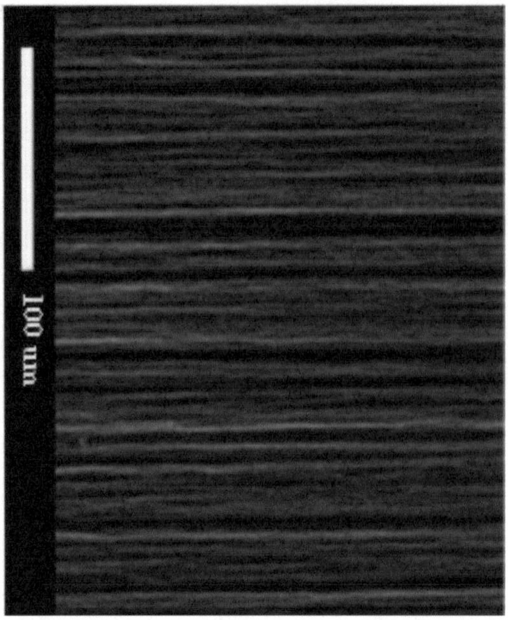

Figure 30. The SEM image of single-walled carbon nanotube (outside diameter: 1-2 nm; inside diameter: 0.8-1.6 nm) (www.us-nano.com).

and cutting force due to higher wettability although MWCNT had lower thermal conductivity than SWCNT [50]. However, they cannot produce effective rolling like other spherical nanoparticles due to their cylinder structures. Therefore, CNTs can only produce limited anti-friction effect [43]. Furthermore, CNTs possess the highest coefficient of thermal conductivity among nanoparticles (**Table 1**), and so their application can be broadened by additionally dispersing other appropriate types of nanoparticles to form hybrid nanofluids [51, 52].

The ZrO_2 nanoparticles are mainly spherical (shown in **Figure 31**). They appear oblique crystal at low temperature and show tetragonal crystal formation at high temperature. ZrO_2 is soluble in sulfuric acid and hydrofluoric acid and has good thermal-chemical stability due to very high melting temperature. When at high temperature, they have good strength and toughness. ZrO_2 nanoparticles have the lowest coefficient of thermal conductivity among nanoparticles (**Table 1**), but their high surface energy and surface activity tend to be adsorbed onto the machined surface establishing a layer of self-healing lubrication film on the friction pair surface and achieving good lubrication effect [43].

The nanodiamond or ND exhibits cubic structure, and the morphology of nanoparticles is spherical or flaky (shown in **Figure 32**). ND presents very high hardness (HV = 98 GPa) which is superior to that of workpiece materials. Moreover, the ND exhibits extremely large elasticity modulus (980 GPa) with a compressive strength of about 13 GPa. With a size less than 1 μm, ND has attracted remarkable scientific attention due to their excellent mechanical and optical properties, high surface areas, and tunable surface structures. Due to unique properties, the excellent performances will certainly influence the lubrication performance in term of reducing

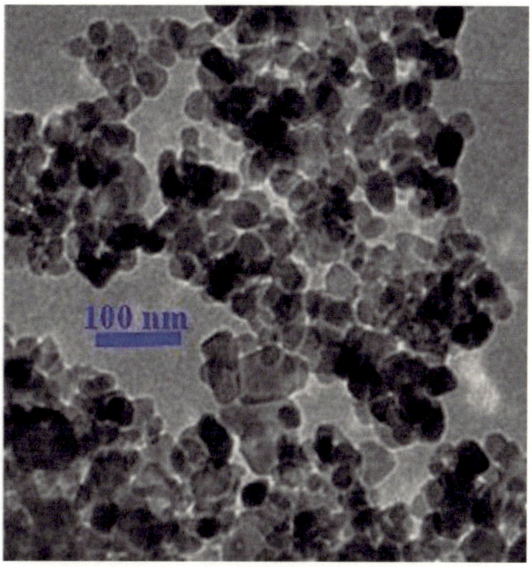

Figure 31. The SEM image of ZrO_2 nanoparticles (40 nm) (www.us-nano.com).

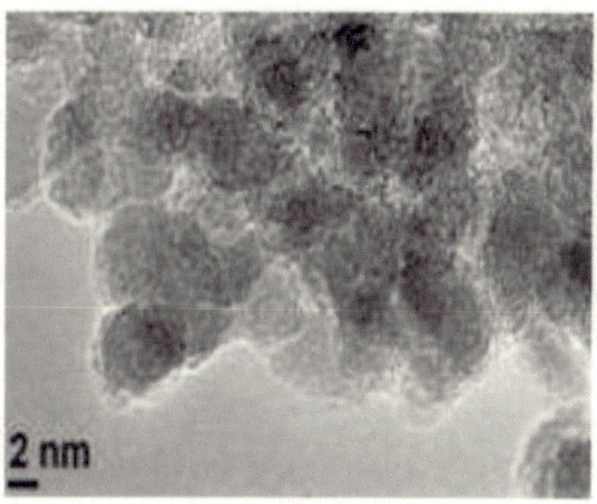

Figure 32. The SEM image of ND nanoparticles (50 nm) (ndp.diamonds).

friction coefficient, cutting forces, cutting temperature [53–56]. Interestingly, NDs can serve as abrasive grains and take part in cutting processes under high pressure like grinding process [43]; therefore, if the size of NDs becomes larger, they may deteriorate the surface quality.

3.3. The size of nanoparticles

The size of nanoparticles should not be avoided because it not only influences on the performance of nanofluids and cutting processes but also contributes significant amount of nanofluid cost. In markets, the smaller the grain size of nanoparticles, the higher their cost will be. **Figure 33** shows the relationship between the NP size and its cost, and it can be seen that the

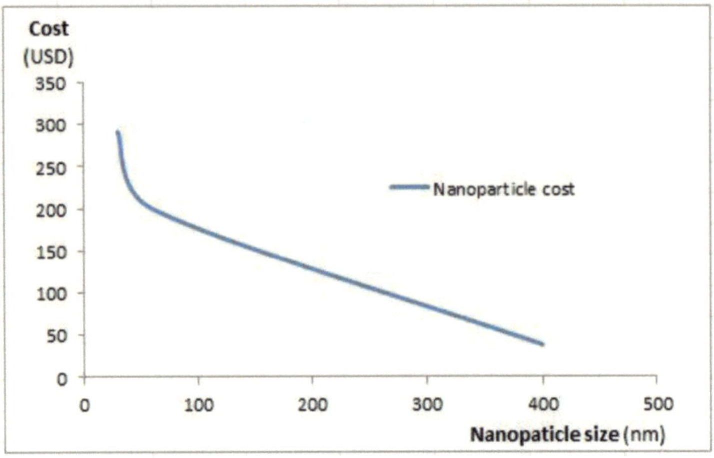

Figure 33. The relationship between 500 g of MoS_2 nanoparticle size and its cost.

grain size of nanoparticles strongly influences the cost of NPs. Hence, the NP size is definitely needed to optimize while remaining the good performance of nanofluids and the reasonable manufacturing cost [58, 59]. The experimental study on nanolubricant of nanographite (0.1 vol%) was carried out with diffrent particle sizes 5 μm, 450 nm, and 55 nm [60].

From **Figure 34**, the friction coefficient of nanolubricant of the disc specimen as a function of the applied normal force exhibits the much lower values compared to microlubricant and raw mineral lubricant. In this test, the fluid with the smallest NP size 55 nm shows the lowest friction coefficient and reaches stable state when increasing the applied normal force. Moreover, the microfluid of the microparticle size 5 μm shows the highest friction coefficient, which is also higher than that of pure mineral lubricant.

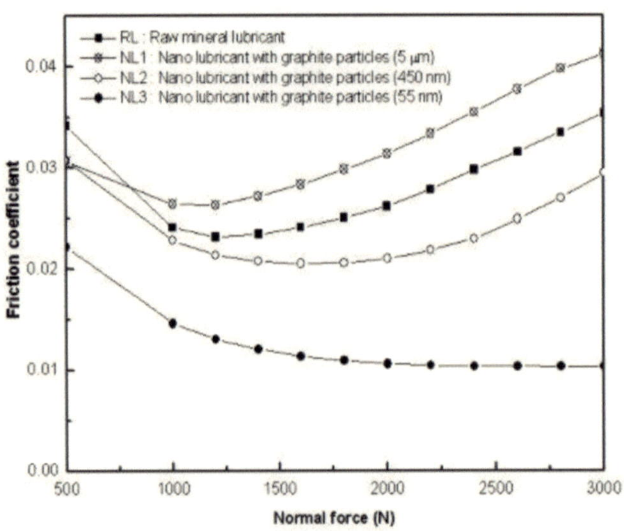

Figure 34. Friction coefficients of the disc specimen as a function of the applied normal force at different particle sizes [60].

The MQL grinding process of hardened AISI 52100 steel with Al_2O_3 nanofluids was done to investigate the effect of size of nanoparticles at grinding wheel speed =0.05 m/s and the grinding depth = 10 μm [61]. The grinding temperature was reduced with the nanofluid of smaller NP size (**Figure 35**). When comparing the surface roughness of Al_2O_3 nanofluids with NP size of 40 nm and 80 nm, it is clearly seen that the better surface roughness can be achieved by using NFs with NP size of 40 nm, even in different nanoconcentration (shown in **Figure 36**).

Another MQL grinding process with Al_2O_3 nanofluids and nanodiamond was done to investigate the effect of size of nanoparticles [62]. From **Figures 37** and **38**, MQL grinding process with nanofluids exhibits better surface roughness and reduces grinding forces when compared to those of dry and pure MQL grinding. Considered the NP size among nanofluids, the ND with smaller size 30 nm gives the best grinding performance in term of surface roughness and grinding forces.

Overall, the nano/microparticle sizes have the strong effects on cutting performance. Nanofluids exhibit better machining performance than microfluids. The smaller the NP size is, the better surface quality will be. However, the cost of NPs rises with smaller size, and so the appropriate nanoconcentration in fluid will be the key parameter affecting the application of nanofluids in machining practice.

3.4. The nanoparticle concentration

The nanoparticle concentration has attained a significant attention of many researchers because it influences on the performance of nanofluids and directly contributes a large fraction of the NF cost. The experimental study on nanolubricant of nanographite with different concentration 0.1% and 0.5% reveals that the lower friction coefficients and average temperature of lubricated surfaces of the specimens can be achieved in case of nanolubricant with larger volume of fraction 0.5% (shown in **Figures 39** and **40**) [60]. The similar observation can be made from **Figure 36** by the comparison of Al_2O_3 nanofluids with three concentrations 1, 3, and 5%. The value of surface roughness decreases as the nanoconcentration rises from 1 to

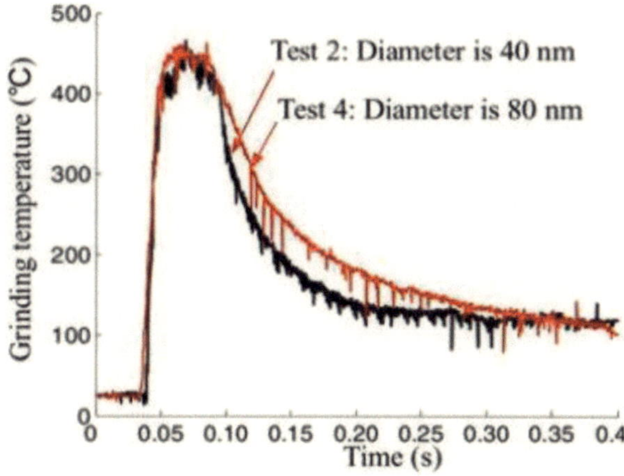

Figure 35. MQL grinding temperature of hardened AISI 52100 steel with Al_2O_3 nanofluids (grinding wheel speed =0.05 m/s; the grinding depth 10 μm) [61].

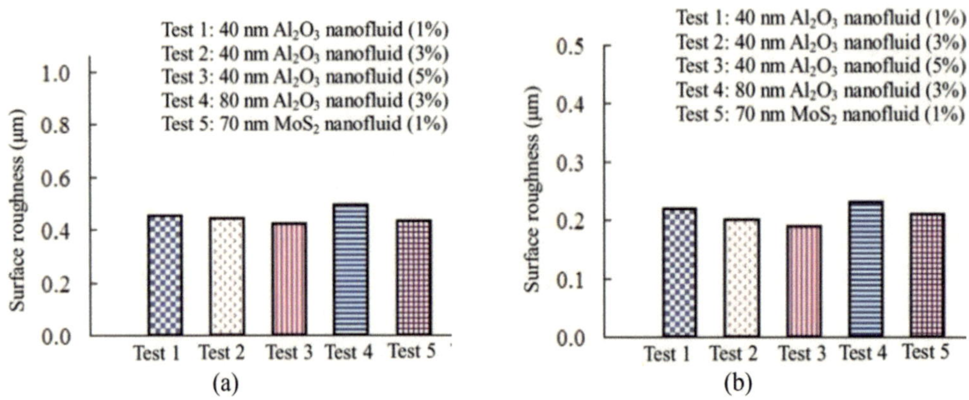

Figure 36. Surface roughness of MQL grinding: (a) across the grinding direction; (b) along the grinding direction [61].

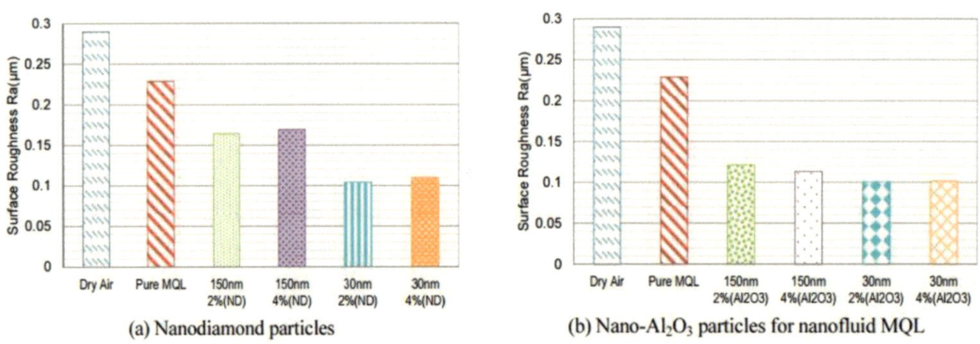

Figure 37. Surface roughness of MQL grinding with different nanofluids [62].

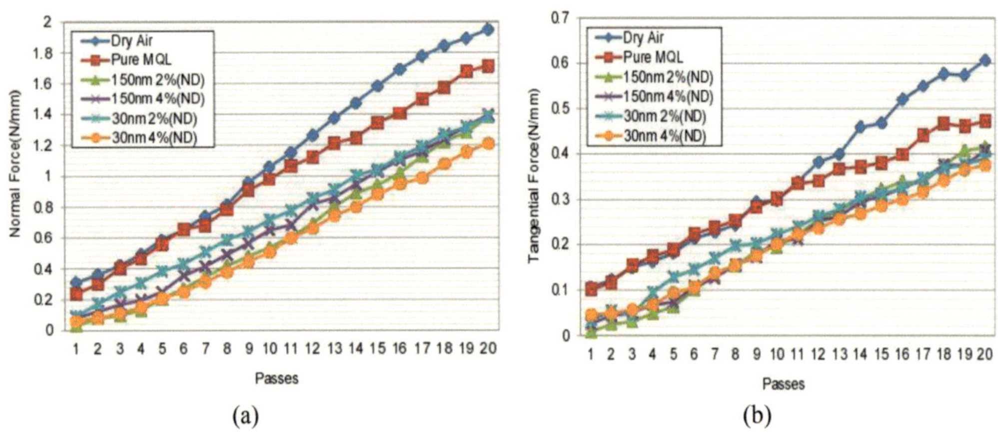

Figure 38. Measured grinding forces; (a) normal direction and (b) tangential direction in the cases of dry, pure MQL and MQL nanofluid with nanodiamond [62].

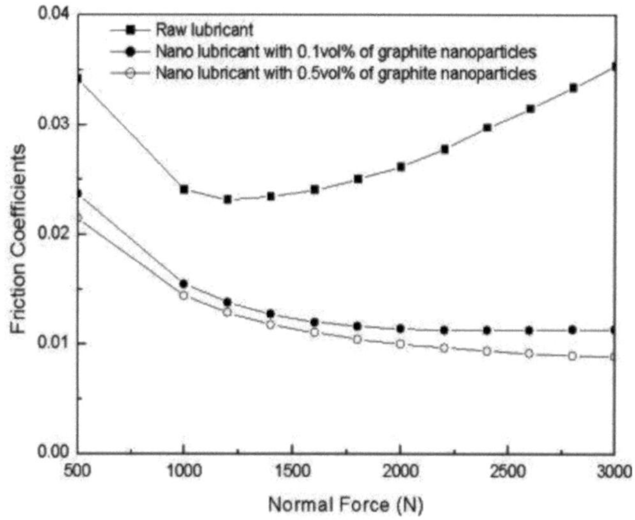

Figure 39. Friction coefficients of the disc specimen as a function of the applied normal force at different nanoconcentration [60].

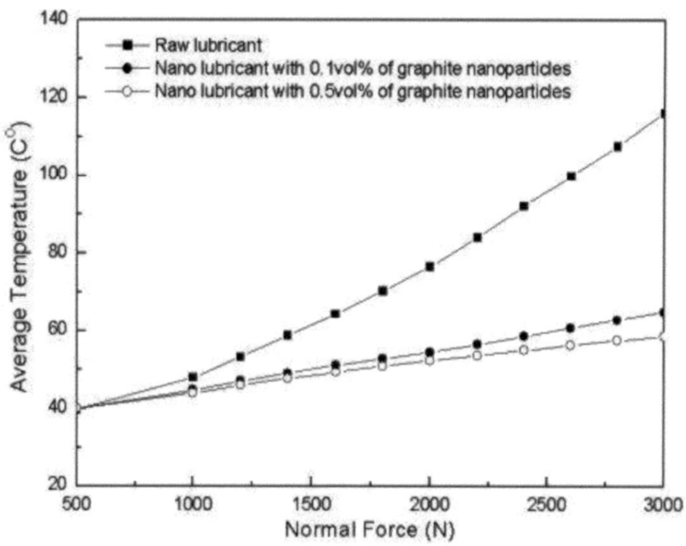

Figure 40. Temperature of lubricated surfaces of the specimens as a function of normal force with different nanoconcentration [60].

5%. In contrast, when rising the volume fraction of nanodiamond in MQL grinding, the little effectiveness on cutting performance can be achieved. From **Figure 37a**, the value of surface roughness increases along with the rise of the concentration of ND.

In summary, the nanoconcentration is necessarily investigated by further research to optimize this key parameter, by which the application of nanofluids together with MQL technique cannot be limited by the cost of nanoparticles. Although the volume fraction of nanoparticles in MQL fluid is relatively small in each experimental research, it will be large when applied to production line in practice.

3.5. Conclusion

The cost-effective and sustainable manufacturing techniques are being given considerable importance these days to ensure economic, societal, and environmental sustainability globally. The basic parameters of nanofluids are discussed in detail, and they give strong effects on their tribological and lubricating characteristics, as well as the cost of NFs. The selection of appropriate nanofluids with suitable parameters is the crucial factor for machining processes. Especially for hard machining materials using geometrically defined cutting edges (hard turning, hard milling, hard drilling, and so on), the effectiveness of nanofluids is evaluated in terms of reducing the friction coefficient, enhancing the cutting performance, ensuring the tool life, and achieving good surface quality. However, very few records can be seen in the literature indicating that more experimental research is needed to find the optimal parameters in order to ensure maximum performance of using nanofluids in machining.

4. The development trend of nanofluids

The successful application of nanofluids in MQL technique using vegetable oils becomes a big step toward sustainable manufacturing. This new trend is well suited to encounter the issue of environmental change. MQL machining under nanofluid condition significantly develops the cutting processes, which generate large amount of heat such as hard turning, hard milling, grinding, and so on. The machining capability and achieved surface integrity of hard cutting operations are much improved and equivalent to those of grinding, but the productivity is higher. They can replace or supplement to some of grinding processes. The progression of hard machining will reach further achievements.

Like the idea coming from composite materials, the use of hybrid nanofluids (two or more different types of nanoparticles suspended in fluids) recently gains much attention due to better performance when compared to nanofluids [63–65], but the ratio of mixing and the types of nanoparticles will be continually studied and proven. On the one hand, the MQCL technique, the novel development of the MQL technique will be the future of modern machining processes. MQCL technique has overcome the cooling effect, the main drawback of MQL method [66, 67]. MQCL technique with nanofluids or hybrid nanofluids definitely brings out novel solutions and develops hard machining and grinding to the new level in the near future. On the other hand, the development of ice jet machining is one of the greatest achievements in the field of modern/advanced machining. It is a nondestructive, nonabrasive, residue-free, and environmentally friendly way of machining [68].

Symbols and Abbreviations

a_p	depth of cut (mm)
v	cutting speed (m/min)

v_w	table speed (m/min)
f	feed rate (mm/rev)
f_t	feed rate (mm/tooth)
γ_o	rake angle (°)
α_o	relief angle (°)
λ_s	inclination angle (°)
k_r	side cutting-edge angle (°)
F_x, F_y, F_z	cutting forces (N)
CNT	carbon nanotube
CNTs	carbon nanotubes
SWCNT	single-walled carbon nanotube
MWCNT	multi-walled carbon nanotube
MQL	minimum quantity lubrication
MQCL	minimum quantity cooling lubrication
SQL	small quantity lubrication
ND	nanodiamond
NF	nanofluid
NFs	nanofluids
NP	nanoparticle
NPs	nanoparticles

Author details

Tran The Long* and Tran Minh Duc

*Address all correspondence to: tranthelong90@gmail.com

Department of Manufacturing Engineering, Mechanical Engineering Faculty, Thai Nguyen University of Technology, Vietnam

References

[1] Wong KV, De Leon O. Review article "applications of Nanofluids: Current and future". Advances in Mechanical Engineering. 2010. DOI: 10.1155/2010/519659

[2] Ekinovica S, Prcanovicb H, Begovic E. Investigation of influence of MQL machining parameters on cutting forces during MQL turning of carbon steel St52-3. Procedia Engineering. 2015;**132**:608-614

[3] Grzesik W. Chapter fifteen – Advanced machining processes. Advanced Machining Processes of Metallic Materials (Second Edition);**2017**:285-397

[4] Lee PH, Nam JS, Li C, et al. An experimental study on micro-grinding process with nanofluid minimum quantity lubrication (MQL). International Journal of Precision Engineering and Manufacturing. 2012;**13**:331-338

[5] Paulo DJ. Machining of Hard Materials. London: Springer-Verlag London Limited; 2011

[6] Obikawa T, Kamata Y, Shinozuka J. High-speed grooving with applying MQL. International Journal of Machine Tools & Manufacture. 2006;**46**:1854-1861

[7] Duc TM, Long TT. Investigation of MQL-employed hard-milling process of S60C steel using coated-cemented carbide tools. Journal of Mechanics Engineering and Automation. 2016;**6**:128-132

[8] Pryazhnikov MI, Minakov AV, Rudyak VY, Guzei DV. Thermal conductivity measurements of nanofluids. International Journal of Heat and Mass Transfer. 2017;**104**:1275-1282

[9] Ali MKA, Xianjun H, Mai L, Qingping C, Turkson RF, Bicheng C. Improving the tribological characteristics of piston ring assembly in automotive engines using Al_2O_3 and TiO_2 nanomaterials as nanolubricant additives. Tribology International. 2016;**103**:540-554

[10] Sheikholeslami M, Hayat T, Alsaedi A. Numerical study for external magnetic source influence on water based nanofluid convective heat transfer. International Journal of Heat and Mass Transfer. 2017;**106**:745-755

[11] Sheikholeslami M, Bhatti MM. Active method for nanofluid heat transfer enhancement by means of EHD. International Journal of Heat and Mass Transfer. 2017;**109**:115-122

[12] Sheikholeslami M, Shehzad SA. Magnetohydrodynamic nanofluid convective flow in a porous enclosure by means of LBM. International Journal of Heat and Mass Transfer. 2017;**113**:796-805

[13] Uysala A, Demirena F, Altana E. Applying minimum quantity lubrication (MQL) method on milling of martensitic stainless steel by using nano MoS2 reinforced vegetable cutting fluid. Procedia - Social and Behavioral Sciences. 2015;**195**:2742-2747

[14] He H-B, Li H-Y, Yang J, Zhang X-Y, Yue Q-B, Jiang X, Lyu S-k. A study on major factors influencing dry cutting temperature of AISI 304 stainless steel. International Journal of Precision Engineering and Manufacturing. 2017;**18**:1387-1392

[15] Hadad MJ, Sadeghi B. Minimum quantity lubrication-MQL turning of AISI 4140 steel alloy. Journal of Cleaner Production. 2013;**54**:332-343

[16] Sinha MK, Madarkar R, Ghosh S, Venkateswara Rao P. Application of eco-friendly nanofluids during grinding of Inconel 718 through small quantity lubrication. Journal of Cleaner Production. 2017;**141**:1359-1375

[17] Lia B, Lia C, Zhang Y, Wang Y, Jia D, Yanga M, Zhang N, Wu Q, Han Z, Sun K. Heat transfer performance of MQL grinding with different nanofluids for Ni-based alloys using vegetable oil. Journal of Cleaner Production. 2017;**154**:1-11

[18] Wang Y, Li C, Zhang Y, Li B, Yang M, Zhang X, Guo S, Liu G. Experimental evaluation of the lubrication properties of the wheel/workpiece interface in MQL grinding with different nanofluids. Tribology International. 2016;**99**:198-210

[19] Minh DT, The LT, Bao NT. Performance of Al_2O_3 Nanofuid in minimum quantity lubrication in hard milling of $60Si_2Mn$ steel using cemented carbide tools. Advances in Mechanical Engineering. 2017;**9**:1-9

[20] Vi H, Tuan NM, Long TT. Investigation of Nanofluids as potential cutting fluids in gear Hobbing processes of AISI 4118 steel. Journal of Mechanics Engineering and Automation. 2016;**6**:180-185

[21] Sheikholeslami M. Influence of Lorentz forces on nanofluid flow in a porous cylinder considering Darcy model. Journal of Molecular Liquids. 2016;**225**:903-912

[22] Sheikholeslami M, Shamlooei M, Moradi R. Fe3O4-Ethylene glycol nanofluid forced convection inside a porous enclosure in existence of Coulomb force. Journal of Molecular Liquids. 2018;**249**:429-437

[23] Sheikholeslami M. Influence of magnetic field on nanofluid free convection in an open porous cavity by means of Lattice Boltzmann method. Journal of Molecular Liquids. 2017;**234**:364-374

[24] Sheikholeslami M. Magnetohydrodynamic nanofluid forced convection in a porous lid driven cubic cavity using Lattice Boltzmann method. Journal of Molecular Liquids. 2017;**231**:555-565

[25] Sheikholeslami M. Numerical investigation of nanofluid free convection under the influence of electric field in a porous enclosure. Journal of Molecular Liquids. 2018;**249**: 1212-1221

[26] Sheikholeslami M. CuO-water nanofluid flow due to magnetic field inside a porous media considering Brownian motion. Journal of Molecular Liquids. 2018;**249**:921-929

[27] Sheikholeslami M. Numerical investigation for CuO-H2O nanofluid flow in a porous channel with magnetic field using mesoscopic method. Journal of Molecular Liquids. 2018;**249**:739-746

[28] Sheikholeslami M, Seyednezhad M. Simulation of nanofluid flow and natural convection in a porous media under the influence of electric field using CVFEM. International Journal of Heat and Mass Transfer. 2018;**120**:772-781

[29] Sheikholeslami M, Shehzad SA. Numerical analysis of $Fe_3O4–H_2O$ nanofluid flow in permeable media under the effect of external magnetic source. International Journal of Heat and Mass Transfer. 2018;**118**:182-192

[30] Sheikholeslami M, Sadoughi MK. Simulation of CuO-water nanofluid heat transfer enhancement in presence of melting surface. International Journal of Heat and Mass Transfer. 2018;**116**:909-919

[31] Sheikholeslami M, Zeeshan A. Analysis of flow and heat transfer in water based nanofluid due to magnetic field in a porous enclosure with constant heat flux using CVFEM. Computer Methods in Applied Mechanics and Engineering. 2017;**320**:68-81

[32] Sheikholeslami M, Shamlooei M. $Fe_3O_4–H_2O$ nanofluid natural convection in presence of thermal radiation. International Journal of Hydrogen Energy. 2017;**42**:5708-5718

[33] Sheikholeslami M, Shehzad SA. Magnetohydrodynamic nanofluid convection in a porous enclosure considering heat flux boundary condition. International Journal of Heat and Mass Transfer. 2017;**106**:1261-1269

[34] Sheikholeslami M, Shamlooei M. $Fe_3O_4–H_2O$ nanofluid natural convection in presence of thermal radiation. International Journal of Hydrogen Energy. 2017;**42**:5708-5718

[35] Lee CG, Hwang YJ, Choi YM, Lee JK, Choi C, Oh JM. A study on the tribological characteristics of graphite nano lubricants. International Journal of Precision Engineering and Manufacturing. 2009;**10**(1):85-90

[36] Wang Y, Li C, Zhang Y, Yang M, Li B, Jia D, Hou Y, Mao C. Experimental evaluation of the lubrication properties of the wheel/workpiece interface in minimum quantity lubrication (MQL) grinding using different types of vegetable oils. Journal of Cleaner Production. 2016;**127**:487-499

[37] Kuram E, Ozcelik B, Cetin MH, Demirbas E, Askin S. Effects of blended vegetable-based cutting fluids with extreme pressure on tool wear and force components in turning of Al 7075-T6. Lubrication Science. 2013;**25**(1):39-52

[38] Debnath S, Reddy MM, Yi QS. Environmental friendly cutting fluids and cooling techniques in machining: A review. Journal of Cleaner Production. 2014;**83**:33-47

[39] Wang JG, Zhang JZ. On formation and breakup of boundary lubricating layer. Lubrication Engineering. 2005;**6**:4-8

[40] Sinha MK, Madarkar R, Ghosh S, Venkateswara Rao P. Application of eco-friendly nanofluids during grinding of Inconel 718 through small quantity lubrication. Journal of Cleaner Production. 2017;**141**:1359-1375

[41] Abdalla HS, Patel S. The performance and oxidation stability of sustainable metalworking fluid derived from vegetable extracts. Proceedings of the Institution of Mechanical Engineers, Part B: Journal of Engineering Manufacture. 2006;**220**:2027-2040

[42] Baumgart P, Canzi G, Hanashiro T, Doezema L, Siniawski MT. Influence of fatty acid additives on the tribological performance of sunflower oil. Lubrication Science. 2010; **22**(9):393-403

[43] Wang Y, Li C, Zhang Y, Li B, Yang M, Zhang X, Guo S, Liu G. Experimental evaluation of the lubrication properties of the wheel/workpiece interface in MQL grinding with different nanofluids. Tribology International. 2016;**99**:198-210

[44] Kato H, Komai K. Tribofilm formation and mild wear by tribo-sintering of nanometer-sized oxide particles on rubbing steel surfaces. Wear. 2007;**262**:36-41

[45] Huang HD, Tu JP, Gan LP, Li CZ. An investigation on tribological properties of graphite nanosheets as oil additive. Wear. 2006;**261**:140-144

[46] Luo T, Wein X, Huang X, Huang L, Yang F. Tribological properties of Al_2O_3 nanoparticles as lubricating oil additives. Ceramics International. 2014;**40**:7143-7149

[47] Ge XY, Xia YQ, Cao ZF. Tribological properties and insulation effect of nanometer TiO_2 and nanometer SiO_2 as additives in grease. Tribology International. 2015;**92**:454-461

[48] Yu HL, Xu Y, Shi PJ, Wang HM, Wei M, Zhao KK, Xu BS. Microstructure, mechanical properties and tribological behavior of tribofilm generated from natural serpentine mineral powders as lubricant additive. Wear. 2013;**297**(1):802-810

[49] Yu B, Liu ZL, Ma CB, Sun JJ, Liu WM, Zhou F. Ionic liquid modified multi-walled carbon nanotubes as lubricant additive. Tribology International. 2015;**81**:38-42

[50] Samuel J et al. Graphene colloidal suspensions as high performance semisynthetic metal-working fluids. Journal of Physical Chemistry C. 2011;**115**(8):3410-3415

[51] Huang WT et al.. Robust design of using MWCNTs in minimum quantity lubrication. In: Applied Mechanics and Materials. 2014. pp. 11-21

[52] Huang WT et al. A combined minimum quantity lubrication and MWCNT cutting fluid approach for SKD 11 end milling. International Journal of Advanced Manufacturing Technology. 2016;**84**(5-8):1697-1704

[53] Nam JS, Lee P-H, Lee SW. Experimental characterization of micro-drilling process using nanofluid minimum quantity lubrication. International Journal of Machine Tools and Manufacture. 2011;**51**(7):649-652

[54] Lee P-H, Nam JS, Li C, Lee SW. An experimental study on micro-grinding process with Nanofluid minimum quantity lubrication (MQL). International Journal of Precision Engineering and Manufacturing. 2012;**13**:331-338

[55] Nam JS, Kim DH, Lee SW. A parametric analysis on microdrilling process with nano-fluid minimum quantity lubrication. In: ASME 2013 International Manufacturing Science

and Engineering Conference Collocated with the 41st North American Manufacturing Research Conference. 2013. pp. V002T04A016-V002T04A016

[56] Kim DH, Lee P-H, Kim JS, Moon H, Lee SW. Experimental study on micro end-milling process of Ti-6AL-4V using nanofluid minimum quantity lubrication (MQL). In: ASME 2014 International Manufacturing Science and Engineering Conference Collocated with the JSME 2014 International Conference on Materials and Processing and the 42nd North American Manufacturing Research Conference, American Society of Mechanical Engineers. 2014. pp. V001T03A015-V001T03A015

[57] Minh DT, The LT, Bao NT. Performance of Al_2O_3 nanofluids in minimum quantity lubrication in hard milling of $60Si_2Mn$ steel using cemented carbide tools. Advances in Mechanical Engineering. 2017;**9**:1-9

[58] Lee CG, Hwang YJ, Choi YM, Park MC, Kim KM, Lee JK, Choi C, Oh JM. The comparison of lubrication characteristics on the effects of the size and shape of the nanoparticles dispersed in Nano lubricants. Proceedings of the KSTLE Spring Conference. 2008:140-145

[59] Lee CG, Hwang YJ, Choi YM, Lee JK, Choi C, Oh JM. A study on the Tribological characteristics of graphite Nano lubricants. International Journal of Precision Engineering and Manufacturing. 2009;**10**:85-90

[60] Hwang Y, Lee C, Choi Y, Cheong S, Kim D, Lee K, Lee1 J, Kim SH. Effect of the size and morphology of particles dispersed in nano-oil on friction performance between rotating discs. Journal of Mechanical Science and Technology. 2011;**25**(11):2853-2857

[61] Mao C, Zhang J, Huang Y, Zou H, Huang X, Zhou Z. Investigation on the effect of Nanofluid parameters on MQL grinding. Materials and Manufacturing Processes. 2013;**28**:436-442

[62] Lee P-H, Nam JS, Li C, Lee SW. An experimental study on micro-grinding process with Nanofluid minimum quantity lubrication (MQL). International Journal of Precision Engineering and Manufacturing. 2012;**13**:331-338

[63] Hamzah MH, Sidik NAC, Ken TL, Mamat R, Najafi G. Factors affecting the performance of hybrid nanofluids: A comprehensive review. International Journal of Heat and Mass Transfer. 2017;**115 (**630-646

[64] Leong KY, Ahmad KZK, Ong HC, Ghazali MJ, Baharum A. Synthesis and thermal conductivity characteristic of hybrid nanofluids – A review. Renewable and Sustainable Energy Reviews. 2017;**75**:868-878

[65] Sharma AK, Singh RK, Dixit AR, Tiwari AK. Novel uses of alumina-MoS2 hybrid nanoparticle enriched cutting fluid in hard turning of AISI 304 steel. Journal of Manufacturing Processes. 2017;**30**:467-482

[66] Maruda RW, Krolczyk GM, Feldshtein E, Nieslony P, Pusavec F. Tool wear characterizations in finish turning of AISI 1045 carbon steel for MQCL conditions. Wear. 2017;**372-373**:54-67

[67] Maruda RW, Krolczyk GM, Niesłony P, Krolczyk JB, Legutko S. Chip formation zone analysis during the turning of austenitic stainless steel 316L under MQCL cooling condition. Procedia Engineering. 2016;**149**:297-304

[68] Duc TM, Dong PQ, Long TT, Chien TQ. Manufacturing research of vortex tube applied to minimum quantity lubrication for hard milling of SKD 11 tool steel. Journal of Science and Technology. 2017;**173**(13):171-176

Chapter 8

Precisely Addressed (DNA Gene) Spray Microfluidic Chip Technology

Jian-Chiun Liou

Additional information is available at the end of the chapter

http://dx.doi.org/10.5772/intechopen.74611

Abstract

This study is the subject of precisely addressed (DNA Gene) spray microfluidic chip. It is a special chip (ASIC) designed to spray liquid medical wisdom DNA gene sequencing system technology transfer fabric onto the glass slide. Thermal bubble liquid bead generation, it produces a very large thrust bubble in a short time to launch the liquid. It forms micro-droplets. It is in the biomedical micro-beads quantitative, it uses the address spray liquid crystal structure of the cavity. It uses the principle of ink-jet printer cloth DNA liquid onto glass slides. It does DNA sorting for each style. Bubble inkjet technology is in the inkjet head position on the wall with heating electrodes. It is by pulsing the selected heating element by electrical pulses. It produces ink droplets on the inkjet head. It is heated to a certain temperature after the electrode. It makes the droplet a tiny bubble and explodes. It is then discharged through the heating chamber through the inkjet head. It is attached to the substrate surface. It is printed on the amount of ink droplets depends on the temperature control of the heating device.

Keywords: DNA gene, microfluidic chip, addressed

1. Introduction

Introduction and application of inkjet technology, inkjet technology is involved in electronics, electrical, mechanical, materials, chemical, physics and other fields of expertise. It is highly integrated technology. Inkjet technology is in industrial applications with high-speed operation, quiet, non-contact and computer-controlled features. It does not require the use of photomask, process simplification, material utilization, low cost and more environmentally friendly. Jetting technology is a widely used technology [1–12]. The principle of ink-jet technology is mainly that the liquid in the chamber is squeezed by an Actuator so that the liquid is squeezed by the pressure to be ejected through the spray hole. It is because each jetting hole in the ink jetting device has an independent jetting chamber, an actuator and a micron-sized jetting hole. Spray chamber can be filled with a certain amount of liquid [13–18]. When the actuator squeezes the liquid in the chamber and then passes through the micron-sized orifice, the fixed-size and uniform-size micrometer-sized liquid droplets can be jetted, as shown in **Figures 1** and **2**. This study has established a complete jet technology. For example, from the droplet nucleation and flow simulation, spray chip design, spray device package, corrosion head design, back pressure control technology, micro-droplet detection technology … and so on. It can be the most complete jet design for the technology, assembly and testing. It currently except in the relevant technology transfer to the industry outside. It also uses this technology in the display, power machinery, bio-chip, printed circuit board (PCB) and other industries in the development of new technologies.

It generates surface tension suction. It pulls the new liquid to replenish the liquid ejection area. Thermal bubble jet technology is constructed from such integrated cycle technology programs and materials.

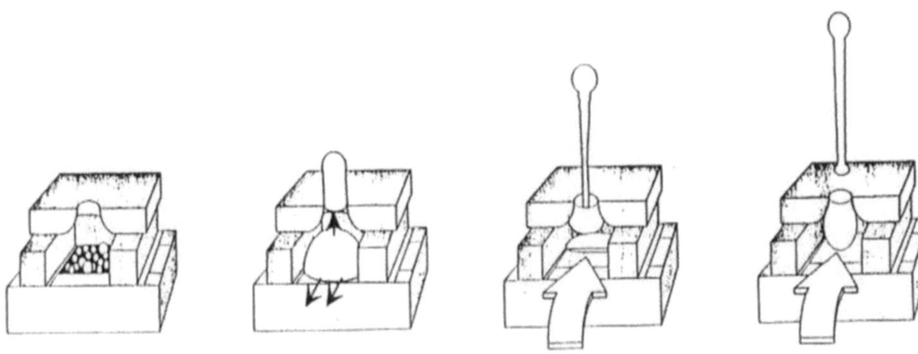

Figure 1. Spray device spray diagram (top jetting).

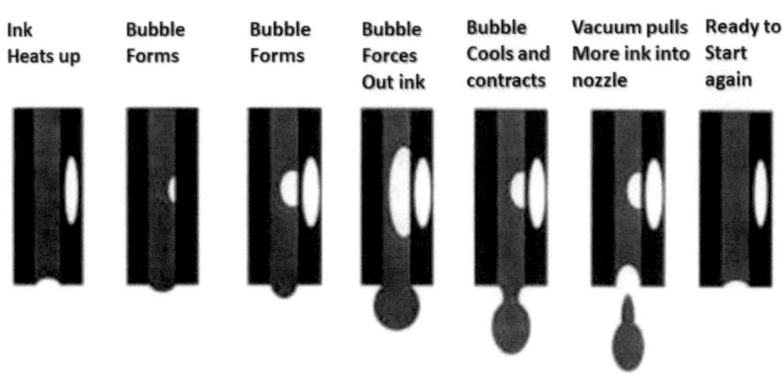

Figure 2. Spray device spray diagram (side jetting).

http://dx.doi.org/10.5772/intechopen.74611

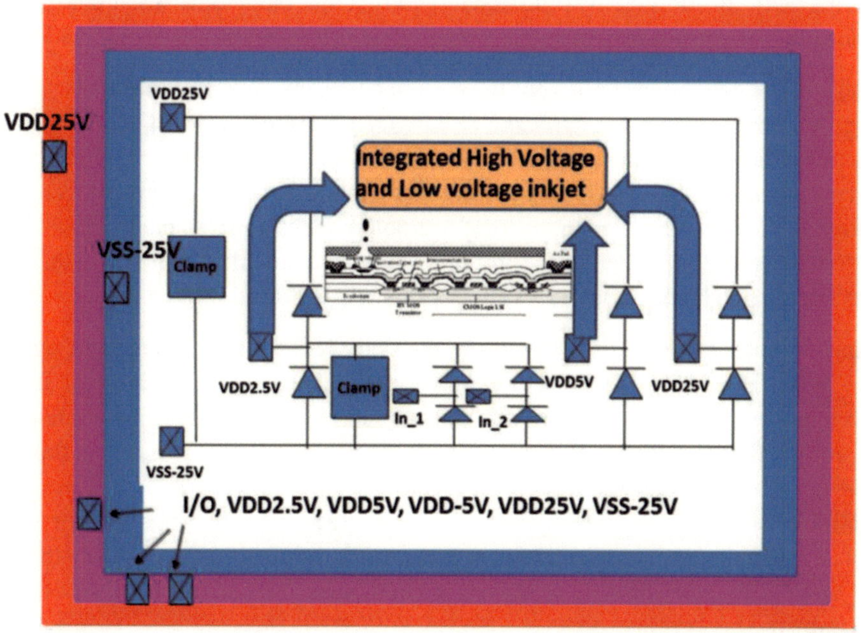

Figure 3. Monolithic DNA liquid jet integrated ESD multi-level output.

Jet dispensing wafer technology includes high-voltage drive, low-voltage logic signal processing, and micro-electromechanical structure array. As shown in **Figure 3**. It drops a small drop of liquid onto the paper. It combines imaging with very small droplets, only 50–60 μ (micron, one millionth of a meter). It is thinner than the human hair (70 μ). The position of the drip ink is precisely finely moved by the ink head. Each inch can be up to 1000 points or more. In this research, HV-ESD Clamp was integrated into the design and fabrication of monolithic inkjet chips for multi-level output integrated circuits. It contains a high-voltage drive power device array, low-voltage logic circuits, and micro-electromechanical components array structure integrated in a silicon substrate process. It is resistant to HBM (±4 kV), MM (±400 V) protection circuit testing.

2. DNA gene spray microfluidic chip technology

Biomedical wafers have many advantages such as trace detection, accurate quantification, automatic operation and rapid parallel processing. Compared with traditional biomedical testing, biomedicine wafers have great advantages and so far, many breakthroughs have been made Technological development. But it cannot be denied that biomedical chips are also facing many technical challenges that need to be overcome by scientists in different fields.

As the design and manufacture of biomedical wafers are cross-cutting projects that involve the operation of microfluidic systems, biomedical reactions and optical signal detection, they are quite different from the highly specialized professional division of labor in many traditional engineering fields. In addition to professional requirements, biomedical wafer design and manufacturing, attaches great importance to cross-cutting technology and communication skills.

The design of the chip often starts from the system or encapsulation level and starts to reverse thinking. Whether the team has sufficient cross-communication between the fields except the experts in each field and jointly solve the problems derived from each other is often the relationship between chip design and The key to success or failure Therefore, different from the training requirements of traditional engineering technology, instead it is able to cross the field of micro-electromechanical systems, microfluidics systems, biomedical technology and optoelectronic technology and other fields, it is very important in the development of biomedical wafer.

When the amount of sample used to reduce, then faced with a sharp decline in signal detection problems. Increasing the signal strength or improving the sensitivity of the sensing device are two ways to solve the above problems. In terms of increasing signal intensity, there are currently artificial methods of replicating biomedical molecules to increase the weight of specimens, such as polymerization and per-chain reaction techniques; for those molecules that cannot be artificially increased, the number of markers for their markers or sensitivity, or to focus molecules in the detection area for detection. In enhancing the sensitivity of the sensing device, more sensitive new sensing technologies are also the focus of development besides reducing background noise.

In this project, the hot bubble liquid bead is generated by heating the liquid by using a micro-heating wire and generating bubbles of great thrust in a short time to push the liquid out to form micro-droplets, as shown in **Figures 3** and **4** below. In the quantification of biomedical microbeads, the integrity and cleanliness of the beaded pellets are often quantitatively

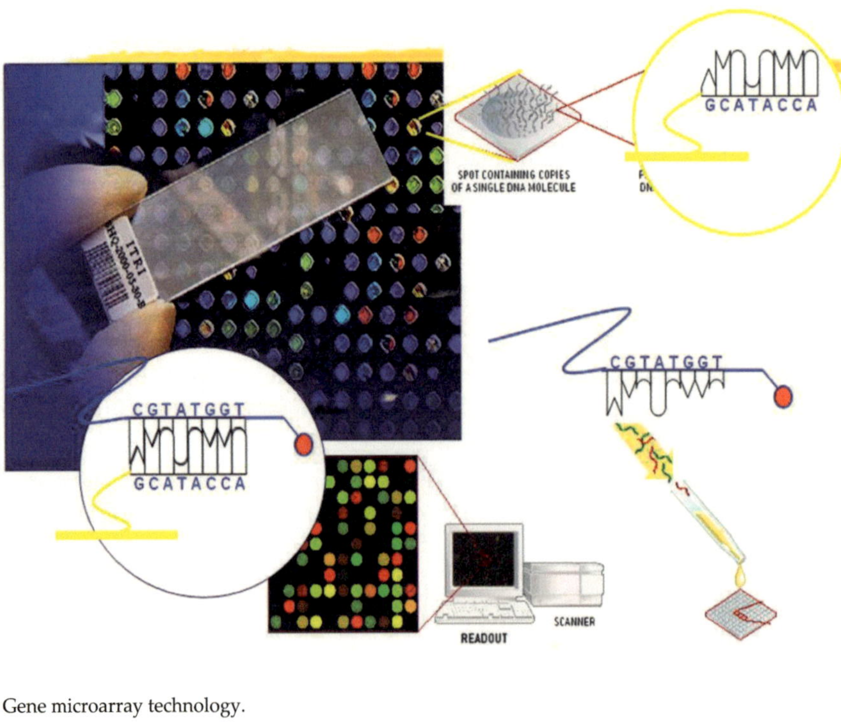

Figure 4. Gene microarray technology.

accurate or not. The way in which the foamed beads are produced will produce a series of beaded beads, that is, in addition to the main beaded beads, Subsequent satellite beads have a considerable degree of impact on accurate quantitation. The program has a complete set of main bead, quantitative determination of liquid beads, in line with biomedical specimens of liquid bead quantitative requirements, this study used to create disease detection and DNA sequencing of DNA microarray chip.

DNA micro-array, bio-chip or DNA chip combination of molecular, materials, motors and other fields of products. This system can be divided into DNA chip, protein chip, cell chip, tissue chip. However, only the first two are commonly used, in which a DNA chip can detect changes in the concentration of mRNA in a cell. The biggest advantage of gene microarrays is the ability to quickly and accurately provide a large number of genetic tests and to understand the performance of these genes. This is a considerable help in understanding the transmission of genes.

The main body of the chip is the use of glass, nylon, silicone, ceramic and other materials. Above there are probes made of oligonucleotides or proteins (DNA chip and protein chip, respectively). Its main principle is that these probes will adsorb their corresponding DNA or RNA species. People want to learn more about these genes or proteins. Currently there are three ways to probe the technology, one is the synthesis spotting method, the first synthesis of oligonucleotides or oligopeptides, and then the way the pressure on the board. The second is to take the light-guided approach, so that oligonucleotide or oligopeptide slowly in the main plate synthesis. The third is similar to the inkjet printer operating mechanism, the oligonucleotide or oligopeptide materials Yiyi sprayed on the board gradually formed.

Gene microarrays have several uses: piecing together an unknown DNA sequence and understanding the performance of a gene. Introduce a relatively simple application. Piece together the DNA sequence of the sample.

First, the DNA of the sample is shattered one by one, and the double-stranded DNA is heated to single-stranded helical DNA. It is a long piece of DNA into a small short DNA sequence. It is the only ATGC single-stranded sequence. It attaches a special sequence of DNA Xs to each small piece of DNA (for example, five A at the beginning). It has a small platform with short pins that correspond to the DNA sequence of the X sequence. To the above example, is to pick five T. In this way, these short DNA sequences are adsorbed onto these short needles.

The system begins to release the same nucleotide one by one and then the next step. For example, it frees C to see what needle DNA will react to it. This is a reaction to give a light. It can shine a bright needle. It can be learned that the first nucleotide of these DNA sequences on these needles is G. The system is constantly repeating this step (of course, to release different kinds of nucleotides) to know the DNA sequence on this needle. Then go through a jigsaw puzzle-like step (using a computer as an aid) to spell out an entire sequence of DNA.

3. Cell detection

In cancer cell detection, gene microarray technology also has an important place. It can be used to compare the difference between normal cells and cancer cells. It can further find out which genes are more in cancer cells for further analysis. The method is to mark normal cells and cancer cells in different colors, then break them up and spread on the wafer because the DNA corresponding to the nucleotide probes on the wafer will remain on the plate. It looks like a colored plaid below.

Take the example on the right, red represents cancer cells and green represents normal cells. Next, quantify the color of the statistics and calculate the ratio between them. It will find that there is significantly more red in certain areas than in green, or significantly more in green than in red. So biologists can take this to know more or less cancer cells than the average number of genes. Suppose today we find that the X gene is much more numerous than normal cells tomorrow. Biologists speculate that the X gene may be a factor in cancer. To test this hypothesis, a biologist can make a cell that has an excess of the X gene and use it to produce the protein. It can see if he has any cancer cell features. With this method we can find out the genes associated with cancer cells. This treatment of cancer cells, whether there is much to be considered for help or prevention.

Gene microarray technology is a multi-field combination of technologies as shown in **Figure 4**. It helps us better understand or crack the genetic code. The combination of technologies in many areas may be more competitive than traditional single-area research. It can solve some difficult problems in a single area.

A biological genome refers to all genetic material on the biological chromosome. Its size is often expressed as the number of base pairs. A proteome refers to the entire protein product of a gene and its performance. Many genes produce more than one type of protein after transcription and translation. There are two main reasons for this: First, a single gene undergoes alternative splicing when it is transcribed into mRNA. It leads to different combinations of exons. It thus produces different proteins. The other is that the translated protein is modified so that the protein has more different structures and functions. Proteasomes are more complicated than the genomes. In addition, a mutation of a single nucleotide (A, T, C, G) may occur at about 100–300 bases per billion of the 3 billion bases in the human genome. It thus makes the DNA sequence change, known as single nucleotide polymorphism (single nucleotide polymorphism, SNP). Therefore, the structure of DNA between individuals is very similar. The difference is very large at the micro level. It leads to significant differences among individuals. After decoding the human genome, it was found that the sequences among different individuals were very similar. It is only 0.1% different. These slight differences determine each person's height, color, size and other aspects of the difference. It also determines the different characteristics of our body. It is easy to suffer certain diseases.

We use different restriction enzymes to process the DNA. It is unique because of each individual's DNA sequence. DNA is cut into fragments of varying lengths. It is different after electrophoresis analysis. This treatment with restriction enzymes gives rise to polymorphism in length of different DNA fragments. It is called restriction fragment length polymorphisms (RFLPs). It is different from everyone's RFLPs, just like our fingerprints. Therefore, RFLPs are also called DNA fingerprinting. RFLPs technology has been widely used in the identification of criminal cases and genetic diseases.

http://dx.doi.org/10.5772/intechopen.74611

RFLPs technology can also be applied to genetic screening. Check if embryos, fetuses or newborns have a genetic disorder. It is for quick medication. The relatives have genetic disease, but their own asymptomatic people. It can also be used by gene scanning to determine if there is a mutant gene or is likely to develop the disease. In addition, DNA Test is the most accurate way to detect a gene. It directly analyzes DNA mutations. Due to PCR (polymerase chain reaction) technology matures. DNA testing requires only a few cells to complete. It applies to one or a few mutations caused by the disease. Gene scanning and DNA detection techniques are most commonly used to test for sickle-cell disease.

Sickle-shaped erythrocyte disease is a general term for a series of hereditary diseases caused by hemoglobin S (HbS). Sickle-anemia is just one of them. It arises from the point mutation in the beta-Hemoglobin gene. It causes heme β to be changed from normal glutamic acid to valine during translation. It is negatively charged by glutamic acid. Valine is electrically neutral. It makes the hemoglobin structure change. It leads to the formation of elongated needles in red blood cells. It causes red blood cells to deform, usually in severe cases. It is more minor growth and development.

Sickle-shaped anemia test in two ways: First, amplified by PCR of DNA sequences, MstII restriction enzyme cutting. It is run by gel electrophoresis. Its normal sequence is cleaved by the restriction enzyme. The molecular weight is smaller and runs farther. Abnormal sequence cannot be cut by restriction enzyme, its molecular weight is larger, runs nearer. It can then check for mutations in the gene. The other is the use of artificial synthesis of two DNA test strips (one for the normal sequence "GGACTCCTC" and the other for the abnormal sequence "GGACACCTC"). It binds to normal or mutated DNA sequences amplified by PCR. Change in color to identify if there are mutations in the gene.

The human genome has about 3 billion nitrogenous bases. It contains only 5% of the DNA sequence. The rest of the sequence is mostly called junk DNA. There are more than 3 million repeats in the head. The number of these repetitive sequences varies from person to person. We call this variable number of tandem repeats (VNTRs). These repetitive sequences are ubiquitous in the genome. Its function is not yet fully understood. It is currently known to have more than ten diseases because of the excess number of repetitions in this area. For example, it is fragile X syndrome, Huntington syndrome, and the like. This area is also the main source of information for identifying paternity.

The genetic variation between people is only one ten thousandth. The main source of variation among individuals is the number of VNTR repeats. The number of repetitions per person varies with VNTR. When the restriction enzyme cuts the region, different individuals will produce fragments of varying lengths and numbers. We call this restriction fragment length polymorphism (RFLP). These cut-off fragments, it is different molecular weights. It has different mobility. It uses electrophoresis to separate DNA of different lengths. It goes through blotting, probe hybridization and other processes. It will show the thickness of the stripes. These stripes are called DNA fingerprinting. This was developed in 1984 by Dr. Alec Jeffreys in the United Kingdom.

It is passed on to offspring as VNTR follows Mendel's laws of inheritance. Every pair of homologous chromosomes is in each human body. It is one from the father, one from the mother. Each person cut the fragment should be at least one of both parents the same. As long as it compares to the three-way RFLP, it can confirm the parent-child relationship.

In recent years, due to the development of polymerase chain reaction (PCR) technology. DNA identification technology uses PCR to amplify the sample DNA collected. It selects mini-satellites DNA with shorter repeat fragments in the VNTR as the target of the assay. Single-parent paternity tests can also be confirmed using maternally inherited mitochondrial DNA or paternally inherited Y chromosomes. Sperm contained in the mitochondria body is very small, most of the mitochondria within the fertilized egg from the egg. It compares the mitochondrial DNA in the offspring cells to determine if the relationship is mother-daughter, mother-daughter, or sibling. The male Y chromosome did not participate in synapsis or gene recombination during meiosis. It therefore determines the relatives of father, son, brother or paternal relative to the Y chromosome. These can be used as a paternity test to determine the relevant evidence.

4. Experiment and results

Micro-injection flow channel design parameters monomer, the impact of the ejection orifice thickness also caused performance is very important, its impact may be the effect of differences in both exit velocity of the droplets and flight direction. It is necessary to know the thickness of the sheet discharge orifice (nozzle plate thickness) affect the size ratio between the length of the two nozzle diameter caused respect. Through computer simulation can be obtained as shown in **Figure 5**. Simulation of DNA in the orifice thickness reduced from 50 to 25 μm single-aisle jet chamber. The results can be seen due to the reduced thickness of the orifice. Droplet flight directionality deteriorated. The reason is that the effect of the thickness of the orifice having rectified so that the flow field velocity with consistent directionality. But too thick so that it will produce additional drag exports slowed down. Orifice having a different orifice diameter should be of different sheet thicknesses. The calculation results are in 200 μs moment, obviously not yet fluid backfill is completed. It causes of reduced thickness orifice, causing the fluid velocity to large-amplitude reduction backfilling, because the droplet from the nozzle holes. Droplet tail flight directionality deteriorated. It has already begun to overflow orifice surface side. Resulting is in resistance, due to surface tension forces led by the great surface. Prevents fluid filled into the interior of the cavity injection phenomenon.

Single injection process in real time, and cannot ensure just above the orifice sheet ejection orifice center is located in the center of the cavity, it is necessary to simulate the effect of off-center position when the spray hole caused by the analysis. The results show that the ejected by the droplet and will not deviate from the center position of the nozzle holes, caused by large droplets tails flying directional radiation deteriorated. The results show the fluid actually increased after the injection backfilling operation speed, almost complete backfill to the surface of the nozzle holes, the reason may be because the orifice is placed away from the channel inlet of the injection cavity side wall surface side so that the fluid this reduces while reducing internal flow field reflux (circulation) from happening; directly back to fill in one

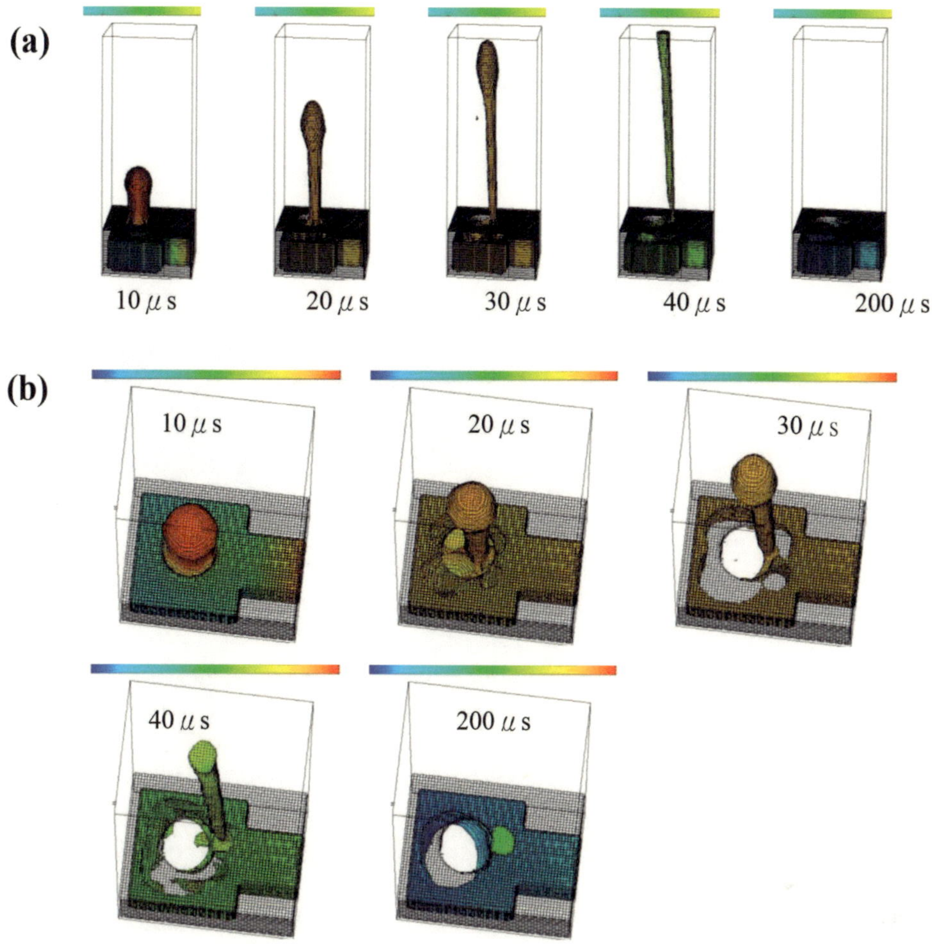

Figure 5. The simulation of DNA was calculated on a three-dimensional top view of (a) a three-dimensional side view (b) at 10, 20, 30, 40 and 200 μs at a single channel ejection cavity with a hole diameter of 60 μm and a thickness of 25 μm at an operating frequency of 5 KHz spray situation.

direction from the channel inlet cavity injection, such as speed XY sectional flow field in an instant 200 μs shown can clearly see resistance to the flow field, it is resulting in increased velocity of the fluid backfill. **Figure 6** is shown Droplet generation control simulation.

Figure 7 is a schematic diagram of a cross-sectional view of a precise overall DNA jet multi-dimensional drive addressing chip. The amount of each single point DNA sequencing (spray volume) is determined by the magnitude of the amplitude power. DNA trace differences can be adjusted. Heater also increases the number of pads as the number of thermal resistance components increases. DNA jetting system costs also increase proportionally. Therefore, Heater thermal resistance components must increase the number of logic multiplexed control circuit. It integrates the related process technology. Making the appropriate driver circuit and selective switching circuit can dramatically reduce the required number of external Pads and reduce manufacturing costs.

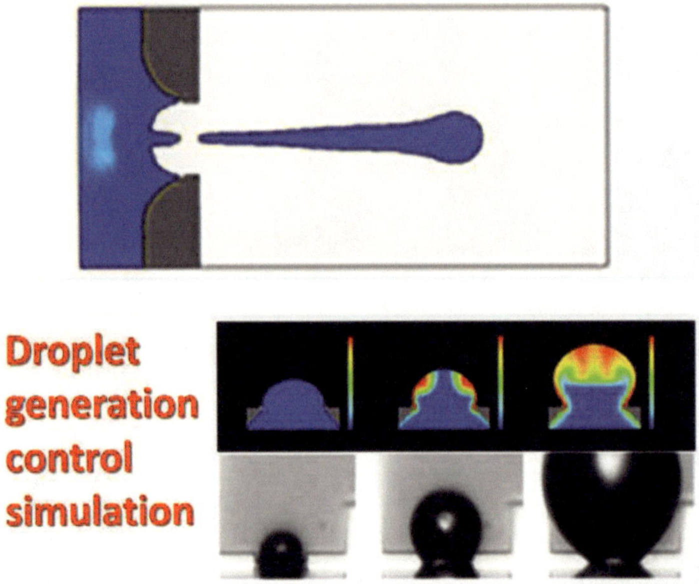

Figure 6. Droplet generation control simulation.

The original voltage source is a group of DC voltage drop. The voltage divider is divided into eight groups by nine resistors, and one group of voltages is selected by three pairs of eight decoders. This circuit architecture can convert the original 5 V logic pulse into the voltage required by the circuit system, and output the required voltage through the PN transistor, Smooth error compensation process, to achieve the amount of liquid DNA is determined by the size

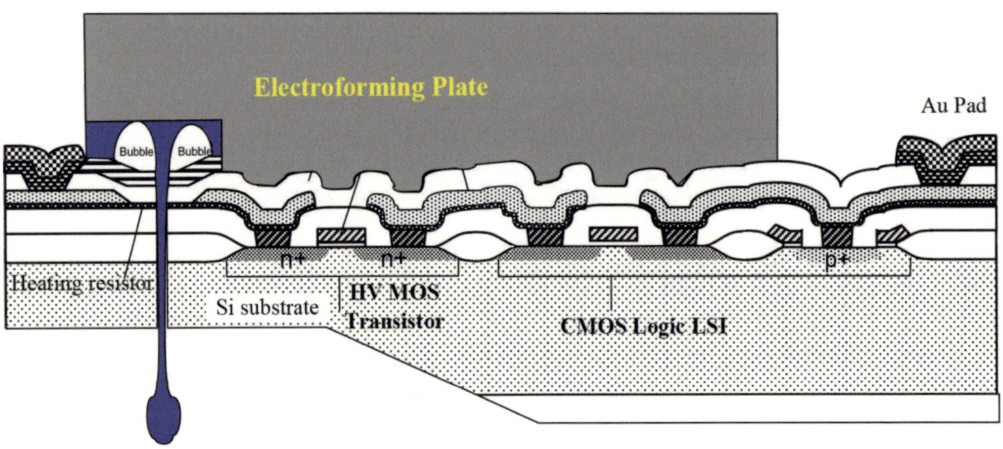

Figure 7. A schematic diagram of a cross-sectional view of a precise overall DNA jet (the merging of multi-dimensional data registration circuit and DNA jet technologies).

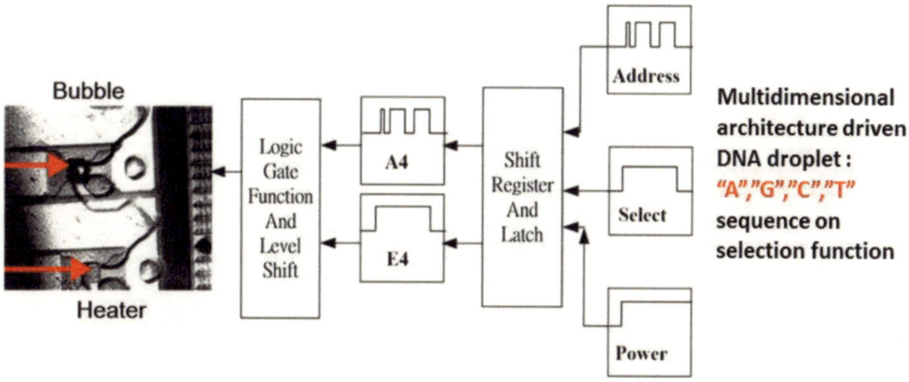

Figure 8. Open pool printhead chip.

of the amplitude and power, DNA trace differences can be adjusted action. For the same DNA sequence of the password, to be able to get different lengths of continuous fragments, and know the end of these fragments corresponding to the A, T, C, G base. Thus, as long as the length of these fragments ordered, the original DNA sequence of the complete base to know. As for the method of obtaining a large number of fragments; after a large amount of DNA to be sequenced is copied first, the DNA is cut at different bases. For example, the restriction will cut off double helical DNA at specific base alignments. The effect of cleaving DNA at specific bases can also be achieved by chemical reactions that remove bases. Also, copy the DNA that used to contain the complete passcode as a template. It is randomly interrupted to interrupt without growth. It produces a variety of pieces of different lengths. It sorts DNA fragments of different lengths. It is a technique called gel electrophoresis. Gel electrophoresis can distinguish small differences in DNA fragments that have been cut off. The rationale is that the flow rate of a molecule fragment in the gel is inversely proportional to the logarithm of the number of base pairs. The shorter the molecular fragment, the faster it will drift in the colloid. Gel electrophoresis on the length of fragment resolution is very high. It is for two DNA fragments that themselves are hundreds of units long. The difference between the length of only one nucleotide can be distinguished.

This systematic phase verification follows the DNA droplet addressing requirements. It is the first plan to send signals to the output thermal resistance module, as shown in **Figure 8**. Each thermal resistance has a corresponding signal and open pool measurement. It allows this signal is generated after the designated system contains quantitative circuits. It can adjust the voltage level of the output circuit to accurately quantify DNA. Adjustable jet emits DNA flow through the system wafer. DNA jetting system wafer characteristics measurement, pulse width modulation (PWM) power corresponding quantitative DNA gene-matching parameters, high voltage power transistor driver design compensation. It optimizes the parameters of power components to observe the changes of DNA beads under different operating conditions. The DNA liquid tube array and droplets on the carrier is shown in **Figure 9**.

In the biomedical micro-liquid bead quantitative, the integrity and cleanliness of the liquid bead is often about the accuracy of quantitative or not. Hot bubble bead generation, will produce a series of beads. It is in addition to the main liquid beads, there are satellite beads

Figure 9. Multiplexer DNA solution jet part.

followed. It has a considerable degree of impact on accurate quantitation. In this study, a complete master bead was used and the beads were quantitatively accurate. It meets the biomedical specimen liquid bead quantitative requirements. The study was used to create DNA microarray chips for disease detection and gene sequencing.

Author details

Jian-Chiun Liou

Address all correspondence to: jcliou@tmu.edu.tw

School of Biomedical Engineering, Taipei Medical University, Taipei, Taiwan

References

[1] Tseng F-G, Kim C-J, Ho C-M. A high-resolution high-frequency monolithic top-shooting microinjector free of satellite drops—Part II fabrication, implementation, and characterization. Journal of Microelectromechanical Systems. October 2002;**11**(5):437

[2] Tseng F-G, Kim C-J, Ho C-M. A high-resolution high-frequency monolithic top-shooting microinjector free of satellite drops—Part I: Concept, design, and model. Journal of Microelectromechanical Systems. October 2002;**11**(5):427

[3] Rembe C, aus der Wiesche S, Beuten M, Hofer EP. Investigations of nonreproducible phenomena in thermal ink jets with real high speed cine photomicrography. In: Proceedings of SPIE Europto Series, Vol. 3409, Electronic Imaging: Processing, Printing ,and Publishing in Color. Zurich; 1998

[4] Ashkin A, Dziedzic JM. Observation of optical resonances of dielectric spheres by light scattering. Applied Optics. 1981;**20**(10):1803-1814

[5] Lin H-B, Huston AL, Justus BL, Campillo AJ. Some characteristics of a droplet whispering-gallery-mode laser. Optics Letters. 1986;**11**(10):614-616

[6] Hill SC, Leach DH, Chang RK. Third-order sum-frequency generation in droplets: Model with numerical results for third-harmonic generation. JOSA-B. 1993;**10**(1):16-33

[7] Lee J-Y, Park H-C, Jung J-Y, Kwak H-Y. Bubble nucleation on micro line heaters. ASME Journal of Heat Transfer. 2003;**121**:687-692

[8] Park J-H, Oh YS. Investigation to minimize heater burnout in thermal thin film print heads. Microsystem Technologies. Jan. 2005;**11**(1):16-22

[9] Akamatsu H, Matsukubo N, Inokuma Y, Aso T, Shindo HK. Maximum performance of thermal printhead. 23rd International Conference on Digital Printing Technologies. Technical Program and Proceedings. Digital Fabrication; 2007. pp. 142-7

[10] Lindemann T, Ashauer H, Goettsche T, Sandmaier H, Yu Y, Peters RP, Sassano D, Bellone A, Scardovi A, Zengerle R, Koltay P. Bubble jet printhead with integrated polyimide nozzle plate. 18th IEEE International Conference on Micro Electro Mechanical Systems (IEEE Cat. No.05CH37610); 2005. pp. 560-563

[11] Shin SJ, Kuk K, Shin JW, Lee CS, Oh YS, Park SO. Thermal design modifications to improve firing frequency of back shooting inkjet printhead. Sensors and Actuators, A: Physical. Sept. 2004;**A114**(2-3):387-391

[12] Shin SJ, Kuk K, Shin JW, Lee CS, Oh YS, Park SO. Firing frequency improvement of back shooting inkjet printhead by thermal management. Transducers '03. 12th International Conference on Solid-State Sensors, Actuators and Microsystems. Digest of Technical Papers (Cat. No.03TH8664). vol. 1; 2003. pp. 380-383

[13] Liou J-C, Liu C-H, Chen C-J. A bubble-jet MEMS printhead integrated with CMOS demultiplexer. 2004 IEEE/SEMI Advanced Semiconductor Manufacturing Conference and Workshop (IEEE Cat. No.04CH37530); 2004. pp. 413-419

[14] Nayve R, Fujii M, Fukugawa A, Takeuchi T, Murata M, Yamada Y, Koyanagi M. High-resolution long-array thermal ink jet printhead fabricated by anisotropic wet etching and deep Si RIE. Journal of Microelectromechanical Systems. Oct. 2004;**13**(5):814-821

[15] Liou J-C, Chen C-J, Chuang M-H. Lower power supply thermal bubble printhead chip with MEMS technology increasing thermal energy effect. Final Program and Proceedings of IS&T's NIP20: International Conference on Digital Printing Technologies. 2004. pp. 863-866

[16] Murata M, Kataoka M, Nayve R, Fukugawa A, Ueda Y, Mihara T, Fujii M, Iwamori T. High resolution long array thermal ink jet printhead with on-chip LSI heater plate and micromachined Si channel plate. IEICE Transactions on Electronics. Dec. 2001;**E84-C**(12):1792-1800

[17] Liou J-C, Tseng F-G. Reduced 30% scanning time 3D multiplexer integrated circuit applied to large array format 20KHZ frequency inkjet print heads. DTIP 2007, Stresa, Italy, 25-27 April 2007. pp. 106-111

[18] Liou J-C, Tseng F-G. Three-dimensional architecture of multiplexing data registration integrated circuit for large-array ink jet Printhead. Journal of Imaging Science and Technology. 2008;**52**(1):1-1-7

Chapter 9

Thermal Transport and Challenges on Nanofluids Performance

José Jaime Taha-Tijerina

Additional information is available at the end of the chapter

http://dx.doi.org/10.5772/intechopen.72505

Abstract

Progress in technology and industrial developments demands the efficient and successful energy utilization and its management in a greater extent. Conventional heat-transfer fluids (HTFs) such as water, ethylene glycol, oils and other fluids are typically low-efficiency heat dissipation fluids. Thermal management is a key factor in diverse applications where these fluids can be used, such as in automotive, microelectronics, energy storage, medical, and nuclear cooling among others. Furthermore, the miniaturization and high efficiency of devices in these fields demand successful heat management and energy-efficient materials. The advent of nanofluids could successfully address the low thermal efficiency of HTFs since nanofluids have shown many interesting properties, and the distinctive features offering extraordinary potential for many applications. Nanofluids are engineered by homogeneously suspending nanostructures with average sizes below 100 nm within conventional fluids. This chapter aims to focus on a detail description of the thermal transport behavior, challenges and implications that involve the development and use of HTFs under the influence of atomistic-scale structures and industrial applications. Multifunctional characteristics of these nanofluids, nanostructures variables and features are discussed in this chapter; the mechanisms that promote these effects on the improvement of nanofluids thermal transport performance and the broad range of current and future applications will be included.

Keywords: nanofluids, thermal conductivity, nanoparticles, challenges, thermal transport

1. General overview

With global population rapid growth and industrialization surge, resources have reached unprecedented levels. Energy has been pointed out as the most important issue facing Humanity in the next years [1]. Energy management becomes crucial for meeting the rising needs of mankind [2]. Nowadays, with increasing pressure of globalized markets and companies' profit race, a dramatic search to obtain proper materials performance, optimizing components and devices designs, improving efficiencies, reducing tools wear, materials consumption and pollution, and obtaining the most possible revenue. In addition to issues regarding materials scrap, maintenance and components wear among others, a hot topic in industry is the heat dissipation. Among diverse forms of energy used, over 70% is produced in or through the form of heat [2]. Heat transfer is a crucial area of research and study in thermal engineering. Heat is transferred either to input energy into a system or to remove the energy produced in a system. Hence, reducing energy loss and intensifying heat transfer processes are becoming paramount tasks to be addressed. Therefore, thermal management plays a vital factor concerning devices, machinery or apparatuses performance; thermal transport role has been subjected to countless investigations and is under the scope of the operational useful life of these components and devices. Being this an opportunity area for successful heat management and energy efficient fluid-based heat transfer systems, with aid of reinforced materials.

Nanotechnology is a science that deals with diverse characteristics and properties of materials at a nanometric level (1 nm = 10^{-9} m). Recently, diverse techniques, equipment, and instrumentations have been devised, as well as various relevant and interesting characteristics and properties of these materials were sorted out. Hence, with aid of nanotechnology, with novel developments linking electronic, optical, mechanical, and magnetic properties, industrial devices have emerged, and this trend is certainly continuing in this century. Cooling of electric, electronic and mechanical devices has been a hot topic in today's fast-growing technologies. The heat required to be dissipated from systems is continually increasing due to industrial and economic trends to miniaturize designs, make better use of resources, obtain more power output, develop and use more environmentally friendly materials, and obtain more profits.

2. Introduction to nanofluids

Nanofluids are a new generation of HTFs with anomalous behavior, engineered by homogeneously suspending nanometer-sized materials or structures within conventional fluids. In other words, nanofluids are nanoscale colloidal suspensions containing condensed nanomaterials. Nanofluids have been found to possess enhanced thermo-physical properties such as thermal conductivity, thermal diffusivity, viscosity, and convective heat transfer coefficients compared to those of base fluids such as water (DiW), ethylene glycol (EG) or oils. They have demonstrated great potential applications in many fields such as microelectronics, transportation, industrial cooling, magnetic sealing, reducing pollution, space and defense, energy storage, air conditioning, power transmission systems, medical therapy and diagnosis, antibacterial activity nanodrug delivery, fuel cells, components and tools wear, friction reduction and nuclear systems cooling, etc. [3–8]. Among diverse techniques to cool down or maintain certain temperature in these systems, the use of fins, vanes or radiators as well as forced air/fluids through cooling channels are being used, even though these are costly. Diverse machinery and devices use inexpensive conventional HTFs to intensify heat dissipation. However, the inherent limitation of these fluids is the relatively low thermal conductivity; water for instance, is roughly three orders of magnitude less conductive than copper or aluminum (**Table 1**). What these conventional fluids lack in thermal conductivity however, is compensated by their ability to flow.

http://dx.doi.org/10.5772/intechopen.72505

Material		Thermal conductivity (W/m K)	Reference
Conventional fluids	Water (DiW)	~0.598–0.609	[9–11]
	Ethylene glycol (EG)	~0.251	[12–14]
	Engine oil	~0.145	[12, 14]
	Mineral oil	~0.115	[3, 15]
	Kerosene	~0.139	[16]
	R141b refrigerant	~0.089	[17]
Carbon structures	Single wall nanotubes (SWCNTs)	3000–6000	[18–21]
	Multi wall nanotubes (MWCNTs)	~3000	[22, 23]
	carbon (diamond)	900–2320	[24, 25]
	Carbon (graphite)	119–165	[24]
	Graphene	~3000	[4]
	Graphite	130–2000	[26]
Metallic solids	Aluminum	237	[27]
	Copper	398	[27]
	Gold	315	[27]
	Silver	424	[27]
Nonmetallic solids	Alumina (Al_2O_3)	31–41	[26, 28]
	Aluminum nitride (AlN)	319–550	[29]
	Boron nitride (h-BN)	~300	[30, 31]
	Boron nitride nanotubes (BNNTs)	~600–960	[32–34]
	Cobalt oxide (Co_3O_4)	12.8	[35]
	Copper oxide (CuO)	76.5	[36]
	Molybdenum disulfide (MoS_2)	34.5 ± 4	[37]
	Silicon carbide (SiC)	148–270	[27, 38]
	Silicon oxide (SiO_2)	1.4–12	[39–42]
	Titania (TiO_2)	8.4–11.2	[13, 39, 42]
	Tungsten disulfide (WS_2)	32–124	[43, 44]
	Zinc oxide (ZnO)	13–29	[13, 39]
	Zirconia (ZrO_2)	2.2	[45]

Table 1. Typical thermal conductivities for diverse conventional fluids and solid materials.

The main mechanism for heat transfer in fluids is convection; its efficacy mostly depends on the thermo-physical properties of the conventional fluids. Furthermore, if the thermal conductivity of conventional fluids were enhanced, it would be much more effective. Hence, since the solid materials possess several orders of higher thermal conductivities, compared with that of conventional fluids, an idea to introduce conducting particles to fluids was considered. Among diverse particles geometry, different particle shapes occur naturally or are engineered for specific applications, as shown in **Figure 1**.

Heat transfer using fluids is a very complicated phenomenon, and various factors such as fluid stability, composition, viscosity, surface charge, interface, and morphology of the dispersed nanostructures influence the observed results [3, 6, 46–59]. Optimization and high efficiency of components and devices have gained great importance since these factors play a crucial role in diverse fields. Solid materials such as metals, CNTs, oxide/nitride/carbide ceramics, semiconductors, and composite materials having higher thermal conductivity can be homogeneously suspended and stabilized within conventional fluids, resulting in better thermal transport performance composite fluids. Nevertheless, improvement in thermal conductivity cannot be achieved by just increasing the solid filler concentration because each system presents a threshold, in which beyond a certain limit, increasing the filler fraction will also increase the viscosity, which will adversely affect the fluid properties and performance.

Most early studies used suspensions of millimeter or micrometer-sized particles, which led to countless problems, such as a tendency to rapidly sediment, unless flow rate is increased; not only losing the improvements in thermal conductivity, but also forming sludge sediments, increasing the thermal resistance and impairing the heat transfer capacity of the conventional fluids. Furthermore, fluids of this scale size could have considerably larger pressure drops [60–64], thus making flow through small channels much more difficult since diverse parameters are critical for device performance, such as morphology and stability of nanostructures, fluids composition, viscosity, fast sedimentation, channels clogging, erosion or wear, among others, which are often very serious for systems consisting of small channels [3, 65–69].

A revolution in the field of HTFs arose with the advent of nanofluids (NFs), a term introduced by Prof. Choi's research group in the late 1990s at Argonne National Lab [68]. The first investigations were performed by Masuda et al. [69] for Al_2O_3 nanoparticles within water, and by

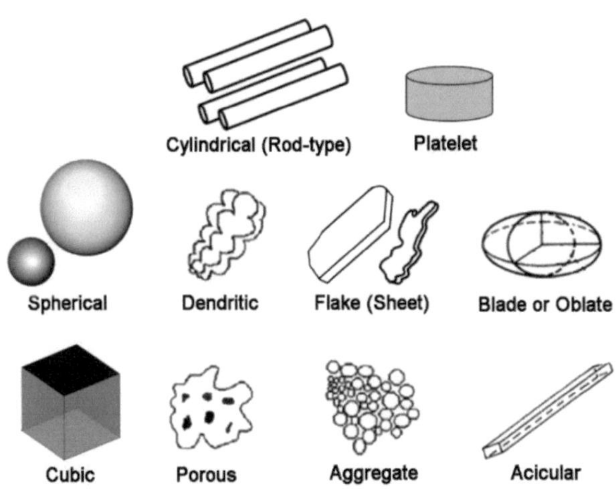

Figure 1. Diverse particle shapes and geometries.

Choi-Eastman group [68] for copper nanostructures dispersed in water as well. Nanofluids research has been exploited and novel developments have been able to fulfill industrial necessities. This research area has been increasing through time, starting at 5 publications in 2003, reaching up to more than 2100 publications by 2017, according to scientific search engine "sciencedirect.com" and keyword Nanofluids. Nanofluids are a novel class of stable heat-transfer suspensions which are engineered containing homogeneously dispersed solid nanofillers. Compared to micro- or millifluids, nanofluids tend to be more stable, since nanofillers possess unique properties, such as large surface area to volume ratio, as well as dimension-dependent physical properties, which make nanostructures better and more stably dispersed in conventional fluids. Nevertheless, some limitations of the effective incorporation of nanostructures within conventional HTFs are dispersion and solubility, because these tend to aggregate and sediment over time. In some cases, additives or surfactants are used to stabilize the nanostructures within the fluids, even though the surfactants could affect and diminish the thermal conductivity of the nanofluids, since surfactants introduce defects at the interfaces [70, 71]. Therefore, one of the main advantages of nanofluids is that they can be specially engineered to optimally fulfill specific objectives, such as enhanced thermal conductivity, a higher thermal energy storage capacity, higher heat transfer coefficients, a better temperature stabilization and less pressure drop, among others. Hence, search for new nanofillers which can get high thermal conductivities at lower filler fractions is important [3, 72].

It has been demonstrated that nanofluids for heat transfer applications have provided better thermal performance than conventional fluids [3, 12, 73, 74]. Hence, the advent of nanofluid-based heat transfer systems can make compact designs with high efficient thermal, physical and electrical performance for instruments and devices. Experiments on convection heat transfer of nanofluids were conducted by several research groups [75–77], showing significant improvements in heat transfer rates of nanofluids. Meanwhile, the thermal conductivity enhancement of nanofluids show a temperature-dependent characteristic and increase of enhancement with rising temperature, which makes the nanofluids more suitable for applications at elevated temperatures [3, 6, 78–81]. Additionally, previous research has shown that nanofluids display better performance in their thermo-physical properties, such as thermal conductivity, thermal diffusivity, viscosity, friction, etc., compared to conventional fluids [3, 4, 82–87]. Hence, nanofluids could be used for aforesaid engineering applications.

3. Synthesis and preparation of nanofluids

The manipulation of matter on the nanometer scale has become a central focus from both fundamental and technological perspectives. Unique, unpredictable, and highly intriguing physical, electrical, mechanical, optical and magnetic phenomena result from the confinement of matter into nanoscale features. Morphology control in nanostructures has become a key issue in the preparation of electronic or mechanical nanodevices and functional materials [88]. A wide variety of combinations of nanostructures and conventional fluids can be used to synthesize and prepare stable nanofluids for diverse applications. Nanofluids could be manufactured by two methods. The first step method is a process in which, simultaneously, nanostructures are made and dispersed within the base fluid [89–91]. This method avoids diverse processes such as particles drying, storage, handling, and dispersion; so, the agglomeration of nanoparticles is minimized; therefore, stability of nanofluids is improved [89, 90]. In most experimental studies, nanofluids are synthesized in a two-step process [3, 6, 56, 92–96], which is a classic synthesis method of nanofluids. Various nanostructures such as nanofibers, nanotubes, nanosheets, among other nanomaterials used in this technique are initially produced by mechanical comminuting, chemical reaction, inner gas condensation or decomposition of organic complex [97–99] and finally obtained as dry powder. Then, it is followed by the second step in which the as-produced nanostructures are homogeneously dispersed into base conventional fluids through mechanical agitation (stirring) or ultrasonication [6, 97–102]. Furthermore, this process is an economic method to produce nanofluids at large scale, since nanostructures synthesis techniques are readily scaled up to mass production levels. To obtain a good stability and homogeneous dispersion of nanostructures within a fluid, sonication process is used to speed dissolution by breaking intermolecular interactions. The main disadvantage of this method is that due to the high surface area and surface attractively, the nanostructures tend to agglomerate. The nanostructures agglomeration in the fluid results in decreasing the thermal conductivity performance and increasing the settlement and clogging of microchannels. Therefore, to reduce these effects, surfactants or additives are widely used to stabilize nanostructures within the fluids.

3.1. Nanofluids: Variables and features

The concept and strategies of this work have a significant departure from diverse investigations in thermal management applications. Even though there have been several investigations on HTFs with nanoparticles reinforcement, there is still a room for great opportunities to continue developments and understand the implications and effects of this technology. Diverse challenges regarding nanoparticles effect on thermal transport and energetic performance as well as nanofluids industrialization have been studied (**Figure 2**). The heat transfer enhancement in nanofluids has been attributed to many variables, which are presented in the following section.

3.1.1. Brownian motion

Researchers have found that Brownian motion, which is the random movement of particles (**Figure 3**), is a key mechanism for the anomalous increase in the heat transport of nanofluids [55, 78, 103–108]. Brownian motion tends to move the particles from higher concentration areas to the lower concentration areas. Research on fillers motion caused by temperature gradient was studied by Koo and Kleinstreuer [103]; it was shown that the Brownian motion has more impact on the thermal properties of nanofluids than to the effects of a temperature gradient. Aminfar and Motallebzadeh [108] investigated the concentration distribution and velocity field of nanoparticles on water/Al_2O_3 nanofluid in a pipe. It was observed that the Brownian forces have most impact on the nanoparticles filler fraction distribution and the velocity field when compared to other forces such as thermophoretic and gravitational forces. Brownian motion only exists when the particles within the fluid are extremely small and, as the size of the particles gets larger, Brownian motion effects are reduced.

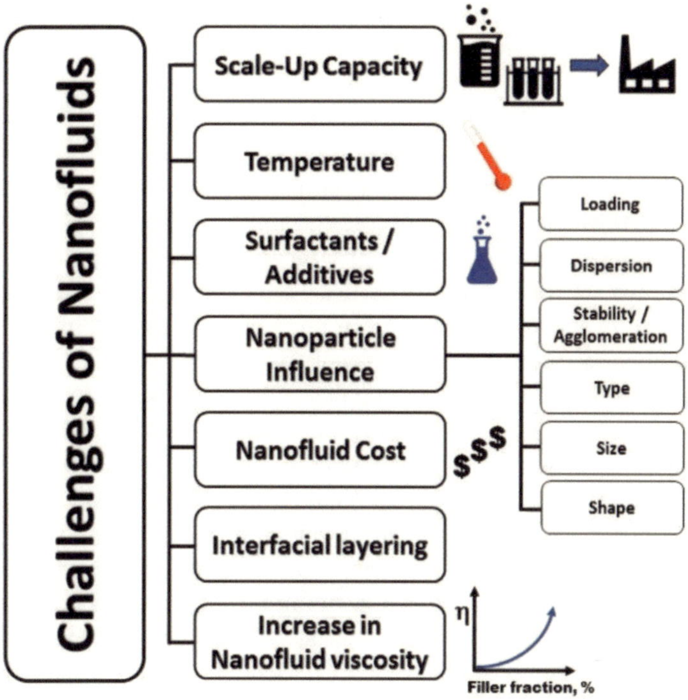

Figure 2. Common challenges of nanofluid developments.

3.1.2. Nanofiller size

Diverse studies have found that as nanostructures are reduced in size, the effective thermal conductivity of the nanofluid increases [13, 109–114]. As the nanoparticle size is reduced, Brownian motion is induced. Also, lighter and smaller nanoparticles are better at resisting sedimentation, one of the biggest technical challenges in experimenting with nanofluids [77]. Li and Peterson [109] investigated the thermal properties of Al_2O_3/DiW nanofluids with particle sizes of 36 and 47 nm at various filler fractions. The nanofluid with 36 nm particles improved the effective thermal conductivity enhancement at ~35°C, as varying the filler fraction from 0.5 vol.% up to 6.0 vol.% (~7% to ~28%, respectively). Agarwal et al. [115] presented the impact of Al_2O_3 size of nanoparticles on thermal conductivity and the dynamic viscosity of kerosene-based nanofluid. Thermal conductivity increases from 1.3% to 9.3% over a particle filler fraction of 0.01 vol.% to 0.10 vol.%. Thermal conductivity performance and viscosity

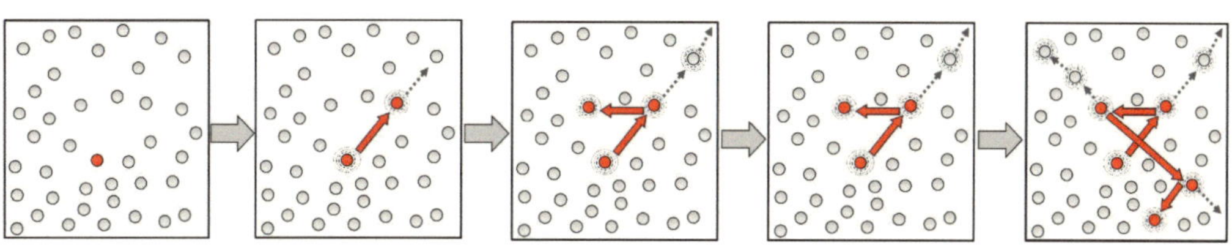

Figure 3. Schematic of Brownian motion of nanoparticles.

were found to be higher for smaller size nanoparticles compared to larger size. At 0.50 vol.%, nanofluid with 13 nm particle size displays 22% improvement in the effective thermal conductivity, compared to only 17% for 50 nm. Beck et al. [116] investigated Al_2O_3/water nanofluids with diverse particle sizes ranging from 8 nm to ~300 nm in diameter; the thermal conductivity enhancement decreases as the particle size decreases below ~50 nm. This behavior was attributed to nanoparticles thermal conductivity, as the particle size becomes small enough to be affected by increased phonon scattering [116]. Teng et al. [112] studied on the effect of particle size, temperature, and weight fraction on the thermal conductivity ratio of Al_2O_3/water nanofluid with filler fraction up to 2.0 wt.% and different Al_2O_3 particle nominal diameters. The results showed a dependence relationship between high thermal conductivity ratios and enhanced sensitivity, and small nanoparticle size and higher temperature. Nguyen et al. [117] studied the heat transfer enhancement and behavior of Al_2O_3/water nanofluid for microprocessors/electronic purposes. It is found that with smaller nanoparticles (36 nm in diameter), nanofluids showed higher convective heat transfer coefficients than with larger ones (47 nm in diameter). From Nguyen's research, thermal transport increased 40% at 6.8 vol.% filler fraction, as compared to water. He et al. [118] studied the heat transfer behavior of TiO_2 water nanofluids with diameters of 95 nm, 145 nm and 210 nm at various filler fractions. For the 95 nm particle size nanofluid, the thermal conductivity showed an increase from 1% to ~5% at 1.0 wt.% and 4.9 wt.%, respectively, compared to water; as filler fraction increased, the thermal conductivity increased as well. It was shown that the effective thermal conductivity decreases as particle size increases. On the other research, Żyła et al. [119] studied the effects of TiN nanoparticles within EG at two different particles size, 20 and 50 nm. It was concluded that the increase in nanoparticle filler fraction led to similar trends for nanofluids properties. At same content, thermal conductivity and surface tension of nanofluids are higher with smaller nanoparticles. Esfahani et al. [120] observed that thermal conductivity of graphene oxide (GO) NFs depends on both particle-size distribution of GO and viscosity of GO NFs. In their work, GO NFs showed enhanced thermal conductivity performance as compared to base water.

There have been a few reports on SiC, CeO_2, and SiO_2 nanoparticles reinforcing conventional fluids that stated a decrease of the effective thermal conductivity with increase in particle size [47, 121–132]. Beck et al. performed studies on water-based ceria (CeO_2) nanofluids [123], which showed an increase in the effective thermal conductivity with an increase of nanoparticle size, although only two particle sizes were studied (12 nm and 74 nm). Silica-ethanol nanofluid was investigated by Hossein Karimi Darvanjooghi and Nasr Esfahany [132] over various particle sizes, temperatures and filler fractions. It was observed that with the increase in particle size and temperature, the thermal conductivity performance of silica–ethanol nanofluid increases for all the investigated filler fractions. 2D nanostructures have also been recently studied; Taha-Tijerina et al. developed a novel nanofluid by adding only 0.10 wt.% of exfoliated hexagonal boron nitride (h-BN) 2D-nanosheets within mineral oil (MO) [30]. Enhancement in thermal conductivity for 0.10 wt.% h-BN/MO nanofluid was more than 75% in comparison with MO. Similarly, Mehrali et al. [133] investigated the effects on thermal transport of graphene-based nanofluids. It was found that at 4.0 vol.%, a significant thermal conductivity enhancement of ~45% was obtained.

The nanostructure size effect is apparently due to interlaying mechanism, van der Waals forces and the micromotions of nanoparticles are stronger when containing lower-sized nanostructures; and the interactions between nanoparticles and fluids are more severe, which results in a stronger energy transmission and a higher thermal transport process in nanofluids. The smaller the particle size, the higher will be the improvement due to the high surface area observed. This trend is observed by most researchers and is supported by two mechanisms: Brownian motion and liquid layering around nanostructures. Hence, nanofillers size is a determinant variable for heat transfer nanofluids, since, as previously stated, its smaller size reduces or avoids critical issues of larger fillers.

3.1.3. Particle shape/surface area

Particle geometry is an important parameter to be considered since it has critical effects on nanofluids performance. There have been several efforts to understand the effects of nanoparticles shape regarding the thermal transport phenomena. It has been observed by several studies that rod-shaped nanoparticles, such as CNTs, remove more heat than spherical nanoparticles [134–139]; this is most probably due to rod-shaped particles' larger aspect ratio (ratio between a particle's surface area to volume) than spherical nanoparticles. For instance, Elias et al. [134] studied various boehmite alumina (γ-AlOOH) nanoparticle shapes (spherical, cylindrical, blades, bricks and platelets) dispersed within EG/water mixtures. The best performance was achieved by the cylindrical-shaped nanoreinforced fluid, followed by bricks, blades, platelets and spherical-shaped nanoparticles, respectively. Thermal conductivity enhancement of cylindrical shape nanoparticles is observed to be ~3% higher than the spherical shape at 1.0 vol.% concentration. On similar study, Kim et al. [140] investigated the effects of particle shape on suspension stability and thermal transport performance of water-based boehmite alumina nanofluids. The thermal conductivities of nanofluids with blade, brick and platelet shaped particles are maximally enhanced up to 16, 28, and 23% at 7.0 vol.%, respectively.

From various studies, it can be concluded that the rod-shaped nanostructures possess higher thermal conductivity performance compared to spherical shape nanoparticles due to larger surface area and rapid heat transport along relatively long distances due to the greater length, usually of the order of micrometers. However, there is some contradiction observed based on studies performed by Xie et al. [141], where thermal conductivity enhancement using spherical (26 nm average) and cylindrical (600 nm average)-shaped SiC nanoparticles suspended within water were evaluated. An enhancement of ~23% at 4.0 vol.% for cylindrical particles was observed, but only 16% increase at 4.2 vol.% for spherical particles was reported. Results showed higher enhancement at larger particle size. These contradictions with other literature could be explained by the severe clustering of nanoparticles having smaller particle size. Although at a certain level clustering may enhance the thermal conductivity, excessive clustering may create an opposite effect, resulting in the sedimentation of nanoparticles [142]. Sudarsana Reddy and Chamkha[139] studied the effects of shape on nanoparticles on natural convection magnetohydrodynamic (MHD) fluids. Results reveal that significant heat transfer enhancement is noticed as the size of nanoparticles decreases. Moreover, the type of the nanoparticles and the type of the base fluid (water/kerosene) also influenced the natural convection heat transfer. In another study, Aaiza et al. [143] investigated energy transport in MHD nanofluids with different nanoparticles shapes such as cylinders, platelets, blades, and bricks. It was observed that elongated structures such as cylinders and platelets result in higher viscosity at the same filler fraction due to structural limitation of rotational and transitional Brownian motion, which resulted in effects on thermal conductivity.

3.1.4. Filler fraction

One of the most extensively considered factor on thermal transport performance, the key variable for nanofluids' improvement, is the nanostructures concentration dispersed within conventional fluids. Filler concentration has been stated by weight and volume percentages in research, reports, papers and patents. Effective thermal conductivity (k_{eff}) among other properties of nanofluids improve with increasing nanoparticles filler fraction [144, 145]. Nevertheless, as the nanoparticle concentration increases, it may no longer be valid to assume well-suspended nanostructures, due to particles agglomeration, sedimentation, stability and increase in viscosity, which could cause other problems such as possible abrasion and clogging of microchannels. It has also been observed that pressure drop increases in diverse conventional fluids as filler concentration is increased [10]. This is why it is more effective to use a very small filler fraction in nanofluids [3, 146–151]. At low filler fractions, nanostructures have more intense Brownian motion at higher temperatures, which can significantly enhance the effective thermal conductivity. But at high volume fractions, nanoparticles have high potential to be agglomerated at high temperatures. Higher concentration of particle shows less stability which leads to the agglomeration process due to increase in number of molecules within the fluid. This causes an increase in weight which cannot be maintained in the suspension by Brownian agitation, and settle out of the suspension [152].

Pang et al. [153] studied the effect of Al_2O_3 and SiO_2 nanoparticles dispersed in methanol at various concentrations, such as 0.01, 0.10 and 0.50 vol.%. Effective thermal conductivity increases with an increase of the nanoparticles volume fraction; for Al_2O_3, the increments were 1, 5 and 11%, respectively, compared to pure fluid and, for SiO_2, the increments were 6, 11 and 16%, respectively, as compared to pure fluid, as well. Arulprakasajothi et al. [151] investigated TiO_2 concentrations of 0.1, 0.25, 0.5 and 0.75% using two step method. It was observed that as concentration increases, the surface area of particle also increases and exchange more heat. The effective thermal conductivity for nanofluid concentrations was increased from 1 to 6%, respectively. On research conducted by Wang et al. [147], 2D-graphene structures with average particle size were 0.5–2.0 μm and thickness of 0.8–1.2 nm. Graphene was dispersed uniformly into base oil without any surfactant by ultrasonic oscillation. Graphene/oil nanofluids' concentration was 0.02, 0.05, 0.1, and 0.2 mg/ml. Thermal conductivity for all concentrations was raised 4, 8, 17 and 25%, as compared to pure oil, respectively. On this same path, according to Taha-Tijerina et al., the superb thermal transport performance of 2D-based nanofluid was observed, in which nanosheets of h-BN within MO showed improvements of ~10% and ~80% at 0.01 wt.% and 0.1 wt.%, respectively, without significant increase of kinematic viscosity [3, 145, 154]. Tiwari et al. [47] investigated CeO_2/water nanofluids and its effects of filler fraction, ranging from 0.5 to 3.0 vol.%, and temperature. The experimental results indicate that the convective heat transfer coefficient increases with increase in nanoparticle filler fraction (up to a threshold-optimum value). It was observed that the increase in particle concentration also increases the fluid viscosity, which should result in an increase in the boundary layer thickness, which overcomes the convective heat transfer coefficient as well. However, significant improvements in thermal conductivity are shown. For instance, at 40°C, the thermal conductivity improvements were significant: 7, 11, 13, 16 and 21% at 0.5, 1.0, 1.5, 2.0 and 3.0 vol.%, respectively. Research by Paul et al. [155] demonstrated that the thermal conductivity of water-based SiC nanofluids could be improved by 12% at only 0.1 vol.%. Studies on mixtures of water and EG-based SiC nanofluids were performed by Timofeeva et al. [121, 156], where nanofluids displayed 1.5–20% thermal conductivity enhancement at different filler fraction and nanoparticle sizes. Ferrouillat et al. [157] estimated that heat transfer of SiO_2/water nanofluid in the particle ranges from 5 wt.% to 34 wt.%, and found an improvement of 10–60% compared to pure water. Lee et al. [158] investigated the thermal conductivity of DI-water-based SiC nanofluids; a ~7% improvement was observed, compared to pure DI-water. Li and Zou [159] prepared homogeneous and stable nanofluids by dispersing SiC nanoparticles within mixtures of ethylene glycol and water. It was observed that thermal conductivity of water/EG-based SiC nanofluids increased with SiC concentrations. Improvements of ~34% at 1.0 vol.% of SiC were achieved. Pang et al. [153] studied the effect of SiO_2 nanoparticles within methanol at various concentrations (0.01, 0.10 and 0.50 vol.%). Thermal conductivity enhancements were 6, 11 and ~16%, respectively, as compared to pure fluid.

3.1.5. Stability/particles agglomeration

A key challenge with nanofluids is that nanoparticles tend to agglomerate due to molecular interactions, such as van der Waals forces [122, 160]. The agglomeration of nanoparticles results not only in the settlement and clogging of microchannels, but also causes the effective surface area to volume ratio to decrease, which impacts the thermal conductivity performance of nanofluids. Nanoparticles agglomeration increases as filler fraction increases, due to closer particles and higher Van der Waals attraction. Similarly, this issue generates other problems such as viscosity increments (**Figure 4**).

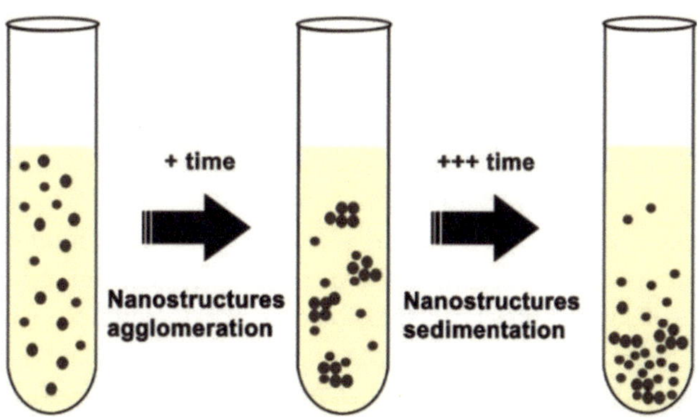

Figure 4. Scheme of nanoparticles sedimentation over time.

The stability of nanofluids is considered one of the critical issues, which should be addressed before any application [48, 145]. Synthesis of nanoparticles and preparation methods of nanofluids play an important role on stability, which effects are observed on nanofluids' thermal transport characteristics. In addition to filler fraction and working temperature, pH has an important role in stability of nanofluids. For instance, Nikkhah et al. observed that by controlling the pH value, the stability of CuO/water nanofluids (50 nm in diameter) can be increased and thus the thermal conductivity performance [161]. Due to nanostructures interlayer adhesion forces, nanoparticles become agglomerated and their settlement can be observed due to gravity forces. Nanostructures sedimentation overcomes one of the major drawbacks of suspensions; nanoparticle aggregates promote settling of particles; hence, the dispersion stability may decay with time. To increase the stability of nanofluids, diverse techniques have been employed, such as extended ultrasonication [162–164]. Ultrasonic vibration is a possible way to break-up cluster formation of nanoparticles and help to scatter the nanostructures within base fluids, so that ultrasonication processes were widely used for nanofluid preparation. Furthermore, to enhance the stability of nanofluids, surfactants or additives are used; nevertheless, these have impacts on thermal characteristics and there could be certain drawbacks by using them.

Timofeeva et al. [122, 165] studied the thermal conductivity and viscosity of Al_2O_3 nanoparticles dispersed in water and EG. It was observed that the main parameters for controlling nanofluids' thermal conductivity enhancement are the geometry, agglomeration state and surface resistance of nanoparticles. Karthikeyan et al. [144] identified that CuO nanoparticles and clusters size have a significant influence on thermal conductivity of water and EG. It was also found that nanoparticles agglomeration is time-dependent; as time elapsed, agglomeration increased, which decreased the thermal conductivity performance. However, some reports show aggregation in water-based Al_2O_3 nanofluids significantly increases the thermal conductivity of the fluid [166], such as research by Shima et al. [167], where an increase in thermal conductivity with particle sizes of average diameters of 2.8–9.5 nm was observed. For 5.5 vol.%, the improvement was 5% and 25%, for 2.8 nm and 9.5 nm, respectively. According to their studies, interfacial resistance, nature, and aspect ratio of agglomerates dictate heat conduction enhancement in nanofluids. Yu et al. observed that stable nanofluids could be able to withstand or maintain no significant variation in thermal conductivity with time. This was observed for EG-based ZnO nanofluids [168] and kerosene-based Fe_3O_4 nanofluids [169].

3.1.6. Surfactants/additives

Surfactants have been widely used to stabilize the nanofillers within conventional fluids, even though these surfactants may affect the nanofluids performance; since surfactants thermal conductivities are generally lower than the base fluids *per se*, the addition ratios of surfactants are generally extremely low to prevent from reducing the thermal conductivity or increasing the viscosity of nanofluids. Surfactants could also introduce defects at the molecular interfaces [170]. The use of surfactants and dispersion agents has shown to be effective providing repulsion between nanoparticles and reducing agglomeration [6, 95, 171]. Additives are also incorporated to materials to enhance their mechanical properties. Nevertheless, the functionality of the surfactants under high temperature is also a big concern, especially for high-temperature applications [39].

Utomo et al. [172] concluded that surfactants in high-loading ratios could reduce the effective thermal conductivity performance of water-based Al_2O_3 and TiO_2 nanofluids. Khairul et al. [173] studied the effects of Al_2O_3 and CuO nanoparticles, as well as sodium dodecyl benzene sulfonate (SDBS) surfactant on viscosity and thermal conductivity of water-based nanofluids. It is observed that increasing the SDBS concentration, thermal conductivity for both systems tend to rapidly decrease, which is attributed to the increased nanostructures aggregation. Murshed et al. [135] observed that low concentration (≤0.02 vol.%) of oleic acid (OA) or a cationic surfactant, cetyl trimethyl ammonium bromide (CTAB), as dispersants could greatly improve the dispersion stability of TiO_2 nanofluids without reducing the thermal conductivity of TiO_2 nanofluids. From other studies, non-ionic surfactants were found to strongly interact with graphite surfaces in case of CNTs stabilization within aqueous suspensions [174]. Quite a few results indicated that surfactants played positive roles in the thermal conductivity. Saleh et al. [175] found that each of the three kinds of surfactants: CTAB, anionic surfactant, sodium dodecyl sulfate (SDS) and nonionic surfactant sorbitan monooleate (Span80) could greatly improve the dispersion behavior and thermal conduction performance of TiO_2 nanofluids. Chen et al. [176] investigated the effects of SiC on saline water, for solar distillation systems. Nanofluids with 0.4 vol.% of SiC were dispersed within saline water, and additionally, polyvinyl pyrrolidone (PVP) dispersant was used (0.02 wt.%) to keep nanoparticles homogeneously dispersed. It was observed that thermal conductivity of seawater/SiC nanofluids improved ~5% compared to pure seawater, which confirms the feasibility of nanofluids application in solar desalination system. Therefore, it can be concluded that the right amount of surfactant can play positive roles in both dispersion and heat conduction performance nanofluids.

3.1.7. *Viscosity*

Among diverse nanofluid properties, viscosity is a paramount parameter. Viscosity describes a fluid's internal resistance to flow. Many parameters affect the nanofluid's viscosity, including the preparation method, base fluid type, operating temperature, nanostructure size and geometry, filler fraction, acidity (pH value), shear rate, usage of additives or surfactants, and particle aggregation and sedimentation [3, 93, 177–181]. It has been demonstrated that the viscosity of nanofluids increases with the nanoparticle volume fraction. This property is troublesome due to lack of understanding of viscosity mechanisms and lack of general mathematical models to predict the viscosity behavior in nanofluids. Nguyen et al. [177] investigated the nanostructures size effect for Al_2O_3 aqueous-based nanofluids and observed that particle size effects on viscosity are more significant for high filler concentrations. Yiamsawas et al. [178] measured the viscosity of Al_2O_3/water nanofluids at high filler concentrations and high temperatures. The filler fraction ranges varied from 1.0 vol.% to 8.0 vol.%, while the temperature evaluation varied between 15°C and 60°C. It was observed that the viscosity decreases with a temperature increase. Nanofluids prepared in higher viscosity base fluids exhibit more enhancement compared to low viscosity base fluids.

Li et al. [179] investigated EG-based nanofluids containing ZnO nanostructures at different concentrations ranging from 1.75 wt.% and 10.5 wt.%. Results showed that viscosity increases with increasing the concentration of ZnO nanoparticles and decreases with temperature. Attari et al. [180] explored the effects of adding TiO_2, Fe_2O_3 and ZnO nanoparticles to crude oil at different filler fractions, ranging from 0.5 to 2.0 wt.% at different temperatures. It was observed that with the increase in nanoparticle loading, the relative viscosity of ZnO-crude oil nanofluid increases. At the mass fraction lower than 1.0 wt.%, the relative viscosity of nanofluid decreases slightly with the increase in temperature and the main factor which can influence the relative viscosity is nanoparticle type. On research by Jeong et al. [181], the viscosity behavior of water-based ZnO nanofluids with two nanoparticle shapes and semi-rectangular and spherical at various filler fractions ranging from 0.05 to 5.0 vol.% was investigated. Their results indicated that the viscosity increased from 5.3 to ~70% with increase in the filler concentrations. Moreover, the enhancement of the viscosity of the nearly rectangular shape nanoparticles was found to be more than 7%, rather than the spherical nanoparticles. Williams et al. [182] studied ZrO_2-water nanofluid for 60 nm particle size, at small filler fractions (0.2–0.9 vol.%) and found a 54% increase in viscosity, when compared to pure water.

3.1.8. Temperature dependence

Thermophoresis or thermodiffusion is an interesting consequence of the Brownian motion of the nanostructures. High energy molecules in a warmer region of a liquid migrate in the direction opposite the temperature gradient to cooler regions; small particles tend to disperse faster in hotter regions and slower in colder regions. Thermophoresis and Brownian motion effects are the mass transfer mechanisms which also influence the convective heat transport performance of nanofluids [78–81, 183–189]. As Michaelides [189] explained, interparticle collisions in the colder regions where the nanostructure concentrations are higher, partly hinder this accumulation and a dynamic equilibrium for nanoparticle concentration is established, with lower concentrations in the hotter regions and higher concentrations in the colder regions. **Figure 5** schematically depicts the differential dispersion and the resulting thermophoresis which shows the effects of the magnitude of the molecular collisions on small particles.

Diverse theoretical and experimental investigations have been developed. For instance, Wang et al. [187] observed the effects of temperature-dependent properties on natural convection

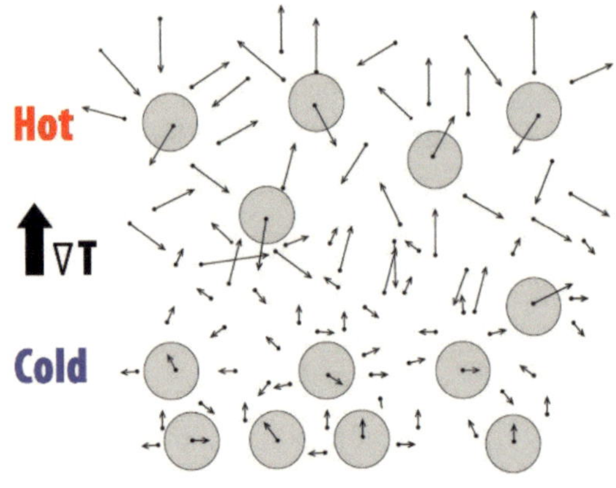

Figure 5. Thermophoretic motion of particles. The Brownian motion brings more particles to the colder region of the system.

of Al_2O_3 nanofluids; lattice Boltzmann (LB) method was considered for their study. It was concluded that for a given nanoparticle filler fraction, the value of the heat transfer enhancement is increased as the temperature increases, and the nanoparticle diameter decreases. On the other hand, Das et al. [78], similarly to Lee et al. [190], observed that thermal conductivity performance of Al_2O_3 and CuO nanofluids has a temperature-dependent influence; they posed motion of reinforced fillers as an important factor for that. Jyothirmayee Aravind and Ramaprabhu [184] observed a temperature dependence on graphene nanosheets reinforcing EG and DiW. It was observed that thermal conductivity performance increases with increasing filler fraction and operating temperature. The thermal conductivity of the base fluids did not show significant improvements as the temperature increases, similar tendency as reported by Jha and Ramaprabhu [191]. An enhancement in thermal conductivity of ~2.4% is observed at 25°C with a very low filler fraction of 0.008 vol.% of the graphene/EG nanofluid; meanwhile, at 50°C, the increment was ~17%. At 0.14 vol.%, the thermal conductivity improvement was 6.5 and 36%, at 25 and 50°C, respectively. On the research conducted by Wen and Ding [192], it was observed that thermal conductivity increases with increasing temperature on the system, showing a nonlinear dependence after temperatures above 30°C.

The influence of temperature on thermal conductivity on 0.5 vol.% and 1.0 vol.% CeO_2-EG nanofluids was studied by Rajan et al. [164]; it was observed that an enhancement in thermal conductivity increases as filler fraction increases. Thermal conductivity enhancements were 5 and 10% for the filler fraction studied. Also, by increasing temperature, the thermal conductivity ratio decreases for both nanofluid concentrations. Li et al. [179] investigated EG-based nanofluids containing ZnO nanoparticles at different filler fractions (1.75–10.5 wt.%. The thermal conductivity increases with increasing the temperature ranging from 15 to 55°C. The thermal conductivity gradually increases with mass fraction and temperature, and it was observed that the growth rate decreases at 15°C in the range of 8.75–10.5 wt.%. It was concluded that thermal conductivity performance depends on filler fraction, and it increases nonlinearly with the mass fraction of nanoparticles. Kandasamy et al. [107] observed that the combined effect of thermophoresis and Brownian motion play a very dominant role on heat transfer in the presence of thermal stratification, mainly due to the nanoparticles geometry and size. Although thermophoresis effect is important in heat transport, there are other characteristics and parameters that may have effects on nanofluids and should be addressed. These effects include the increase in nanofluids viscosity due to the presence of nanoparticles and fluid density variation due to variable volume fraction.

3.1.9. Interfacial layering on the liquid-nanostructure interface

Interfaces are ideal templates for assembling nanoparticles into structures by the nature of the interfaces. Interfacial layering or nanolayer refers to a phenomenon at the liquid-particle interface where liquid molecules are more ordered than those in the conventional fluids; therefore, the interface effect could enhance the thermal conductivity by the layering of the liquid at the solid interface (given that crystalline solids possess much better thermal transport that liquids) [193–195], by which the atomic structure of the liquid layer is significantly more ordered than that of the conventional liquid. At the interfaces, the nanostructures are mobile, and defects could be eliminated [196]. Nanoparticles suspended in base fluids form clusters that create a low thermal resistance path which can enhance the thermal conductivity, according to Saterlie et al. [197]. The ordered structure could have higher thermal conductivity than that of the conventional, therefore an enhancement of the effective thermal conductivity. Various researchers have suggested that there is liquid layering on the nanoparticles, which helps to enhance the heat transfer properties of nanofluids [197–201]. Yu et al. [200] proved the formation of layers by the liquid molecules close to a solid surface, even though the thickness and thermal conductivity of the nanolayers are not well known yet. According to Elis Josna Mary et al. [164], a temperature-dependent, linear variation in thermal conductivity increase with filler fraction was observed, which could be attributed to liquid layering.

4. Nanofluids application fields

Diverse studies on nanofluids have been carried out by many researchers. This section deals with literature review on nanofluids; nanofluids preparation and characterization, thermo-physical properties, as well as nanofluids applications, which lays foundation and basis for further investigations.

4.1. Thermal performance of nanofluids

Heat transfer is classified into various mechanisms, such as thermal conduction, thermal convection and thermal radiation. In diverse fields, thermal transport is a critical parameter to obtain efficient performance of machinery and devices. Heat convection occurs when bulk flow of a fluid (liquid or gas) carries heat along with the flow of matter in the fluid, this process could be "forced," where fluid motion is generated by an external source such as a pump, fan or other mechanical means, or "natural," by density differences in the fluid occurring due to temperature gradients. Radiation heat transfer is the transfer of energy by means of photons in electromagnetic waves in much the same way as electromagnetic light waves transfer light. On the other hand, heat conduction is the direct microscopic exchange of kinetic energy of particles through the boundary between two systems. When an object is at a different temperature from another body or its surroundings, heat flows so that the body and the surroundings reach the same temperature, at which point they are in thermal equilibrium. The thermal conductivity (k) of liquids can be successfully measured if the time taken to measure k is very small so that the convection current does not develop [202].

Diverse techniques have been proposed to measure nanofluids thermal conductivity over the past years. The most common techniques to measure the effective thermal conductivity of nanofluids are the transient hot-wire method [3, 147, 202–207], steady-state method [109, 208–210], cylindrical cell method [211], temperature oscillation method [183, 212, 213], and 3-ω method [40, 214–216] to name some. Eastman et al. reported a 40% enhancement with only 0.40 vol.% of copper oxide (CuO) particles [217], while Choi et al. reported a remarkable 160% increase in thermal conductivity of MWCNTs/engine oil nanofluid at 1.0 vol.% filler fraction of nanotubes [218]. Ilyas et al. [58] study a commercial brand oil (THO) with incorporation of MWCNTs where significant thermal conductivity improvement of ~22 and ~30% was achieved at 35 and 60°C, respectively, at 1.0 wt.%. Hwang et al. [41] investigated the thermal conductivity of

DiW- and EG-based nanofluids reinforced with MWCNTs, CuO and SiO_2. It was observed that thermal conductivity was improved almost linearly as filler fraction increased. For DiW-based systems, the addition of SiO_2, CuO and MWCNTs at 1.0 vol.% filler fraction showed an increase of 3, 5 and ~12%, respectively. Also, CuO/EG nanofluid at 1.0 vol.% showed an increase of ~9%. Wen and Ding et al. [192] also investigated the effects of MWCNTs within DiW, with addition of 0.25 wt.% gum Arabic (GA) dispersant with respect to DiW. For MWCNTs at 0.50 wt.% and 1.0 wt.%, an increase in thermal conductivity enhancement was achieved up to ~30% and ~38% at 25°C, and ~35% and ~80% at 30°C, respectively. It was found that these improvements were slightly higher than the results reported by Liu et al. [14], Assael et al. [219], Wen et al. [220], and Xie et al. [221], but lower than the results showed by Choi et al. [218]. There are diverse factors that cause these discrepancies among the different groups; as mentioned by Sing [55], these discrepancies should rely on the dependency of thermal conductivity is on diverse important factors such as the structure and properties of the CNTs, their aspect ratio, clustering, addition of dispersants, temperature and the experimental errors involved as well.

Nanodiamonds (NDs) dispersed in EG and mineral oil (with addition of OA) were studied by Branson et al. [222]. It was observed that addition of 0.88 vol.% of NDs enhanced the thermal conductivity by ~12%. In MO, for instance, an enhancement of ~6% and ~11% is achieved at NDs loading of 1.0 vol.% and 1.9 vol.%, respectively. According to Branson et al., the differences on enhancement efficiencies are attributable to divergence in thermal boundary resistance at nanoparticle/surfactant interfaces [222]. Research by Khairul et al. [173] on the effects of Al_2O_3 and CuO nanoparticles filler fraction and use of SDBS surfactant on viscosity and thermal conductivity of water-based nanofluids was performed. It was observed that thermal conductivity of the nanofluids increased nonlinearly with increasing nanoparticles filler fraction, with a maximum increase of 10 and 14% for Al_2O_3 and CuO, respectively, at 0.15 wt.%, similar to what Kong et al. found for the maximal enhancement for Al_2O_3 nanofluids [223]. On the field of oxide nanostructures, Yiamsawasd et al. [224] reported a maximum thermal conductivity enhancement of 20% for TiO_2/water nanofluid. Elis Josna Mary et al. [164] investigated CeO_2/EG nanofluid and observed a temperature effect on thermal conductivity rise of 17% and ~11% at 10 and 30°C, respectively. Serebryakova et al. [225] investigated the effects of dispersing Al_2O_3 within EG/water mixtures. It was observed that thermal conductivity performance was improved by 5% at 1.5 vol.%. Mariano et al. [226] estimated thermal conductivity behavior and rheological properties of Co_3O_4/EG nanofluids and obtained thermal conductivity enhancement of 27% at 5.0 wt.%. Aluminum nitrides (AlN) can also find many applications in the heat exchange process. Thermal conductivity performance of AlN/ethanol nanofluid was investigated by Hu et al. [227]. Results showed a 20% increase in the thermal conductivity of ethanol with 4.0 vol.% at room temperature. Furthermore, a strong temperature dependence of the thermal conductivity was observed in this research. Yu et al. [228] thermal conductivity of AlN dispersed in two different conventional fluids, such as EG and propylene glycol (PG), was investigated. It was found a 39 and 40% thermal conductivity improvement for EG and PG, respectively, having the same particle size and nanoparticles filler fraction.

A great improvement on 2D-nanostructure-based nanofluids was obtained by Taha-Tijerina et al. [3], where exfoliated h-BN and graphene were homogeneously dispersed within mineral oil with superb thermal conductivity increase up to ~80% at very low filler fractions (<0.10 wt.%)

with no significant increase in viscosity. Continuing with 2D structures, several research studies have developed graphene-based nanofluids with high nanoparticle stability and significant enhancements [229–235]. Shaikh et al. studied the effect of exfoliated graphite dispersed within PAO oil at various concentrations, ranging from 0.10 vol.% up to 1.0 vol.%. It was observed that addition of 2D-structures improved the thermal conductivity from 18% up to ~130%, respectively [230]. Hadadian et al. [236] prepared highly stable graphene oxide (GO)-based nanofluid. Thermal transport of EG increased by 30% with 0.07 GO mass fraction. Other EG-based nanofluids synthesized by Yu et al. [237, 238] have shown better enhancements of 61 and 86% with GO [237] and graphene nanosheets [238], respectively, at 5.0 vol.% loading.

Diverse theories explain the mechanisms that could affect the behavior of nanofluids; the most accepted being Brownian motion [40, 104, 239, 240], percolation theory [55, 104, 198, 241, 242], micro convection cell model [104, 198, 239–242], and liquid layering theory [55, 104, 193, 198, 242, 243]. **Table 2** shows the influence of oil-based nanofluids on thermal conductivity.

Filler	Type of oil	Nanoparticles morphology	Filler fraction	TC enhancement	Ref.
Al_2O_3	Engine oil	Spherical ~80 nm diam.	0.5 vol.%	~9%	[194]
			1.0 vol.%	~12%	
Al_2O_3	Engine oil	Spherical ~28 nm diam.	5.0 vol.%	~26%	[209]
			7.5 vol.%	~30%	
Al	Engine oil	Spherical ~80 nm diam.	1.0 vol.%	~20%	[194]
			3.0 vol.%	~37%	
AlN	Mineral oil	Spherical ~50 nm diam.	0.05 vol.%	~8%	[244]
CeO_2	Transformer oil @ 50C	Spherical ~3–7 nm diam.	0.7 vol.%	~15%	[203]
CuO	Mineral oil	Spherical ~100 nm diam.	2.5 vol.%	~12%	[245]
			5.0 vol.%	~23%	
Diamond	Mineral oil	Spherical ~<10 nm diam.	1.0 vol.%	~5%	[222]
			1.9 vol.%	~11%	
Graphene	Mineral oil (50°C)	2D-sheets ~500 by 500 nm	0.01 wt.%	~10%	[3]
		~8–10 atomic layer thick	0.10 wt.%	~80%	
Graphene	Heat-transfer oil	2D sheets ~0.5–2.0 μm	0.05 wt.%	8%	[147]
		~0.8–1.2 nm thick	0.10 wt.%	17%	
			0.20 wt.%	25%	
h-BN	Mineral oil (50°C)	2D sheets ~500 by 500 nm	0.01 wt.%	~9%	[3]
		~5 atomic layers thick	0.05 wt.%	~10%	
			0.10 wt.%	~80%	
h-BN	Synthetic fluid	2D sheets ~500 by 500 nm	0.10 wt.%	8%	[6]
		~5 atomic layers thick			

Filler	Type of oil	Nanoparticles morphology	Filler fraction	TC enhancement	Ref.
MWCNT	Engine oil (15 W-40)	Rods ~length: 0.3–10 μm	0.25 vol.%	~10%	[18]
		Diameter: 10–50 nm	0.5 vol.%	~17%	
			1.0 vol.%	~45%	
MWCNT	Mineral oil	Rods ~length: 10–50 μm	0.5 vol.%	~8.5%	[92]
		Diameter: 10–30 nm			
MWCNT	Synthetic PAO oil	Rods ~length: 50 μm	1.0 vol.%	160%	[218]
		Diameter: 25 nm			
MWCNT	Poly-α-olefin (PAO)	Rods ~length: 1–100 μm	1.0 vol.%	~175%	[19]
		Diameter: 20–300 nm			

Notes: if not specified, measurements were conducted at room temperature.

Table 2. Influence of oil-based nanofluids in thermal management.

Similarly, **Table 3** shows the results from diverse investigations on other water-based nanofluids; and various materials and sizes used as reinforced nanoparticles. **Table 4** shows the influence of various nanofluids in thermal management properties as well.

Filler	Nanoparticles morphology	Filler fraction	TC enhancement	Ref.
Al_2O_3	Spherical ~60 nm diam.	5.0 vol.%	~20%	[141]
Al_2O_3	Spherical ~131 nm diam.	1.0 vol.%[1]	~11%	[246]
		4.0 vol.%	~10%	
		4.0 vol.%[1]	~25%	
Al_2O_3	Spherical ~27–56 nm diam.	1.6 vol.%	~10%	[247]
Al_2O_3	Brick ~20 × 40 × 40 nm	7.0 vol.%	16%	[140]
Al_2O_3	Platelet ~15 nm diam., 5 nm thick	7.0 vol.%	28%	[140]
Al_2O_3	Blade ~8 × 15 nm, 5 nm thick	7.0 vol.%	23%	[140]
Au	Spherical ~10–20 nm diam.	0.00026 vol.%	~8%	[11]
Ag	Spherical ~60–80 nm diam.	0.001 vol.%	~5%	[11]
CeO_2	Spherical 74 nm diam.	2.0 vol.%	~9%	[123]
		3.0 vol.%	~14%	
CeO_2	Spherical 30 nm diam.	1.0 vol.%	11%	[47]
		2.0 vol.%	16%	
		3.0 vol.%	21%	
Cu[3]	Spherical ~60–100 nm diam.	1.0 vol.%	~48%	[197]
CuO	Spherical ~36 nm diam.	1.0 vol.%	~12%	[12]
		5.0 vol.%	~60%	

Filler	Nanoparticles morphology	Filler fraction	TC enhancement	Ref.
CuO	Spherical ~24 nm diam.	1.0 vol.%	~14%	[78]
		1.0 vol.%[1]	~29%	
		4.0 vol.%	~15%	
		4.0 vol.%[1]	~36%	
CuO	Spherical ~100 nm diam.	2.5 vol.%	~24%	[245]
		5.0 vol.%	~55%	
		7.5 vol.%	~78%	
CuO	Spherical ~25 nm diam.	0.10 vol.%	~7%	[248]
		0.30 vol.%	~12%	
Graphene	Sheets, 1 μm lateral	0.40 vol.%	~9%	[232]
GO	Sheets, range of 200 nm to 1000 nm	0.01 wt.%	~9%	[120]
		0.10 wt.%	~19%	
Graphene + MWCNTs	G sheets, 1 μm lateral; MWCNTs ~19 nm diam.	0.40 vol.%	~11%	[232]
SiC	Spherical ~26 nm	4.2 vol.%	16%	[141]
SiC	Cylindrical ~600 nm	4.0 vol.%	23%	[141]
SiC	Spherical ~37–110 nm diam.	0.1 vol.%	~12%	[155]
SiO_2	Spherical ~12 nm diam.	1.0 vol.%	~3%	[92]
TiO_2	Spherical ~15 nm diam.	1.0 vol.%	~18%	[135]
		5.0 vol.%	~30%	
TiO_2	Spherical ~95 nm diam.	1.0 wt.%	~1%	[118]
		4.9 wt.%	~5%	
CNTs	Rods ~length: 35 μm	0.01 wt.%	~38%	[249]
	Diam.: 20 nm	0.10 wt.%[2]	~126%	
		0.10 wt.%[2]	~288%	
MWCNTs	Rods ~length: 30 μm	1.0 vol.%	~7%	[221]
	Diam.: 15 nm			
ZnO	Semi-rectangular (90–210 nm)	1.0 vol.%	10%	[181]
		3.0 vol.%	21%	
ZnO	Spherical, 20–40 nm diam.	1.0 vol.%	9%	[181]
		3.0 vol.%	18%	

Notes: if not specified, measurements were conducted at room temperature.

[1]At 50°C.

[2]At 60°C.

[3]With addition of CTAB.

Table 3. Influence of water-based nanofluids in thermal management.

Filler	Conventional fluid	Nanoparticles morphology	Filler fraction	TC enhancement	Ref.
Al_2O_3	Ethylene glycol	Spherical ~10 nm diam.	5.0 vol.%	~18%	[217]
AlN	Ethylene glycol	Spherical ~50 nm diam.	5.0 vol.% 10.0 vol.%	~20% ~40%	[228]
Au	Toluene	Spherical ~10–20 nm diam.	0.011 vol.%	~9%	[11]
CuO	Ethylene glycol	Spherical ~35 nm diam.	4.0 vol.%	~22%	[190]
CuO	Ethylene glycol	Spherical ~23 nm diam.	~15.0 vol.%	~55%	[209]
Fe_3O_4	Kerosene	Spherical ~15 nm diam.	0.50 vol.% 1.0 vol.%	~15% ~34%	[169]
Graphene	Ethylene glycol @ 20°C Ethylene glycol @ 50°C	2D sheets ~500 × 600 nm	0.14 vol.% 0.14 vol.%	6.5% 36%	[184]
h-BN	Stamping lubricant @ 50°C	2D sheets ~500 × 500 nm ~5 atomic layer thick	0.01 wt.% 0.10 wt.%	~25% ~30%	[6]
h-BN	Metal cutting fluid @ 50°C	2D sheets ~500 × 500 nm ~5 atomic layer thick	0.01 wt.% 0.10 wt.%	~14% ~18%	[6]
MWCNT	Ethylene glycol	Rods length: 30 μm Diam.: 15 nm	0.05 vol.% 1.0 vol.%	~7% ~13%	[67]
MWCNT	Ethylene glycol	Rods length: μm range Diam.: ~20–30 nm	0.50 vol.% 1.0 vol.%	~8% ~13%	[14]
SiO_2	Ethanol	Spherical ~23 nm diam.	1.0 vol.%	~5%	[40]
SiO_2	Ethylene glycol	Spherical ~23 nm diam.	1.0 vol.%	~4%	[40]
TiO_2	Ethylene glycol/water (20/80%)	Spherical ~21 nm diam.	4.0 vol.%	~15%	[224]

Notes: if not specified, measurements were conducted at room temperature.

Table 4. Influence of diverse nanofluids in thermal management.

5. Summary

The heat required to be dissipated from systems is continually increasing due to industrial and economic trends to miniaturize designs, make better use of resources, obtain more power output, environmentally friendly materials, and obtain more profits. The present work offers a general overview of the recent research and development on preparation and characterization of nanofluids for thermal management applications, with emphasis on experimental data, variables and features. Nowadays, many technologies search for the highest efficiency mainly for energy savings, particularly on cooling or heat dissipation challenges within devices and machinery components. Many interesting properties of nanofluids have been reported in the past decades. Several efforts have been made trying to homogeneously disperse nanostructures within conventional HTFs to improve their properties. Nanofillers size has positive effects on conventional HTFs performance, i.e., compared to larger dispersed solid particles making flow through microchannels much easier, also since diverse parameters are critical for devices performance, such as morphology and stability of dispersed nanostructures, fluids composition, viscosity, fast sedimentation, channels clogging, erosion, wear, among others, which are often very serious for systems consisting of small channels. It is noted that nature of enhancement in thermal transport with nanoparticles concentration and temperature increment differs from fluid to fluid, which is comprehensible due to many factors such as fluids composition, viscosity, nature of fluids (morphology as well as interaction between fluid and nanofillers), etc.

It is found that factors such as temperature and filler fraction are more sensitive in determining the effective thermal conductivity in low viscosity fluids. The lower the filler concentration, the higher the stability (but lower thermal conductivity improvement), which means that a medium must be found between the two to prevent nanoparticles sedimentation/agglomeration, the free phonon/electron movement is affected by these defects, and hence a surfactant-free stable suspension can provide much better thermal conductivity. Nanofluids stability is a key factor to evaluate the quality of the nanofluids, and is considerably valued in the industrial applications. Additives or surfactants could be used to promote nanoparticles stabilization, but with some main drawbacks such as decrease of thermal conductivity, since surfactants could introduce defects at the solution/particle interfaces. Some nanofluids are currently expensive, partly due to the difficulty in manufacturing either the nanostructures to be afterward dispersed within conventional fluids or the nanofluids by themselves. Optimum layer thickness and filler fraction are important parameters in research of thermal transport, electrical and physical behavior and general aspects of both fundamental and applied characteristics. Mass production of nanostructures could further reduce the costs, and also using low filler fractions is another way to make nanofluids more affordable. Although nanofluids have displayed paramount and exciting potential applications, some vital hinders also exist before regular commercialization and industrialization of nanofluids.

Acknowledgements

Author acknowledges the support from UdeM, Rice University and CONACYT.

Author details

José Jaime Taha-Tijerina

Address all correspondence to: jose.taha@udem.edu

Universidad de Monterrey, Engineering Department, San Pedro Garza García, N. L., México

References

[1] Smalley RE. Future global energy prosperity: The terawatt challenge. MRS Bulletin. 2005;**30**(6):412-417. DOI: 10.1557/mrs2005.124

[2] Wen D, Lin G, Vafaei S, Zhang K. Review of nanofluids for heat transfer applications. Particuology. 2009;**7**(2):141-150. DOI: 10.1016/j.partic.2009.01.007

[3] Taha-Tijerina J, Narayanan TN, Gao G, Rohde M, Tsentalovich DA, Pasquali M, Ajayan PM. Electrically insulating thermal nano-oils using 2D fillers. ACS Nano. 2012;**6**(2):1214-1220. DOI: 10.1021/nn203862p

[4] Balandin AA. Thermal properties of graphene, carbon nanotubes and nanostructured carbon materials. Nature Materials. 2011;**10**:569-581. DOI: 10.1038/nmat3064

[5] Aravind SSJ, Baskar P, Baby TT, Sabareesh RK, Das S, Ramaprabhu S. Investigation of structural stability, dispersion, viscosity, and conductive heat transfer properties of functionalized carbon nanotube based nanofluids. Journal of Physical Chemistry C. 2011;**115**(34):16737-16744. DOI: 10.1021/jp201672p

[6] Taha-Tijerina J, Peña-Paras L, Narayanan TN, Garza L, Lapray C, Gonzalez J, Palacios E, Molina D, García A, Maldonado D, Ajayan PM. Multifunctional nanofluids with 2D nanosheets for thermal and tribological management. Wear. 2013;**302**(1-2):1241-1248. DOI: 10.1016/j.wear.2012.12.010

[7] Mahian O, Kianifar A, Kalogirou SA, Pop I, Wongwises S. A review of the applications of nanofluids in solar energy. International Journal of Heat and Mass Transfer. 2013;**57**(2):582-594. DOI: 10.1016/j.ijheatmasstransfer.2012.10.037

[8] Saha G, Paul MC. Numerical analysis of the heat transfer behaviour of water based Al_2O_3 and TiO_2 nanofluids in a circular pipe under the turbulent flow condition. International Communications in Heat and Mass Transfer. 2014;**56**:96-108. DOI: 10.1016/j.rser.2015.10.042

[9] Kroto HW, Heath JR, O'Brien SC, Curl RF, Smalley RE. C60: Buckminsterfullerene. Nature. 1985;**318**(6042):162-163. DOI: 10.1038/318162a0

[10] Lee J, Mudawar I. Assessment of the effectiveness of nanofluids for single-phase and two-phase heat transfer in micro-channels. International Journal of Heat and Mass Transfer. 2007;**50**(3):452-463. DOI: 10.1016/j.ijheatmasstransfer.2006.08.001

[11] Patel HE, Das SK, Sundararajan T, Nair AS, George B, Pradeep T. Thermal conductivities of naked and monolayer protected metal nanoparticle based nanofluids: Manifestation of anomalous enhancement and chemical effects. Applied Physics Letters. 2003;**83**:2931. DOI: 10.1063/1.1602578

[12] Eastman JA, Choi US, Li S, Thompson LJ, Lee S. Enhanced thermal conductivity through the development of nanofluids. MRS Proceedings. 1996;**457**:3-11. DOI: 10.1557/PROC-457-3

[13] Kim SH, Choi SR, Kim D. Thermal conductivity of metal-oxide nanofluids: Particle size dependence and effect of laser irradiation. Journal of Heat Transfer. 2006;**129**(3):298-307. DOI: 10.1115/1.2427071

[14] Liu M-S, Ching-Cheng Lin M, Huang I-T, Wang C-C. Enhancement of thermal conductivity with carbon nanotube for nanofluids. International Communications in Heat and Mass Transfer. 2005;**32**(9):1202-1210. DOI: 10.1016/j.icheatmasstransfer.2005.05.005

[15] Godfrey D, Herguth WR. Physical and chemical properties of industrial mineral oils affecting lubrication. Lubrication Engineering. 1995;**51**(6):493-496

[16] Iijima S. Helical microtubules of graphitic carbon. Nature. 1991;**354**(6348):56-58. DOI: 10.1038/354056a0

[17] Mahbubul IM, Saidur R, Amalina MA. Influence of particle concentration and temperature on thermal conductivity and viscosity of Al_2O_3/R141b nanorefrigerant. International Communications in Heat and Mass Transfer. 2013;**43**:100-104. DOI: 10.1016/j.icheatmasstransfer.2013.02.004

[18] Marquis FDS, Chibante LPF. Improving the heat transfer of nanofluids and nanolubricants with carbon nanotubes. JOM. 2005;**57**(12):32-43. DOI: 10.1007/s11837-005-0180-4

[19] Berber S, Kwon Y-K, Tománek D. Unusually high thermal conductivity of carbon nanotubes. Physical Review Letters. 2000;**84**(20):4613-4616. DOI: 10.1103/PhysRevLett.84.4613

[20] Che J, Çagin T, Goddard WA. Thermal conductivity of carbon nanotubes. Nanotechnology. 2000;**11**(2):65-69. DOI: 10.1088/0957-4484/11/2/305

[21] Osman MA, Srivastava D. Temperature dependence of the thermal conductivity of single-wall carbon nanotubes. Nanotechnology. 2001;**12**(1):21. DOI: 10.1088/0957-4484/12/1/305

[22] Kim P, Shi L, Majumdar A, PL ME. Thermal transport measurements of individual multiwalled nanotubes. Physical Review Letters. 2001;**87**(21):215502. DOI: 10.1103/PhysRevLett.87.215502

[23] Hong H, Zheng Y, Roy W. Nanomaterials for efficiently lowering the freezing point of anti-freeze coolants. Journal of Nanoscience and Nanotechnology. 2007;**7**(9):3180-3184

[24] Speight JG. Lange's Handbook of Chemistry. 16th ed. New York: McGraw-Hill Inc; 2004. 1608 p

[25] Kang HU, Kim SH, Oh JM. Estimation of thermal conductivity of Nanofluid using experimental effective particle volume. Experimental Heat Transfer. 2006;**19**(3):181-191. DOI: 10.1080/08916150600619281

[26] Callister WD, Rethwisch DG. Materials Science and Engineering: An Introduction. 9th ed. United States: Wiley; 2014. 984 p

[27] Perry RH, Green DW. Perry's Chemical Engineers' Handbook. 8th ed. New York: McGraw-Hill Education; 2007. 2704 p

[28] Dodd AD, Murfin D. Dictionary of Ceramics. 3rd ed. London, UK: Maney Publishing; 1994. 384 p

[29] Slack GA, Tanzilli RA, Pohl RO, Vandersande JW. The intrinsic thermal conductivity of AIN. Journal of Physics and Chemistry of Solids. 1987;**48**(7):641-647. DOI: 10.1016/0022-3697(87)90153-3

[30] Xu Y, Chung DDL. Increasing the thermal conductivity of boron nitride and aluminum nitride particle epoxy-matrix composites by particle surface treatments. Composite Interfaces. 2000;**7**(4):243-256. DOI: 10.1163/156855400750244969

[31] Wang J, Lee CH, Bando Y, Golberg D, Yap YK. Multiwalled boron nitride nanotubes: Growth, properties, and applications. In: B-C-N Nanotubes and Related Nanostructures. New York: Springer; 2009. pp. 23-44. DOI: 10.1007/978-1-4419-0086-9_2

[32] Duclaux L, Nysten B, Issi J-P, Moore AW. Structure and low-temperature thermal conductivity of pyrolytic boron nitride. Physical Review B. 1992;**46**(6):3362-3367. DOI: 10.1103/PhysRevB.46.3362

[33] Lindsay L, Broido DA. Theory of thermal transport in multilayer hexagonal boron nitride and nanotubes. Physical Review B. 2012;**85**(3):35436. DOI: 10.1103/PhysRevB.85.035436

[34] Chang CW, Han W-Q, Zettl A. Thermal conductivity of B–C–N and BN nanotubes. Applied Physics Letters. 2005;**86**(17):173102. DOI: 10.1063/1.1914963

[35] Sahoo P, Djieutedjeu H, Poudeu PFP. Co_3O_4 nanostructures: The effect of synthesis conditions on particles size, magnetism and transport properties. Journal of Materials Chemistry A. 2013;**1**(47):15022. DOI: 10.1039/c3ta13442c

[36] Hwang Y, Lee JK, Lee CH, Jung YM, Cheong SI, Lee CG, Ku BC, Jang SP. Stability and thermal conductivity characteristics of nanofluids. Thermochimica Acta. 2007;**455**(1-2):70-74. DOI: 10.1016/j.tca.2006.11.036

[37] Yan R, Simpson JR, Bertolazzi S, Brivio J, Watson M, Wu X, Kis A, Luo T, Hight Walker AR, Xing HG. Thermal conductivity of monolayer molybdenum disulfide obtained from temperature-dependent Raman spectroscopy. ACS Nano. 2014;**8**(1):986-993. DOI: 10.1021/nn405826k

[38] Li X, Zou C, Lei X, Li W. Stability and enhanced thermal conductivity of ethylene glycol-based SiC nanofluids. International Journal of Heat and Mass Transfer. 2015;**89**:613-619. DOI: 10.1016/j.ijheatmasstransfer.2015.05.096

[39] Yu W, Xie H. A review on Nanofluids: Preparation, stability mechanisms, and applications. Journal of Nanomaterials. 2012;**2012**:1-17. DOI: 10.1155/2012/435873

[40] Wang ZL, Tang DW, Liu S, Zheng XH, Araki N. Thermal-conductivity and thermal-diffusivity measurements of nanofluids by 3ω method and mechanism. International Journal of Thermophysics. 2007;**28**(4):1255-1268. DOI: 10.1007/s10765-007-0254-3

[41] Hwang YJ, Ahn YC, Shin HS, Lee CG, Kim GT, Park HS, Lee JK. Investigation on characteristics of thermal conductivity enhancement of nanofluids. Current Applied Physics. 2006;**6**(6):1068-1071. DOI: 10.1016/j.cap.2005.07.021

[42] Salman SD, Kadhum AAH, Takriff MS, Mohamad AB. Comparative studies on thermal performance of conic cut twist tape inserts with SiO_2 and TiO_2 nanofluids. Journal of Nanomaterials. 2015;**2015**:1-14. DOI: 10.1155/2015/921394

[43] Peimyoo N, Shang J, Yang W, Wang Y, Cong C, Yu T. Thermal conductivity determination of suspended mono- and bilayer WS_2 by Raman spectroscopy. Nano Research. 2014;**8**(4):1210-1221. DOI: 10.1007/s12274-014-0602-0

[44] Pisoni A, Jacimovic J, Gaál R, Náfrádi B, Berger H, Révay Z, Forró L. Anisotropic transport properties of tungsten disulfide. Scripta Materialia. 2016;**114**:48-50. DOI: 10.1016/j.scriptamat.2015.11.028

[45] Iqbal SM, Raj CS, Michael JJ, Irfan AM. A comparative investigation of Al_2O_3/H_2O, SiO_2/H_2O and ZrO_2/H_2O nanofluid for heat transfer applications. Journal of Nanomaterials and Biostructures. 2017;**12**(2):255-264. DOI: 10.1016/j.proeng.2015.01.107

[46] Shima PD, Philip J, Raj B. Synthesis of aqueous and nonaqueous iron oxide nanofluids and study of temperature dependence on thermal conductivity and viscosity. Journal of Physical Chemistry C. 2010;**114**(44):18825-18833. DOI: 10.1021/jp107447q

[47] Tiwari AK, Ghosh P, Sarkar J. Heat transfer and pressure drop characteristics of CeO_2/water nanofluid in plate heat exchanger. Applied Thermal Engineering. 2013;**57**(1-2):24-32. DOI: https://doi.org/10.1016/j.applthermaleng.2013.03.047

[48] Raja M, Vijayan R, Dineshkumar P, Venkatesan M. Review on nanofluids characterization, heat transfer characteristics and applications. Renewable and Sustainable Energy Reviews. 2016;**64**:163-173. DOI: 10.1016/j.rser.2016.05.079

[49] Yazid MNAWM, Sidik NAC, Yahya WJ. Heat and mass transfer characteristics of carbon nanotube nanofluids: A review. Renewable and Sustainable Energy Reviews. 2017;**80**:914-941. DOI: 10.1016/j.rser.2017.05.192

[50] Azmi WH, Sharma KV, Mamat R, Najafi G, Mohamad MS. The enhancement of effective thermal conductivity and effective dynamic viscosity of nanofluids—A review. Renewable and Sustainable Energy Reviews. 2016;**53**:1046-1058. DOI: 10.1016/j.rser.2015.09.081

[51] Koca HD, Doganay S, Turgut A, Tavman IH, Saidur R, Mahbubul IM. Effect of particle size on the viscosity of nanofluids: A review. Renewable and Sustainable Energy Reviews. 2017;**115**:630-646. DOI: 10.1016/j.rser.2017.07.016

[52] Ilyas SU, Pendyala R, Narahari M, Susin L. Stability, rheology and thermal analysis of functionalized alumina-thermal oil-based nanofluids for advanced cooling systems. Energy Conversion and Management. 2017;**142**:215-229. DOI: 10.1016/j.enconman.2017.01.079

[53] Sundar LS, Irurueta GO, Venkata Ramana E, Singh MK, Sousa ACM. Thermal conductivity and viscosity of hybrid nanfluids prepared with magnetic nanodiamond-cobalt oxide (ND-Co_3O_4) nanocomposite. Case Studies in Thermal Engineering. 2016;**7**:66-77. DOI: 10.1016/j.csite.2016.03.001

[54] Agarwal DK, Vaidyanathan A, Sunil Kumar S. Investigation on convective heat transfer behaviour of kerosene-Al_2O_3 nanofluid. Applied Thermal Engineering. 2015;**84**:64-73. DOI: 10.1016/j.applthermaleng.2015.03.054

[55] Singh A. Thermal conductivity of nanofluids. Defence Science Journal. 2008;**58**(5):600-607. DOI: 10.14429/dsj.58.1682

[56] Taha-Tijerina JJ, Narayanan TN, Tiwary CS, Lozano K, Chipara M, Ajayan PM. Nanodiamond-based thermal fluids. ACS Applied Materials & Interfaces. 2014;**6**(7):4778-4785. DOI: 10.1021/am405575t

[57] Bashirnezhad K, Bazri S, Safaei MR, Goodarzi M, Dahari M, Mahian O, Dalkiliça AS, Wongwises S. Viscosity of nanofluids: A review of recent experimental studies. International Communications in Heat and Mass Transfer. 2016;**73**:114-123. DOI: 10.1016/j.icheatmasstransfer.2016.02.005

[58] Ilyas SU, Pendyala R, Narahari M. Stability and thermal analysis of MWCNT-thermal oil-based nanofluids. Colloids and Surfaces A: Physicochemical and Engineering Aspects. 2017;**527**:11-22. DOI: 10.1016/j.colsurfa.2017.05.004

[59] Zhang H, Wang S, Lin Y, Feng M, Wu Q. Stability, thermal conductivity, and rheological properties of controlled reduced graphene oxide dispersed nanofluids. Applied Thermal Engineering. 2017;**119**:132-139. DOI: 10.1016/j.applthermaleng.2017.03.064

[60] Saraaz MM, Hormozi F. Heat transfer, pressure drop and fouling studies of multi-walled carbon nanotube nano-fluids inside a plate heat exchanger. Experimental Thermal and Fluid Science. 2016;**72**:1-11. DOI: 10.1016/j.expthermflusci.2015.11.004

[61] Tiwari AK, Ghosh P, Sarkar J. Performance comparison of the plate heat exchanger using different nanofluids. Experimental Thermal and Fluid Science. 2013;**49**:141-151. DOI: https://doi.org/10.1016/j.expthermflusci.2013.04.012

[62] Hekmatipour F, Akhavan-Behabadi MA, Sajadi B, Fakoor-Pakdaman M. Mixed convection heat transfer and pressure drop characteristics of the copper oxide-heat transfer oil (CuO-HTO) nanofluid in vertical tube. Case Studies in Thermal Engineering. 2017;**110**:248-256. DOI: 10.1016/J.CSITE.2017.09.009

[63] Gupta NK, Tiwari AK, Ghosh SK. Heat transfer mechanisms in heat pipes using nanofluids—A review. Experimental Thermal and Fluid Science. 2018;**90**:84-100. DOI: 10.1016/j.expthermflusci.2017.08.013

[64] Arshad W, Ali HM. Experimental investigation of heat transfer and pressure drop in a straight minichannel heat sink using TiO_2 nanofluid. International Journal of Heat and Mass Transfer. 2017;**110**:248-256. DOI: 10.1016/j.ijheatmasstransfer.2017.03.032

[65] Husain E, Narayanan TN, Taha-Tijerina JJ, Vinod S, Vajtai R, Ajayan PM. Marine corrosion protective coatings of hexagonal boron nitride thin films on stainless steel. ACS Applied Materials & Interfaces. 2013;**5**(10):4129-4135. DOI: 10.1021/am400016y

[66] Baby TT, Sundara R. Synthesis and transport properties of metal oxide decorated Graphene dispersed Nanofluids. Journal of Physical Chemistry. 2011;**115**(17):8527-8533. DOI: 10.1021/jp200273g

[67] Eswaraiah V, Jyothirmayee Aravind SS, Ramaprabhu S. Top down method for synthesis of highly conducting graphene by exfoliation of graphite oxide using focused solar radiation. Journal of Materials Chemistry. 2011;**21**(19):6800. DOI: 10.1039/c1jm10808e

[68] SUS C, Eastman JA. Enhancing thermal conductivity of fluids with nanoparticles. In: International Mechanical Engineering Congress and Exhibition (IMECE). San Francisco: ASME; 1995. p. 196525

[69] Masuda H, Ebata A, Teramae K, Hishinuma N. Alteration of thermal conductivity and viscosity of liquid by dispersing ultra-fine particles (dispersion of y-Al_2O_3, SiO_2 and TiO_2 ultra-fine particles). Netsu Bussei. 1993;**7**(4):227-233

[70] Chen T-C, Chou C-C, Yung T-Y, Tsai K-C, Huang J-Y. Wear behavior of thermally sprayed Zn/15Al, Al and Inconel 625 coatings on carbon steel. Surface and Coatings Technology. 2016;**303**:78-85. DOI: https://doi.org/10.1016/j.surfcoat.2016.03.095

[71] Li J, Du B, Wang F, Yao W, Yao S. The effect of nanoparticle surfactant polarization on trapping depth of vegetable insulating oil-based nanofluids. Physics Letters A. 2016;**380**(4):604-608. DOI: 10.1016/j.physleta.2015.12.008

[72] SUS C. Nanofluids: From vision to reality through research. Journal of Heat Transfer. 2009;**131**(3):33106. DOI: 10.1115/1.3056479

[73] Lee C-G, Hwang Y-J, Choi Y-M, Lee J-K, Choi C, Oh J-M. A study on the tribological characteristics of graphite nano lubricants. International Journal of Precision Engineering and Manufacturing. 2009;**10**(1):85-90. DOI: https://doi.org/10.1007/s12541-009-0013-4

[74] Wu D, Zhu H, Wang L, Liu L. Critical issues in nanofluids preparation, characterization and thermal conductivity. Current Nanoscience. 2009;**5**(1):103-112. DOI: 10.2174/157341309787314548

[75] Das SK, Choi SU, Yu W, Pradeep T. Nanofluids: Science and Technology. Hoboken, NJ, USA: John Wiley & Sons, Inc.; 2007. 416 p

[76] Peterson GP, Li CH. Heat and mass transfer in fluids with nanoparticle suspensions. Advances in Heat Transfer. 2006;**39**:257-376. DOI: 10.1016/S0065-2717(06)39003-X

[77] Prasher R, Bhattacharya P, Phelan PE. Brownian-motion-based convective-conductive model for the effective thermal conductivity of Nanofluids. Journal of Heat Transfer. 2006;**128**(6):588. DOI: 10.1115/1.2188509

[78] Das SK, Putra N, Thiesen P, Roetzel W. Temperature dependence of thermal conductivity enhancement for nanofluids. Journal of Heat Transfer. 2003;**125**(4):567. DOI: 10.1115/1.1571080

[79] Yang B, Han ZH. Temperature-dependent thermal conductivity of nanorod-based nanofluids. Applied Physics Letters. 2006;**89**(8):83111. DOI: 10.1063/1.2338424

[80] Chon CH, Kihm KD, Lee SP, Choi SUS. Empirical correlation finding the role of temperature and particle size for nanofluid (Al_2O_3) thermal conductivity enhancement. Applied Physics Letters. 2005;**87**(15):153107. DOI: 10.1063/1.2093936

[81] Pil Jang S, Choi SUS. Effects of various parameters on nanofluid thermal conductivity. Journal of Heat Transfer. 2007;**129**(5):617. DOI: 10.1115/1.2712475

[82] Zhi C, Xu Y, Bando Y, Golberg D. Highly thermo-conductive fluid with boron nitride nanofillers. ACS Nano. 2011;**5**(8):6571-6577. DOI: 10.1021/nn201946x

[83] Wong KV, De Leon O. Applications of nanofluids: Current and future. Advances in Mechanical Engineering. 2010;**2**:519659. DOI: 10.1155/2010/519659

[84] Hernández Battez A, González R, Viesca JL, Fernández JE, Díaz Fernández JM, Machado A, Chou R, Riba J. CuO, ZrO_2 and ZnO nanoparticles as antiwear additive in oil lubricants. Wear. 2008;**265**(3-4):422-428. DOI: 10.1016/j.wear.2007.11.013

[85] Lee P-H, Nam JS, Li C, Lee SW. An experimental study on micro-grinding process with nanofluid minimum quantity lubrication (MQL). International Journal of Precision Engineering and Manufacturing. 2012;**13**(3):331-338. DOI: 10.1007/s12541-012-0042-2

[86] Chang H, Li ZY, Kao MJ, Huang KD, Wu HM. Tribological property of TiO_2 nanolubricant on piston and cylinder surfaces. Journal of Alloys and Compound. 2010;**495**(2):481-484. DOI: 10.1016/j.jallcom.2009.10.017

[87] Hernández Battez A, Viesca JL, González R, Blanco D, Asedegbega E, Osorio A. Friction reduction properties of a CuO nanolubricant used as lubricant for a NiCrBSi coating. Wear. 2010;**268**(1-2):325-328. DOI: 10.1016/j.wear.2009.08.018

[88] Narayanan TN, Anantharaman MR. Template assisted fabrication of I-D nanostructures of Nickel,Cobalt, Iron Oxide and Carbon nanotubes and a study on their structural, magnetic and nonlinear optical properties for applications [PhD thesis]. Cochin University of Science & Technology; 2009

[89] Li Y, Zhou J, Tung S, Schneider E, Xi S. A review on development of nanofluid preparation and characterization. Powder Technology. 2009 [cited 2015 Nov 3];**196**(2):89-101. DOI: 10.1016/j.powtec.2009.07.025

[90] Munkhbayar B, Tanshen MR, Jeoun J, Chung H, Jeong H. Surfactant-free dispersion of silver nanoparticles into MWCNT-aqueous nanofluids prepared by one-step technique and their thermal characteristics. Ceramics International. 2013;**39**(6):6415-6425. DOI: 10.1016/j.ceramint.2013.01.069

[91] Zhu H, Lin Y, Yin Y. A novel one-step chemical method for preparation of copper nanofluids. Journal of Colloid and Interface Science. 2004;**277**(1):100-103. DOI: 10.1016/j.jcis.2004.04.026

[92] Hwang Y, Park HS, Lee JK, Jung WH. Thermal conductivity and lubrication characteristics of nanofluids. Current Applied Physics. 2006;**6**(suppl. 1):67-71. DOI: 10.1016/j.cap.2006.01.014

[93] Sohel Murshed SM, Tan S-H, Nguyen N-T. Temperature dependence of interfacial properties and viscosity of nanofluids for droplet-based microfluidics. Journal of Physics D: Applied Physics. 2008;**41**(8):85502. DOI: 10.1088/0022-3727/41/8/085502

[94] Li D, Hong B, Fang W, Guo Y, Lin R. Preparation of well-dispersed silver nanoparticles for oil-based nanofluids. Industrial and Engineering Chemistry Research. 2010; **49**(4):1697-1702. DOI: 10.1021/ie901173h

[95] Hwang Y, Lee J-K, Lee J-K, Jeong Y-M, Cheong S, Ahn Y-C, Kim SH. Production and dispersion stability of nanoparticles in nanofluids. Powder Technology. 2008;**186**(2):145-153. DOI: 10.1016/j.powtec.2007.11.020

[96] Ghozatloo A, Shariaty-Niasar M, Rashidi AM. Preparation of nanofluids from functionalized graphene by new alkaline method and study on the thermal conductivity and stability. International Communications in Heat and Mass Transfer. 2013;**42**:89-94. DOI: 10.1016/j.icheatmasstransfer.2012.12.007

[97] Lee K, Hwang Y, Cheong S, Kwon L, Kim S, Lee J. Performance evaluation of nano-lubricants of fullerene nanoparticles in refrigeration mineral oil. Current Applied Physics. 2009;**9**(2 suppl):e128-e131. DOI: 10.1016/j.cap.2008.12.054

[98] Chen M-L, Meng Z-D, Zhu L, Choi J-G, Park C-Y, Lee S-C, Hong D-S, Lee J-G, Jang W-K, Oh W-C. Dispersion stability of metal (oxide)-graphene nanofluids with electrical and thermal properties. Science of Advanced Materials. 2011;**3**(6):887-892. DOI: 10.1166/sam.2011.1213

[99] Murshed SMS, de Castro CAN, Lourenço MJV. Effect of surfactant and nanoparticle clustering on thermal conductivity of aqueous Nanofluids. Journal of Nanofluids. 2012;**1**(2):175-179. DOI: 10.1166/jon.2012.1020

[100] Narayanan TN, Gupta BK, Vithayathil SA, Aburto RR, Mani SA, Taha-Tijerina J, Xie B, Kaipparettu BA, Torti SV, Ajayan PM. Hybrid 2D nanomaterials as dual-mode contrast agents in cellular imaging. Advanced Materials. 2012;**24**(22):2992-2998. DOI: 10.1002/adma.201200706

[101] Contreras JE, Rodriguez EA, Taha-Tijerina J. Nanotechnology applications for electrical transformers—A review. Electric Power Systems Research. 2017;**143**:573-584. DOI: https://doi.org/10.1016/j.epsr.2016.10.058

[102] Sudeep PM, Taha-Tijerina J, Ajayan PM, Narayanan TN, Anantharaman MR. Nanofluids based on fluorinated graphene oxide for efficient thermal management. RSC Advances. 2014;**4**(47):24887-24892. DOI: 10.1039/c4ra00843

[103] Koo J, Kleinstreuer C. Impact analysis of nanoparticle motion mechanisms on the thermal conductivity of nanofluids. International Communications in Heat and Mass Transfer. 2005;**32**(9):1111-1118. DOI: 10.1016/j.icheatmasstransfer.2005.05.014

[104] Kleinstreuer C, Feng Y. Experimental and theoretical studies of nanofluid thermal conductivity enhancement: A review. Nanoscale Research Letters. 2011;**6**(1):229. DOI: 10.1186/1556-276X-6-229

[105] Malvandi A, Heysiattalab S, Ganji DD. Thermophoresis and Brownian motion effects on heat transfer enhancement at film boiling of nanofluids over a vertical cylinder. Journal of Molecular Liquids. 2016;**216**:503-509. DOI: 10.1016/j.molliq.2016.01.030

[106] Shukla KN, Koller TM, Rausch MH, Fröba AP. Effective thermal conductivity of nanofluids—A new model taking into consideration Brownian motion. International Journal of Heat and Mass Transfer. 2016;**99**:532-540. DOI: 10.1016/j.ijheatmasstransfer.2016.03.129

[107] Kandasamy R, Muhaimin I, Mohamad R. Thermophoresis and Brownian motion effects on MHD boundary-layer flow of a nanofluid in the presence of thermal stratification due to solar radiation. International Journal of Mechanical Sciences. 2013;**70**:146-154. DOI: 10.1016/j.ijmecsci.2013.03.007

[108] Aminfar H, Motallebzadeh R. Investigation of the velocity field and nanoparticle concentration distribution of nanofluid using Lagrangian-Eulerian approach. Journal of Dispersion Science and Technology. 2012;**33**(1):155-163. DOI: 10.1080/0193 2691.2010.528336

[109] Li CH, Peterson GP. The effect of particle size on the effective thermal conductivity of $Al2O_3$-water nanofluids. Journal of Applied Physics. 2007;**101**(4):44312. DOI: 10.1063/ 1.2436472

[110] Mintsa HA, Roy G, Nguyen CT, Doucet D. New temperature dependent thermal conductivity data for water-based nanofluids. International Journal of Thermal. 2009 [cited 2015 Nov 6];**48**(2):363-371. DOI: 10.1016/j.ijthermalsci.2008.03.009

[111] Anoop KB, Sundararajan T, Das SK. Effect of particle size on the convective heat transfer in nanofluid in the developing region. International Journal of Heat and Mass Transfer. 2009;**52**(9-10):2189-2195. DOI: 10.1016/j.ijheatmasstransfer.2007.11.063

[112] Teng T-P, Hung Y-H, Teng T-C, Mo H-E, Hsu H-G. The effect of alumina/water nanofluid particle size on thermal conductivity. Applied Thermal Engineering. 2010;**30**(14-15):2213-2218. DOI: 10.1016/j.applthermaleng.2010.05.036

[113] Abbasian Arani AA, Amani J. Experimental investigation of diameter effect on heat transfer performance and pressure drop of TiO_2–water nanofluid. Experimental Thermal and Fluid Science. 2013;**44**:520-533. DOI: 10.1016/j.expthermflusci.2012.08.014

[114] Kwek D, Crivoi A, Duan F. Effects of temperature and particle size on the thermal property measurements of Al_2O_3 –water Nanofluids. Journal of Chemical & Engineering Data. 2010;**55**(12):5690-5695. DOI: 10.1021/je1006407

[115] Agarwal DK, Vaidyanathan A, Sunil Kumar S. Synthesis and characterization of kerosene–alumina nanofluids. Applied Thermal Engineering. 2013;**60**(1-2):275-284. DOI: 10.1016/ j.applthermaleng.2013.06.049

[116] Beck MP, Yuan Y, Warrier P, Teja AS. The effect of particle size on the thermal conductivity of alumina nanofluids. Journal of Nanoparticle Research. 2009;**11**(5):1129-1136. DOI: 10.1007/s11051-008-9500-2

[117] Nguyen CT, Roy G, Gauthier C, Galanis N. Heat transfer enhancement using Al_2O_3–water nanofluid for an electronic liquid cooling system. Applied Thermal Engineering. 2007;**27**(8-9):1501-1506. DOI: 10.1016/j.applthermaleng.2006.09.028

[118] He Y, Jin Y, Chen H, Ding Y, Cang D, Lu H. Heat transfer and flow behaviour of aqueous suspensions of TiO_2 nanoparticles (nanofluids) flowing upward through a vertical pipe. International Journal of Heat and Mass Transfer. 2007;**50**(11-12):2272-2281. DOI: 10.1016/j.ijheatmasstransfer.2006.10.024

[119] Żyła G, Fal J, Estellé P. Thermophysical and dielectric profiles of ethylene glycol based titanium nitride (TiN–EG) nanofluids with various size of particles. International Journal of Heat and Mass Transfer. 2017;**113**:1189-1199. DOI: 10.1016/j.ijheatmasstransfer.2017.06.032

[120] Esfahani MR, Languri EM, Nunna MR. Effect of particle size and viscosity on thermal conductivity enhancement of graphene oxide nanofluid. International Journal of Heat and Mass Transfer. 2016;**76**:308-315. DOI: 10.1016/j.icheatmasstransfer.2016.06.006

[121] Singh D, Timofeeva E, Yu W, Routbort J, France D, Smith D, Lopez-Cepero JM. An investigation of silicon carbide-water nanofluid for heat transfer applications. Journal of Applied Physics. 2009;**105**(6):64306. DOI: 10.1063/1.3082094

[122] Timofeeva EV, Gavrilov AN, JM MC, Tolmachev YV, Sprunt S, Lopatina LM, Selinger JV. Thermal conductivity and particle agglomeration in alumina nanofluids: Experiment and theory. Physical Review. E, Statistical, Nonlinear, and Soft Matter Physics. 2007;**76**(6 Pt 1):61203. DOI: 10.1103/PhysRevE.76.061203

[123] Beck MP, Yuan Y, Warrier P, Teja AS. The thermal conductivity of aqueous nanofluids containing ceria nanoparticles. Journal of Applied Physics. 2010;**107**(6):66101. DOI: 10.1063/1.3330506

[124] Suresh T, Uthayakumar G, Srinivasan R. Synthesis and characterization of CeO_2/water nanofluids and its thermal properties. 2016 International Conference on Energy Efficient Technologies for Sustainability (ICEETS). IEEE; 2016. p. 202-205. DOI: 10.1109/ICEETS.2016.7582926

[125] Sharafeldin MA, Gróf G. Experimental investigation of flat plate solar collector using CeO_2-water nanofluid. Energy Conversion and Management. 2018;**155**:32-41. DOI: https://doi.org/10.1016/j.enconman.2017.10.070

[126] Amiri M, Movahedirad S, Manteghi F. Thermal conductivity of water and ethylene glycol nanofluids containing new modified surface SiO_2-Cu nanoparticles: Experimental and modeling. Applied Thermal Engineering. 2016;**108**:48-53. DOI: 10.1016/j.applthermaleng.2016.07.091

[127] Abdolbaqi MK, Sidik NAC, Rahim MFA, Mamat R, Azmi WH, Yazid MNAWM, Najafi G. Experimental investigation and development of new correlation for thermal conductivity and viscosity of BioGlycol/water based SiO_2 nanofluids. International

Communications in Heat and Mass Transfer. 2016;**77**:54-63. DOI: 10.1016/j.icheatmasstransfer.2016.07.001

[128] Nabil MF, Azmi WH, Abdul Hamid K, Mamat R, Hagos FY. An experimental study on the thermal conductivity and dynamic viscosity of TiO_2-SiO_2 nanofluids in water: Ethylene glycol mixture. International Communications in Heat and Mass Transfer. 2017;**86**:181-189. DOI: 10.1016/j.icheatmasstransfer.2017.05.024

[129] Minea AA. Hybrid nanofluids based on Al_2O_3, TiO_2 and SiO_2: Numerical evaluation of different approaches. International Journal of Heat and Mass Transfer. 2017;**104**:852-860. DOI: 10.1016/j.ijheatmasstransfer.2016.09.012

[130] Guo Y, Zhang T, Zhang D, Wang Q. Experimental investigation of thermal and electrical conductivity of silicon oxide nanofluids in ethylene glycol/water mixture. International Journal of Heat and Mass Transfer. 2018;**117**:280-286. DOI: 10.1016/j.ijheatmasstransfer.2017.09.091

[131] Zhu BJ, Zhao WL, Li DD, Li JK. Effect of volume fraction and temperature on thermal conductivity of SiO_2 nanofluids. Advanced Materials Research. 2011;**306-307**:1178-1181. DOI: 10.4028/www.scientific.net/AMR.306-307.1178

[132] Hossein Karimi Darvanjooghi M, Nasr Esfahany M. Experimental investigation of the effect of nanoparticle size on thermal conductivity of in-situ prepared silica–ethanol nanofluid. International Communications in Heat and Mass Transfer. 2016;**77**:148-154. DOI: 10.1016/j.icheatmasstransfer.2016.08.001

[133] Mehrali M, Sadeghinezhad E, Akhiani AR, Tahan Latibari S, Talebian S, Dolatshahi-Pirouz A, Metselaar HSC, Mehrali M. An ecofriendly graphene-based nanofluid for heat transfer applications. Journal of Cleaner Production. 2016;**137**:555-566. DOI: 10.1016/j.jclepro.2016.07.136

[134] Elias MM, Miqdad M, Mahbubul IM, Saidur R, Kamalisarvestani M, Sohel MR, Hepbasli A, Rahim NA, Amalina MA. Effect of nanoparticle shape on the heat transfer and thermodynamic performance of a shell and tube heat exchanger. International Communications in Heat and Mass Transfer. 2013;**44**:93-99. DOI: 10.1016/j.icheatmasstransfer.2013.03.014

[135] Murshed SMS, Leong KC, Yang C. Enhanced thermal conductivity of TiO_2—Water based nanofluids. International Journal of Thermal Sciences. 2005;**44**(4):367-373. DOI: 10.1016/j.ijthermalsci.2004.12.005

[136] Zyła G, Fal J. Experimental studies on viscosity, thermal and electrical conductivity of aluminum nitride-ethylene glycol (AlN-EG) nanofluids. Thermochimica Acta. 2016;**637**:11-16. DOI: https://doi.org/10.1016/j.tca.2016.05.006

[137] Khan I. Shape effects of MoS_2 nanoparticles on MHD slip flow of molybdenum disulphide nanofluid in a porous medium. Journal of Molecular Liquids. 2017;**233**:442-451. DOI: 10.1016/J.MOLLIQ.2017.03.009

[138] Arani AAA, Sadripour S, Kermani S. Nanoparticle shape effects on thermal-hydraulic performance of boehmite alumina nanofluids in a sinusoidal–wavy mini-channel with

phase shift and variable wavelength. International Journal of Mechanical Sciences. 2017;**128-129**:550-563. DOI: 10.1016/j.ijmecsci.2017.05.030

[139] Sudarsana Reddy P, Chamkha AJ. Influence of size, shape, type of nanoparticles, type and temperature of the base fluid on natural convection MHD of nanofluids. Alexandria Engineering Journal. 2016;**55**:331-341. DOI: 10.1016/j.aej.2016.01.027

[140] Kim HJ, Lee SH, Lee JH, Jang SP. Effect of particle shape on suspension stability and thermal conductivities of water-based bohemite alumina nanofluids. Energy. 2015;**90**: 1290-1297. DOI: 10.1016/j.energy.2015.06.084

[141] Xie H, Wang J, Xi T, Liu Y. Thermal conductivity of suspensions containing nano-sized SiC particles. International Journal of Thermophysics. 2002;**23**(2):571-580. DOI: 10.1023/A:1015121805842

[142] Prasher R, Phelan PE, Bhattacharya P. Effect of aggregation kinetics on the thermal conductivity of nanoscale colloidal solutions (nanofluid). Nano Letters. 2006;**6**(7):1529-1534. DOI: 10.1021/nl060992s

[143] Aaiza G, Khan I, Shafie S. Energy transfer in mixed convection MHD flow of nanofluid containing different shapes of nanoparticles in a channel filled with saturated porous medium. Nanoscale Research Letters. 2015;**10**(1):490. DOI: 10.1186/s11671-015-1144-4

[144] Karthikeyan NR, Philip J, Raj B. Effect of clustering on the thermal conductivity of nanofluids. Materials Chemistry and Physics. 2008;**109**(1):50-55. DOI: 10.1016/j.matchemphys.2007.10.029

[145] Taha-Tijerina J, Peña-Parás L, Maldonado-Cortés D. 2D-based nanofluids: Materials evaluation and performance. In: Nayak PK, editor. Two-Dimensional Materials—Synthesis, Characterization and Potential Applications. Rijeka, Croatia: InTech; 2016. pp. 153-198. DOI: 10.5772/64760

[146] Peña-Parás L, Taha-Tijerina J, García A, Maldonado D, González JA, Molina D, Palacios E, Cantú P. Antiwear and extreme pressure properties of Nanofluids for industrial applications. Tribology Transactions. 2014;**57**(6):1072-1076. DOI: 10.1080/10402004.2014.933937

[147] Wang N, Xu G, Li S, Zhang X. Thermal properties and solar collection characteristics of oil-based nanofluids with low graphene concentration. Energy Procedia. 2017;**105**:194-199. DOI: 10.1016/j.egypro.2017.03.301

[148] Hejazian M, Moraveji MK. A comparative analysis of single and two-phase models of turbulent convective heat transfer in a tube for TiO_2 nanofluid with CFD. Numerical Heat Transfer, Part A: Applications. 2013;**63**(10):795-806. DOI: 10.1080/10407782.2013.756759

[149] Abbassi Y, Talebi M, Shirani AS, Khorsandi J. Experimental investigation of TiO_2/water nanofluid effects on heat transfer characteristics of a vertical annulus with non-uniform heat flux in non-radiation environment. Annals of Nuclear Energy. 2014;**69**:7-13. DOI: 10.1016/j.anucene.2014.01.033

[150] Barzegarian R, Moraveji MK, Aloueyan A. Experimental investigation on heat transfer characteristics and pressure drop of BPHE (brazed plate heat exchanger) using TiO_2–water nanofluid. Experimental Thermal and Fluid Science. 2016;**74**(74):11-18. DOI: 10.1016/j.expthermflusci.2015.11.018

[151] Arulprakasajothi M, Elangovan K, HemaChandra Reddy K, Suresh S. Heat transfer study of water-based nanofluids containing titanium oxide nanoparticles. Materials Today. 2015;**2**:3648-3655. DOI: 10.1016/j.matpr.2015.07.123

[152] Krishnam M, Bose S, Das C. Boron nitride (BN) nanofluids as cooling agent in thermal management system (TMS). Applied Thermal Engineering. 2016;**106**:951-958. DOI: 10.1016/j.applthermaleng.2016.06.099

[153] Pang C, Jung J-Y, Lee JW, Kang YT. Thermal conductivity measurement of methanol-based nanofluids with $Al2O_3$ and SiO_2 nanoparticles. International Journal of Heat and Mass Transfer. 2012;**55**(21-22):5597-5602. DOI: 10.1016/j.ijheatmasstransfer.2012.05.048

[154] Mohan M, Thomas S, Taha-Tijerina J, Narayanan TN, Sobhan CB, Ajayan PM. Heat transfer studies in thermally conducting and electrically insulating nano-oils in a natural circulation loop. In: ASME International Mechanical Engineering Congress & Exposition (IMECE 2013). Volume 6B: Energy ASME; 2013, IMECE 2013-64285. V06BT07A040. DOI: 10.1115/IMECE2013-64285

[155] Manna O, Singh SK, Paul G. Enhanced thermal conductivity of nano-SiC dispersed water based nanofluid. Bulletin of Materials Science. 2012;**35**(5):707-712. DOI: 10.1007/s12034-012-0366-7

[156] Timofeeva EV, Yu W, France DM, Singh D, Routbort JL. Base fluid and temperature effects on the heat transfer characteristics of SiC in ethylene glycol/H_2O and H_2O nanofluids. Journal of Applied Physics. 2011;**109**(1):14914. DOI: 10.1063/1.3524274

[157] Ferrouillat S, Bontemps A, Ribeiro JP, Gruss JA, Soriano O. Hydraulic and heat transfer study of SiO_2/water nanofluids in horizontal tubes with imposed wall temperature boundary conditions. International Journal of Heat and Fluid Flow. 2011;**32**(2):424-439. DOI: 10.1016/j.ijheatfluidflow.2011.01.003

[158] Lee SW, Park SD, Kang S, Bang IC, Kim JH. Investigation of viscosity and thermal conductivity of SiC nanofluids for heat transfer applications. International Journal of Heat and Mass Transfer. 2011;**54**(1-3):433-438. DOI: 10.1016/j.ijheatmasstransfer.2010.09.026

[159] Li X, Zou C. Thermo-physical properties of water and ethylene glycol mixture based SiC nanofluids: An experimental investigation. International Journal of Heat and Mass Transfer. 2016;**101**:412-417. DOI: 10.1016/j.ijheatmasstransfer.2016.05.089

[160] Nasiri A, Shariaty-Niasar M, Rashidi A, Amrollahi A, Khodafarin R. Effect of dispersion method on thermal conductivity and stability of nanofluid. Experimental Thermal and Fluid Science. 2011;**35**(4):717-723. DOI: 10.1016/j.expthermflusci.2011.01.006

[161] Nikkhah V, Sarafraz MM, Hormozi F, Peyghambarzadeh SM. Particulate fouling of CuO-water nanofluid at isothermal diffusive condition inside the conventional heat exchanger-experimental and modeling. Experimental Thermal and Fluid Science. 2014;**60**:83-95. DOI: 10.1016/j.expthermflusci.2014.08.009

[162] Taha-Tijerina J, Venkataramani D, Aichele CP, Tiwary CS, Smay JE, Mathkar A, Chang P, Ajayan PM. Quantification of the particle size and stability of graphene oxide in a

variety of solvents. Particle and Particle Systems Characterization. 2015;**32**(3):334-339. DOI: 10.1002/ppsc.201400099

[163] Yang L, Du K, Zhang X. A theoretical investigation of thermal conductivity of nanofluids with particles in cylindrical shape by anisotropy analysis. Powder Technology. 2017; **314**:328-338. DOI: https://doi.org/10.1016/j.powtec.2016.09.032

[164] Elis Josna Mary E, Suganthi KS, Manikandan S, Anusha N, Rajan KS. Cerium oxide-ethylene glycol nanofluids with improved transport properties: Preparation and elucidation of mechanism. Journal of the Taiwan Institute of Chemical Engineers. 2015;**49**:183-191. DOI: 10.1016/j.jtice.2014.10.027

[165] Timofeeva EV, Routbort JL, Singh D. Particle shape effects on thermophysical properties of alumina nanofluids. Journal of Applied Physics. 2009;**106**(1):14304. DOI: 10.1063/1.3155999

[166] Gharagozloo PE, Goodson KE. Aggregate fractal dimensions and thermal conduction in nanofluids. Journal of Applied Physics. 2010;**108**(7):74309. DOI: 10.1063/1.3481423

[167] Shima PD, Philip J, Raj B. Role of microconvection induced by Brownian motion of nanoparticles in the enhanced thermal conductivity of stable nanofluids. Appl Phys Lett. 2009;**94**(22):223101. DOI: 10.1063/1.3147855

[168] Yu W, Xie H, Chen L, Li Y. Investigation of thermal conductivity and viscosity of ethylene glycol based ZnO nanofluid. Thermochimica Acta. 2009;**491**(1-2):92-96. DOI: 10.1016/j.tca.2009.03.007

[169] Yu W, Xie H, Chen L, Li Y. Enhancement of thermal conductivity of kerosene-based Fe_3O_4 nanofluids prepared via phase-transfer method. Colloids and Surfaces A: Physicochemical and Engineering Aspects. 2010;**355**(1-3, 113):109. DOI: 10.1016/j.colsurfa.2009.11.044

[170] Xie H, Chen L. Review on the preparation and thermal performances of carbon nanotube contained nanofluids. Journal of Chemical & Engineering Data. 2011;**56**(4):1030-1041. DOI: 10.1021/je101026j

[171] Gulzar M, Masjuki H, Varman M, Kalam M, Mufti RA, Zulkifli N, Yunus R, Zahid R. Improving the AW/EP ability of chemically modified palm oil by adding CuO and MoS_2 nanoparticles. Tribology International. 2015;**88**:271-279. DOI: 10.1016/j.triboint.2015.03.035

[172] Utomo AT, Poth H, Robbins PT, Pacek AW. Experimental and theoretical studies of thermal conductivity, viscosity and heat transfer coefficient of titania and alumina nanofluids. International Journal of Heat and Mass Transfer. 2012;**55**(25-26):7772-7781. DOI: 10.1016/j.ijheatmasstransfer.2012.08.003

[173] Khairul MA, Shah K, Doroodchi E, Azizian R, Moghtaderi B. Effects of surfactant on stability and thermo-physical properties of metal oxide nanofluids. International Journal of Heat and Mass Transfer. 2016;**98**:778-787. DOI: 10.1016/j.ijheatmasstransfer.2016.03.079

[174] Findenegg GH, Pasucha B, Strunk H. Adsorption of non-ionic surfactants from aqueous solutions on graphite: Adsorption isotherms and calorimetric enthalpies of displacement for C_8E_4 and related compounds. Colloids and Surfaces. 1989;**37**:223-233. DOI: 10.1016/0166-6622(89)80121-0

[175] Saleh R, Putra N, Wibowo RE, Septiadi WN, Prakoso SP. Titanium dioxide nanofluids for heat transfer applications. Experimental Thermal and Fluid Science. 2014;**52**:19-29. DOI: 10.1016/j.expthermflusci.2013.08.018

[176] Chen W, Zou C, Li X, Li L. Experimental investigation of SiC nanofluids for solar distillation system: Stability, optical properties and thermal conductivity with saline water-based fluid. International Journal of Heat and Mass Transfer. 2017;**107**:264-270. DOI: 10.1016/j.ijheatmasstransfer.2016.11.048

[177] Nguyen CT, Desgranges F, Roy G, Galanis N, Maré T, Boucher S, Angue Mintsa H. Temperature and particle-size dependent viscosity data for water-based nanofluids—Hysteresis phenomenon. International Journal of Heat and Fluid Flow. 2007;**28**(6):1492-1506. DOI: 10.1016/j.ijheatfluidflow.2007.02.004

[178] Yiamsawas T, Mahian O, Dalkilic AS, Kaewnai S, Wongwises S. Experimental studies on the viscosity of TiO_2 and Al_2O_3 nanoparticles suspended in a mixture of ethylene glycol and water for high temperature applications. Applied Energy. 2013;**111**:40-45. DOI: 10.1016/j.apenergy.2013.04.068

[179] Li H, Wang L, He Y, Hu Y, Zhu J, Jiang B. Experimental investigation of thermal conductivity and viscosity of ethylene glycol based ZnO nanofluids. Applied Thermal Engineering. 2014;**88**:363-368. DOI: 10.1016/j.applthermaleng.2014.10.071

[180] Attari H, Derakhshanfard F, Darvanjooghi MHK. Effect of temperature and mass fraction on viscosity of crude oil-based nanofluids containing oxide nanoparticles. International Communications in Heat and Mass Transfer. 2017;**82**:103-113. DOI: 10.1016/j.icheatmasstransfer.2017.02.007

[181] Jeong J, Li C, Kwon Y, Lee J, Kim SH, Yun R. Particle shape effect on the viscosity and thermal conductivity of ZnO nanofluids. International Journal of Refrigeration. 2013;**36**(8):2233-2241. DOI: 10.1016/j.ijrefrig.2013.07.024

[182] Williams W, Buongiorno J, Hu L-W. Experimental investigation of turbulent convective heat transfer and pressure loss of alumina/water and zirconia/water nanoparticle colloids (nanofluids) in horizontal tubes. Journal of Heat Transfer. 2008;**130**(4):42412. DOI: 10.1115/1.2818775

[183] Bhattacharya P, Nara S, Vijayan P, Tang T, Lai W, Phelan PE, Prasher RS, Song DW, Wang J. Characterization of the temperature oscillation technique to measure the thermal conductivity of fluids. International Journal of Heat and Mass Transfer. 2006;**49**(17-18):2950-2956. DOI: 10.1016/j.ijheatmasstransfer.2006.02.023

[184] Jyothirmayee Aravind SS, Ramaprabhu S. Surfactant free graphene nanosheets based nanofluids by in-situ reduction of alkaline graphite oxide suspensions. Journal of Applied Physics. 2011;**110**(12):124326. DOI: 10.1063/1.3671613

[185] Paul G, Philip J, Raj B, Das PK, Manna I. Synthesis, characterization, and thermal property measurement of nano-Al95Zn05 dispersed nanofluid prepared by a two-step process. International Journal of Heat and Mass Transfer. 2011;**54**(15-16):3783-3788. DOI: 10.1016/j.ijheatmasstransfer.2011.02.044

[186] Kalidasan K, Velkennedy R, Rajesh Kanna P. Natural convection heat transfer enhancement using nanofluid and time-variant temperature on the square enclosure with diagonally constructed twin adiabatic blocks. Applied Thermal Engineering. 2016;**92**(92): 219-235. DOI: 10.1016/j.applthermaleng.2015.09.077

[187] Wang L, Shi B, Chai Z. Effects of temperature-dependent properties on natural convection of nanofluids in a partially heated cubic enclosure. Applied Thermal Engineering. 2017;**128**:204-213. DOI: 10.1016/j.applthermaleng.2017.09.006

[188] Kuznetsov AV, Nield DA. Natural convective boundary-layer flow of a nanofluid past a vertical plate. International Journal of Thermal Sciences. 2010;**49**(2):243-247. DOI: 10.1016/j.ijthermalsci.2009.07.015

[189] Michaelides EE. Transport properties of nanofluids. A critical review. Journal of Non-Equilibrium Thermodynamics. 2013;**38**(1):1-79. DOI: 10.1515/jnetdy-2012-0023

[190] Lee S, Choi SU-S, Li S, Eastman JA. Measuring thermal conductivity of fluids containing oxide nanoparticles. Journal of Heat Transfer. 1999;**121**(2):280. DOI: 10.1115/1.2825978

[191] Jha N, Ramaprabhu S. Thermal conductivity studies of metal dispersed multiwalled carbon nanotubes in water and ethylene glycol based nanofluids. Journal of Applied Physics. 2009;**106**(8):84317. DOI: 10.1063/1.3240307

[192] Wen D, Ding Y. Effective thermal conductivity of aqueous suspensions of carbon nanotubes (carbon nanotube nanofluids). Journal of Thermophysics and Heat Transfer. 2004; **18**(4):481-485. DOI: 10.2514/1.9934

[193] Henderson JR, van Swol F. On the interface between a fluid and a planar wall. Molecular Physics. 2007;**51**(4):991-1010. DOI: 10.1080/00268978400100651

[194] Murshed SMS, Leong KC, Yang C. Investigations of thermal conductivity and viscosity of nanofluids. International Journal of Thermal Sciences. 2008;**47**(5):560-588. DOI: 10.1016/j.ijthermalsci.2007.05.004

[195] Rana P, Dhanai R, Kumar L. MHD slip flow and heat transfer of Al_2O_3-water nanofluid over a horizontal shrinking cylinder using Buongiorno's model: Effect of nanolayer and nanoparticle diameter. Advanced Powder Technology. 2017;**28**(7):1727-1738. DOI: 10.1016/j.apt.2017.04.010

[196] Du K. Self-assembly of nanoparticles at liquid-liquid interfaces [PhD thesis]. 2010

[197] Saterlie M, Sahin H, Kavlicoglu B, Liu Y, Graeve O. Particle size effects in the thermal conductivity enhancement of copper-based nanofluids. Nanoscale Research Letters. 2011;**6**(1):217. DOI: 10.1186/1556-276X-6-217

[198] Ding Y, Chen H, Wang L, Yang C-Y, He Y, Yang W, Lee WP, Zhang L, Huo R. Heat transfer intensification using nanofluids. KONA Powder and Particle Journal. 2007;**25**:23-38. DOI: 10.14356/kona.2007006

[199] Li L, Zhang Y, Ma H, Yang M. Molecular dynamics simulation of effect of liquid layering around the nanoparticle on the enhanced thermal conductivity of nanofluids. Journal of Nanoparticle Research. 2009;**12**(3):811-821. DOI: 10.1007/s11051-009-9728-5

[200] Yu C-J, Richter AG, Datta A, Durbin MK, Dutta P. Observation of molecular layering in thin liquid films using X-ray reflectivity. Physical Review Letters. 1999;**82**(11):2326-2329. DOI: 10.1103/PhysRevLett.82.2326

[201] Ren Y, Xie H, Cai A. Effective thermal conductivity of nanofluids containing spherical nanoparticles. Journal of Physics D: Applied Physics. 2005;**38**(21):3958-3961. DOI: 10.1088/0022-3727/38/21/019

[202] Duangthongsuk W, Wongwises S. Measurement of temperature-dependent thermal conductivity and viscosity of TiO_2-water nanofluids. Experimental Thermal and Fluid Science. 2009;**33**(4):706-714. DOI: 10.1016/j.expthermflusci.2009.01.005

[203] Wang B, Wang X, Lou W, Hao J. Thermal conductivity and rheological properties of graphite/oil nanofluids. Colloids and Surfaces A: Physicochemical and Engineering Aspects. 2012;**414**:125-131. DOI: 10.1016/j.colsurfa.2012.08.008

[204] Kole M, Dey TK. Investigation of thermal conductivity, viscosity, and electrical conductivity of graphene based nanofluids. Journal of Applied Physics. 2013;**113**(8):84307. DOI: 10.1063/1.4793581

[205] Botha SS, Ndungu P, Bladergroen BJ. Physicochemical properties of oil-based nanofluids containing hybrid structures of silver nanoparticles supported on silica. Industrial & Engineering Chemistry Research. 2011;**50**(6):3071-3077. DOI: 10.1021/ie101088x

[206] Yan X, Jiang Y, Jiang M, Hu Z, Hong H. ZnO Nanorod based nanofluids. Journal of Nanofluids. 2013;**2**(1):63-68. DOI: 10.1016/j.cossms.2014.02.002

[207] Hamid KA, Azmi WH, Nabil MF, Mamat R, Sharma KV. Experimental investigation of thermal conductivity and dynamic viscosity on nanoparticle mixture ratios of TiO_2-SiO_2 nanofluids. International Journal of Heat and Mass Transfer. 2018;**116**:1143-1152. DOI: 10.1016/j.ijheatmasstransfer.2017.09.087

[208] Glory J, Bonetti M, Helezen M, Mayne-L'Hermite M, Reynaud C. Thermal and electrical conductivities of water-based nanofluids prepared with long multiwalled carbon nanotubes. Journal of Applied Physics. 2008;**103**(9):94309. DOI: https://doi.org/10.1063/1.2908229

[209] Wang X, Xu XS, Choi SU. Thermal conductivity of nanoparticle-fluid mixture. Journal of Thermophysics and Heat Transfer. 1999;**13**(4):474-480. DOI: 10.2514/2.6486

[210] Challoner AR, Powell RW. Thermal conductivities of liquids: New determinations for seven liquids and appraisal of existing values. Proceedings of the Royal Society A: Mathematical, Physical and Engineering Science. 1956;**238**(1212):90-106. DOI: 10.1098/rspa.1956.0205

[211] Kurt H, Kayfeci M. Prediction of thermal conductivity of ethylene glycol–water solutions by using artificial neural networks. Applied Energy. 2009;**86**(10):2244-2248. DOI: 10.1016/j.apenergy.2008.12.020

[212] Santucci A, Verdini L, Verdini PG. Data-acquisition system for measurement of thermal diffusivity and propagation properties of thermal waves by a non-steady-state method. Review of Scientific Instruments. 1986;**57**(8):1627. DOI: 10.1063/1.1138541

[213] Czarnetzki W, Roetzel W. Temperature oscillation techniques for simultaneous measurement of thermal diffusivity and conductivity. International Journal of Thermophysics. 1995;**16**(2):413-422. DOI: 10.1007/BF01441907

[214] Cahill DG. Thermal conductivity measurement from 30 to 750 K: The 3ω method. The Review of Scientific Instruments. 1990;**61**(2):802. DOI: 10.1063/1.1141498

[215] Oh D-W, Jain A, Eaton JK, Goodson KE, Lee JS. Thermal conductivity measurement and sedimentation detection of aluminum oxide nanofluids by using the 3ω method. International Journal of Heat and Fluid Flow. 2008;**29**(5):1456-1461. DOI: 10.1016/j.ijheatfluidflow.2008.04.007

[216] Wang H, Sen M. Analysis of the 3-omega method for thermal conductivity measurement. International Journal of Heat and Mass Transfer. 2009;**52**(7-8):2102-2109. DOI: 10.1016/j.ijheatmasstransfer.2008.10.020

[217] Eastman JA, Choi SUS, Li S, Yu W, Thompson LJ. Anomalously increased effective thermal conductivities of ethylene glycol-based nanofluids containing copper nanoparticles. Applied Physics Letters. 2001;**78**(6):718. DOI: 10.1063/1.1341218

[218] Choi SUS, Zhang ZG, Yu W, Lockwood FE, Grulke EA. Anomalous thermal conductivity enhancement in nanotube suspensions. Applied Physics Letters. 2001;**79**(14):2252. DOI: 10.1063/1.1408272

[219] Assael MJ, Chen C-F, Metaxa I, Wakeham WA. Thermal conductivity of suspensions of carbon nanotubes in water. International Journal of Thermophysics. 2004;**25**(4):971-985. DOI: 10.1023/B:IJOT.0000038494.22494.04

[220] Ding Y, Alias H, Wen D, Williams RA. Heat transfer of aqueous suspensions of carbon nanotubes (CNT nanofluids). International Journal of Heat and Mass Transfer. 2006;**49**(1-2):240-250. DOI: 10.1016/j.ijheatmasstransfer.2005.07.009

[221] Xie H, Lee H, Youn W, Choi M. Nanofluids containing multiwalled carbon nanotubes and their enhanced thermal conductivities. Journal of Applied Physics. 2003;**94**(8):4967. DOI: 10.1063/1.1613374

[222] Branson BT, Beauchamp PS, Beam JC, Lukehart CM, Davidson JL. Nanodiamond nanofluids for enhanced thermal conductivity. ACS Nano. 2013;**7**(4):3183-3189. DOI: 10.1021/nn305664x

[223] Kong D, Yang H, Yang Y, Wei S, Wang J, Cheng B. Dispersion behavior and stabilization mechanism of alumina powders in silica sol. Materials Letters. 2004;**58**(27-28):3503-3508. DOI: 10.1016/j.matlet.2004.06.060

[224] Yiamsawasd T, Dalkilic AS, Wongwises S. Measurement of the thermal conductivity of titania and alumina nanofluids. Thermochimica Acta. 2012;**545**:48-56. DOI: 10.1016/j.tca.2012.06.026

[225] Serebryakova MA, Dimov SV, Bardakhanov SP, Novopashin SA. Thermal conductivity, viscosity and rheology of a suspension based on Al_2O_3 nanoparticles and mixture of 90% ethylene glycol and 10% water. International Journal of Heat and Mass Transfer. 2015;**83**:187-191. DOI: 10.1016/j.ijheatmasstransfer.2014.12.002

[226] Mariano A, Pastoriza-Gallego MJ, Lugo L, Mussari L, Piñeiro MM. Co_3O_4 ethylene glycol-based nanofluids: Thermal conductivity, viscosity and high pressure density. International Journal of Heat and Mass Transfer. 2015;**85**:54-60. DOI: 10.1016/j.ijheatmasstransfer.2015.01.061

[227] Hu P, Shan W-L, Yu F, Chen Z-S. Thermal conductivity of AlN–ethanol nanofluids. International Journal of Thermophysics. 2008;**29**(6):1968-1973. DOI: 10.1007/s10765-008-0529-3

[228] Yu W, Xie H, Li Y, Chen L. Experimental investigation on thermal conductivity and viscosity of aluminum nitride nanofluid. Particuology. 2011;**9**(2):187-191. DOI: 10.1016/j.partic.2010.05.014

[229] Yu W, Xie H, Chen W. Experimental investigation on thermal conductivity of nanofluids containing graphene oxide nanosheets. Journal of Applied Physics. 2010;**107**(9):94317. DOI: 10.1063/1.3372733

[230] Shaikh S, Lafdi K, Ponnappan R. Thermal conductivity improvement in carbon nanoparticle doped PAO oil: An experimental study. Journal of Applied Physics. 2007;**101**(6):64302. DOI: 10.1063/1.2710337

[231] Yu W, Xie H, Chen L, Zhu Z, Zhao J, Zhang Z. Graphene based silicone thermal greases. Physics Letters A. 2014;**378**(3):207-211. DOI: 10.1016/j.physleta.2013.10.017

[232] Aravind SSJ, Ramaprabhu S. Graphene–multiwalled carbon nanotube-based nanofluids for improved heat dissipation. RSC Advances. 2013;**3**(13):4199. DOI: 10.1039/c3ra22653k

[233] Zhang W, Zhou M, Zhu H, Tian Y, Wang K, Wei J, Ji F, Li X, Li Z, Zhang P, Wu D. Tribological properties of oleic acid-modified graphene as lubricant oil additives. Journal of Physics D: Applied Physics. 2011;**44**(20):205303. DOI: 10.1088/0022-3727/44/20/205303

[234] Baby TT, Ramaprabhu S. Enhanced convective heat transfer using graphene dispersed nanofluids. Nanoscale Research Letters. 2011;**6**(1):289. DOI: 10.1186/1556-276X-6-289

[235] Sadeghinezhad E, Mehrali M, Saidur R, Mehrali M, Tahan Latibari S, Akhiani AR, Metselaar HSC. A comprehensive review on graphene nanofluids: Recent research, development and applications. Energy Conversion and Management. 2016;**111**:466-487. DOI: 10.1016/j.enconman.2016.01.004

[236] Hadadian M, Goharshadi EK, Youssefi A. Electrical conductivity, thermal conductivity, and rheological properties of graphene oxide-based nanofluids. Journal of Nanoparticle Research. 2014;**16**(12):2788. DOI: 10.1007/s11051-014-2788-1

[237] Yu W, Xie H, Bao D. Enhanced thermal conductivities of nanofluids containing graphene oxide nanosheets. Nanotechnology. 2010;**21**(5):55705. DOI: 10.1088/0957-4484/21/5/055705

[238] Yu W, Xie H, Wang X, Wang X. Significant thermal conductivity enhancement for nanofluids containing graphene nanosheets. Physics Letters A. 2011;**375**(10):1323-1328. DOI: 10.1016/j.physleta.2011.01.040

[239] Patel HE, Sundararajan T, Pradeep T, Dasgupta A, Dasgupta N, Das SK. A micro-convection model for thermal conductivity of nanofluids. Pramana. 2005;**65**(5):863-869. DOI: 10.1007/BF02704086

[240] Kumar DH, Patel HE, VRR K, Sundararajan T, Pradeep T, Das SK. Model for heat conduction in nanofluids. Physical Review Letters. 2004;**93**(14):144301. DOI: 10.1103/PhysRevLett.93.144301

[241] Venkata Sastry NN, Bhunia A, Sundararajan T, Das SK. Predicting the effective thermal conductivity of carbon nanotube based nanofluids. Nanotechnology. 2008;**19**(5):55704. DOI: 10.1088/0957-4484/19/05/055704

[242] Wang X-Q, Mujumdar AS. Heat transfer characteristics of nanofluids: A review. Interational Journal of Thermal Sciences. 2007;**46**(1):1-19. DOI: 10.1016/j.ijthermalsci.2006.06.010

[243] Keblinski P, Phillpot S, Choi SU, Eastman J. Mechanisms of heat flow in suspensions of nano-sized particles (nanofluids). International Journal of Heat and Mass Transfer. 2002;**45**(4):855-863

[244] Choi C, Yoo HS, Oh JM. Preparation and heat transfer properties of nanoparticle-in-transformer oil dispersions as advanced energy-efficient coolants. Current Applied Physics. 2008;**8**(6):710-712. DOI: 10.1016/j.cap.2007.04.060

[245] Xuan Y, Li Q. Heat transfer enhancement of nanofluids. International Journal of Heat and Fluid Flow. 2000;**21**(1):58-64. DOI: 10.1016/S0142-727X(99)00067-3

[246] Putra N, Roetzel W, Das SK. Natural convection of nano-fluids. Heat and Mass Transfer. 2002;**39**(8-9):775-784. DOI: 10.1007/s00231-002-0382-z

[247] Wen D, Ding Y. Experimental investigation into convective heat transfer of nanofluids at the entrance region under laminar flow conditions. International Journal of Heat and Mass Transfer. 2004;**47**(24):5181-5188. DOI: 10.1016/j.ijheatmasstransfer.2004.07.012

[248] Lee D, Kim J-W, Kim BG. A new parameter to control heat transport in nanofluids: Surface charge state of the particle in suspension. Journal of Physical Chemistry B. 2006;**110**(9):4323-4328. DOI: 10.1021/jp057225m

[249] Walvekar R, Faris IA, Khalid M. Thermal conductivity of carbon nanotube nanofluid-experimental and theoretical study. Heat Transfer Research. 2012;**41**(2):145-163. DOI: 10.1002/htj.20405

Chapter 10

Magnetite Molybdenum Disulphide Nanofluid of Grade Two: A Generalized Model with Caputo-Fabrizio Derivative

Farhad Ali, Madeha Gohar, Ilyas Khan,
Nadeem Ahmad Sheikh, Syed Aftab Alam Jan and
Muhammad Saqib

Additional information is available at the end of the chapter

http://dx.doi.org/10.5772/intechopen.72863

Abstract

Heat and mass transfer analysis in magnetite molybdenum disulphide nanofluid of grade two is studied. MoS_2 powder with each particle of nanosize is dissolved in engine oil chosen as base fluid. A generalized form of grade-two model is considered with fractional order derivatives of Caputo and Fabrizio. The fluid over vertically oscillating plate is subjected to isothermal temperate and species concentration. The problem is modeled in terms of partial differential equations with sufficient initial conditions and boundary conditions. Fractional form of Laplace transform is used and exact solutions in closed form are determined for velocity field, temperature and concentration distributions. These solutions are then plotted for embedded parameters and discussed. Results for the physical quantities of interest (skin friction coefficient, Nusselt number and Sherwood number) are computed in tables. Results obtained in this work are compared with some published results from the open literature.

Keywords: Caputo-Fabrizio approach, MHD, nanofluid, generalized second-grade fluid, exact solutions

1. Introduction

The idea of fractional order calculus is as old as traditional order calculus. The pioneering systematic studies are devoted to Riemann-Liouville and Leibniz [1]. The subject is growing day by day and its applications have been utilized in different fields, for example, viscoelasticity, bioengineering, biophysics and mechatronics [2]. The applications of non-integer order calculus have also been encountered in different areas of science despite mathematics and physics drastically [3–5]. In fluid dynamics, the fractional order calculus has been broadly used to describe the viscoelastic behaviour of the material. Viscoelasticity of a material is defined it deforms evince both viscous and elastic behaviour via storage of mechanical energy and simultaneous behaviour. Mainardi [6] examined the connections among fractional calculus, wave motion and viscoelasticity. It is increasingly seen as an efficient tool through which useful generalization of physical concepts can be obtained. Hayat et al. [7] studied the periodic unidirectional flows of a viscoelastic fluid with the Maxwell model (fractional). Qi and Jin [8] analyzed the unsteady rotating flows of viscoelastic fluid with the fractional Maxwell model between coaxial cylinders. Many other researchers used the idea of fractional calculus and published quite number of research papers in some reputable journals [9–11].

Several versions of fractional derivatives are now available in the literature; however, the widely used derivatives are the Riemann-Liouville fractional derivatives and Caputo/fractional derivative [12, 13]. However, the researchers were facing quite number of difficulties in using them. For example, the Riemann-Liouville derivative of a constant is not zero and the Laplace transform of Riemann-Liouville derivative contains terms without physical significance. Though the Caputo fractional derivative has eliminated the short fall of Riemann-Liouville derivative, its kernal has singularity point. Ali et al. [14] reported the conjugate effect of heat and mass transfer on time fraction convective flow of Brinkman type fluid using the Caputo approach. Shahid et al. [15] investigated the approach of Caputo fractional derivatives to study the magnetohyrodynamic (MHD) flow past over an oscillating vertical plate along with heat and mass transfer. Recently, Caputo and Fabrizio (CF) have initiated a fractional derivative with no singular kernel [16]. However, Shah and Khan [17] analyzed that heat transfer analysis in a grade-two fluid over an oscillating vertical plate by using CF derivatives. Ali et al. [18] studied the application of CF derivative to MHD free convection flow of generalized Walter's-B fluid model. Recently, Sheikh et al. [19] applied CF derivatives to MHD flow of a regular second-grade fluid together with radiative heat transfer.

However, the idea of fractional calculus is very new in nanoscience, particularly in nanofluid also called smart fluid [20]. In this study, we have applied the fractional calculus idea more exactly, the idea of CF derivatives to a subclass of differential type fluid known as the second-grade fluid with suspended nanoparticles in spherical shape of molybdenum disulphide (MoS_2). Generally, the purpose of nanoparticles when dropped in regular fluid/base fluid/host fluid is to enhance the thermal conductivity of the host fluid. The inclusion of nanomaterial not only increases the thermal conductivity but also increases the base fluid viscosity (Wu et al. [21], Wang et al. [22], Garg et al. [23] and Lee et al. [24]). For this purpose, several types of nanomaterials, such as carbides, oxides and iron, and so on, are available in the market with their specific usage/characteristics and applications. For example, nanomaterial can be used as a nanolubricants, friction reductant, anti-wear agent and additive to tribological performance.

Oxides, such as copper (CuO_2) and titanium oxides (TiO_2), can be used as an additive to lubricants. The combustion of fossil fuels produces injurious gases (CO and NO) that cause air pollution and global warming. To save natural recourses and produce environment-friendly products, currently, nanomaterials are used to enhance the fuel efficiency of the oils [25].

Among the different types of nanomaterial, there is one called molybdenum disulphide nanomaterial MoS_2, used very rarely in nanofluid studies. Although MoS_2 nanoparticles are not focused more, they have several interesting and useful applications. Applications of molybdenum disulphide can be seen in MoS_2-based lubricants such as two-stroke engines, for example, motorcycle engines, automotive CV and universal joints, bicycle coaster brakes, bullets and ski waxes [26]. Moreover, the MoS_2 has a very high boiling point and many researchers have investigated it as a lubricant The first theoretical study on MoS_2-based nanofluid was performed by Shafie et al. [27], where they studied the shape effect of MoS_2 nanoparticles of four different shapes (platelet, cylinder, brick and blade) in convective flow of fluid in a channel filled with saturated porous medium.

By keeping in mind the importance of MoS_2 nanoparticles, this chapter studies the joint analysis of heat and mass transfer in magnetite molybdenum disulphide viscoelastic nanofluid of grade two. The concept of fractional calculus has been used in formulating the generalized model of grade-two fluid. MoS_2 nanoparticles of spherical shapes have been used in engine oil chosen as base fluid. The problem is formulated in fractional form and Laplace transform together. CF derivatives have been used for finding the exact solution of the problem. Results are obtained in tabular and graphical forms and discussed for rheological parameters.

2. Solution of the problem

Let us consider heat and mass transfer analysis in magnetite molybdenum disulphide nanofluid of grade two with viscosity and elasticity effects. MoS_2 nanoparticles in powder form of spherical shape are dissolved in engine oil chosen as base fluid. MoS_2 nanofluid is taken over an infinite plate placed in xy-plane. The plate is chosen in vertical direction along x-axis, and y-axis is transverse to the plate. Electrically conducting fluid in the presence of uniform magnetic B_0 is considered which is taken normal to the flow direction. Magnetic Reynolds number is chosen very small so that induced magnetic field can be neglected. Before the time start, both the fluid and plate are stationary with ambient temperature T_∞ and ambient concentration C_∞. At time $t = 0^+$, both the plate and fluid starts to oscillate in its own direction with constant amplitude U and frequency ω. Schematic diagram is shown in **Figure 1**.

Under these assumptions, the problem is governed by the following system of differential equations:

$$\rho_{nf}\frac{\partial u}{\partial t} = \mu_{nf}\frac{\partial^2 u}{\partial y^2} + \alpha_1\frac{\partial^3 u}{\partial t\partial y^2} - \sigma_{nf}B_0^2 u + g(\rho\beta_T)_{nf}(T - T_\infty) + g(\rho\beta_C)_{nf}(C - C_\infty), \tag{1}$$

$$(\rho c_p)_{nf}\frac{\partial T}{\partial t} = k_{nf}\frac{\partial^2 T}{\partial y^2}, \tag{2}$$

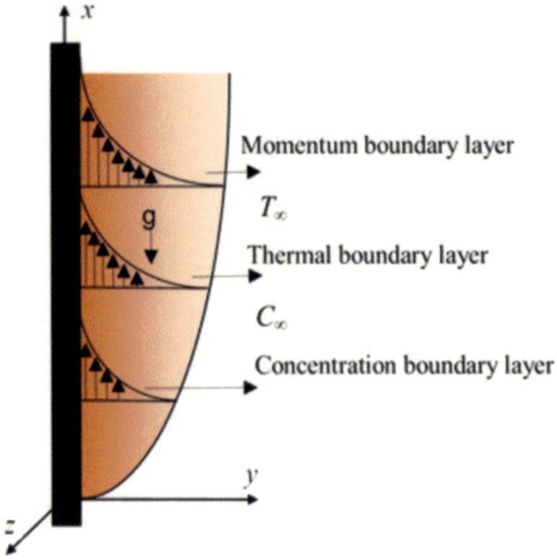

Figure 1. Schematic Diagram of the flow.

$$\frac{\partial C}{\partial t} = D_{nf}\frac{\partial^2 C}{\partial y^2}, \tag{3}$$

where $\rho_{nf}, \sigma_{nf}, \mu_{nf}(\beta_T)_{nf}, (\beta_C)_{nf}, k_{nf}, (\rho C_p)_{nf}, D_{nf}$ are the density, electrical conductivity, viscosity, thermal expansion coefficient, coefficient of concentration, thermal conductivity, heat capacity and mass diffusivity of nanofluid. α_1 shows second two parameters, and g denotes acceleration due to gravity.

The appropriate initial and boundary conditions are

$$\begin{array}{llll} u(y,0)=0 & T(y,0)=T_\infty & C(y,0)=C_\infty & y>0, \\ u(0,t)=UH(t)\cos\omega t & T(0,t)=T_w & C(0,t)=C_w & t>0, \\ u(\infty,t)=0 & T(\infty,t)=T_\infty & C(\infty,t)=C_\infty & as\ y\to\infty, \quad t>0. \end{array} \tag{4}$$

For nanofluids, the expressions for $\rho_{nf}, \ \mu_{nf}, \rho\beta_{nf}, \ (\rho c_p)_{nf}$ are given by:

$$\mu_{nf} = \frac{\mu_f}{(1-\phi)^{2.5}}, \quad \rho_{nf} = (1-\phi)\rho_f + \phi\rho_s, \quad (\rho\beta_T)_{nf} = (1-\phi)(\rho\beta)_f + \phi(\rho\beta)_s,$$

$$(\rho\beta_C)_{nf} = (1-\phi)(\rho\beta)_f + \phi(\rho\beta)_s, (\rho c_p)_{nf} = (1-\phi)(\rho c_p)_f$$
$$+\phi(\rho c_p)_s, \quad \sigma_{nf} = \sigma_f\left(1+\frac{3(\sigma-1)\phi}{(\sigma+2)-(\sigma-1)\phi}\right),$$

$$\sigma = \frac{\sigma_s}{\sigma_f}, \frac{k_{nf}}{k_f} = \frac{(k_s+2k_f)-2\phi(k_f-k_s)}{(k_s+2k_f)+\phi(k_f-k_s)}, \quad D_{nf} = (1-\phi)D_f. \tag{4a}$$

http://dx.doi.org/10.5772/intechopen.72863

Model	$\rho\left(kgm^{-3}\right)$	$C_p\left(Jkg^{-1}K^{-1}\right)$	$C_p\left(Jkg^{-1}K^{-1}\right)$	$\beta \times 10^{-5}\left(K^{-1}\right)$
Engine oil	863	2048	0.1404	0.00007
MoS_2	5.06×10^3	397.21	904.4	2.8424

Table 1. Numerical values of thermophysical properties.

where ϕ describes the volume fraction of nanoparticles. The subscripts s and f stands for solid nanoparticles and base fluid, respectively. The numerical values of physical properties of nanoparticle and base fluid are mentioned in **Table 1**.

Introducing the following dimensional less variables

$$v = \frac{u}{U}, \ \xi = \frac{U}{\nu}y, \ \tau = \frac{U^2}{\nu}t, \ \theta = \frac{T - T_\infty}{T_w - T_\infty}, \quad \Phi = \frac{C - C_\infty}{C_w - C_\infty}.$$

into Eqs. (1)–(4), we get

$$\frac{\partial v}{\partial \tau} = \frac{1}{\mathrm{Re}}\frac{\partial^2 v}{\partial \xi^2} + \frac{\beta}{a_1}\frac{\partial^3 v}{\partial \tau \partial \xi^2} - M_1 v + Gr\phi_2\theta + Gm\phi_3\Phi, \tag{5}$$

$$\frac{\mathrm{Pr}\phi_4}{\lambda_{nf}}\frac{\partial \theta}{\partial \tau} = \frac{\partial^2 \theta}{\partial \xi^2}, \tag{6}$$

$$\frac{\partial \Phi}{\partial \tau} = \frac{1}{a_4}\frac{\partial^2 \Phi}{\partial \xi^2}. \tag{7}$$

$$\begin{array}{lll} v(\xi, 0) = 0 & \theta(\xi, 0) = 0 & \Phi(\xi, 0) = 0 \\ v(0, \tau) = \cos\omega\tau & \theta(0, \tau) = 1 & \Phi(0, \tau) = 1 \\ v(\infty, \tau) = 0 & \theta(\infty, \tau) = 0 & \Phi(\infty, \tau) = 0, \end{array} \tag{8}$$

where

$$\mathrm{Re} = (1 - \phi)^{2.5}a_1, \qquad M_1 = \frac{M\phi_1}{a_1}, \qquad \phi_1 = 1 + \frac{3(\sigma - 1)\phi}{(\sigma + 2) - (\sigma - 1)},$$

$$\phi_2 = (1 - \phi)\rho_f + \phi\rho_s\frac{\beta_{Ts}}{\beta_{Tf}}, \quad \phi_3 = (1 - \phi)\rho_f + \phi\rho_s\frac{\beta_{Cs}}{\beta_{Cf}},$$

$$\phi_4 = (1 - \phi) + \phi\frac{(\rho c_p)_s}{(\rho c_p)_f}, \qquad \lambda_{nf} = \frac{k_{nf}}{k_f}, \quad a_1 = (1 - \phi) + \phi\frac{\rho_s}{\rho_f}, \qquad a_4 = \frac{Sc}{(1 - \phi)},$$

where Re is the Reynolds number, $\beta = \frac{\alpha_1 U^2}{\rho_f \nu^2}$ is the non-dimensional second-grade parameter, $M = \frac{\sigma_f B_0^2 \nu}{\rho_f U^2}$ shows the Hartmann number (magnetic parameter), $Gr = \frac{g\nu\beta_{Tf}}{U^3}(T_w - T_\infty)$ is the thermal Grashof number, $Gm = \frac{g\nu\beta_{Cf}}{U^3}(C_w - C_\infty)$ is the mass Grashof number, $\Pr = \frac{(\mu c_p)_f}{k}$ is the Prandtl number and $Sc = \frac{\nu}{D_f}$ is the Schmidt number.

3. Exact solution

In order to develop the generalized second-grade nanofluid model, we replace the partial derivative with respect to τ by CF fractional operator of order α, and Eqs. (5)–(7) can be written as

$$D_\tau^\alpha v(\xi,\tau) = \frac{1}{\text{Re}}\frac{\partial^2 v}{\partial \xi^2} + \frac{\beta}{a_1} D_\tau^\alpha \frac{\partial^3 v}{\partial \tau \partial \xi^2} - M_1 v + Gr\phi_2\theta + Gm\phi_3\Phi, \tag{9}$$

$$\frac{\Pr\phi_4}{\lambda_{nf}} D_\tau^\alpha \theta = \frac{\partial^2 \theta}{\partial \xi^2}, \tag{10}$$

$$a_4 D_\tau^\alpha \Phi = \frac{\partial^2 \Phi}{\partial \xi^2}, \tag{11}$$

where D_τ^α (.) is known as Caputo-Fabrizio time fractional operator and is defined as:

$$D_\tau^\alpha f(\tau) = \frac{1}{1-\alpha}\int_0^\tau \exp\left(\frac{-\alpha(\tau - t)}{1-\alpha}\right) f'''(\tau)dt; \qquad \text{for} \quad 0 < \alpha < 1 \tag{12}$$

Applying Laplace transform to Eqs. (9)–(11) and using the corresponding initial conditions from Eq. (8), we have:

$$\frac{d^2\overline{\theta}(\xi,q)}{d\xi^2} - \frac{b_1 q}{q+\gamma_1}\overline{\theta}(\xi,q) = 0, \tag{13}$$

$$\frac{d^2\overline{\Phi}(\xi,q)}{d\xi^2} - \frac{a_2 q}{q+\gamma_1}\overline{\Phi}(\xi,q) = 0, \tag{14}$$

$$\frac{d^2\overline{v}(\xi,q)}{d\xi^2} - \frac{M_4(q+M_3)}{(q+d_1)}\overline{v}(\xi,q) = -Gr\phi_2\overline{\theta}(\xi,q) - Gm\phi_3\overline{\Phi}(\xi,q). \tag{15}$$

where $b_1 = \frac{\Pr\phi_4}{\lambda_{nf}}\gamma_0$, $\gamma_1 = \alpha\gamma_0$, $a_2 = a_4\gamma_0$, $M_4 = \frac{M_1+\gamma_0}{d_0}$, $M_3 = \frac{M_1\gamma_1}{M_1+\gamma_0}$, $d_1 = \frac{a_1\gamma_1}{a_1+\text{Re}\beta\gamma_0}$, $\gamma_0 = \frac{1}{1-\alpha}$ and $d_0 = \frac{a_1+\text{Re}\beta\gamma_0}{\text{Re}a_1}$.

Boundary conditions are transformed to:

$$\overline{\theta}(0,q) = \frac{1}{q}, \quad \overline{\theta}(\infty,q) = 0, \tag{16}$$

$$\overline{\Phi}(0,q) = \frac{1}{q}, \quad \overline{\Phi}(\infty,q) = 0, \tag{17}$$

http://dx.doi.org/10.5772/intechopen.72863

$$\overline{v}(0,q) = \frac{q}{q^2+\omega^2}, \quad \overline{v}(\infty,q) = 0. \tag{18}$$

Upon solving Eqs. (13)–(15) and using the boundary conditions from Eqs. (16)–(18), we get:

$$\overline{\theta}(\xi,q) = \frac{1}{q}\exp\left(-\xi\sqrt{\frac{b_1 q}{q+\gamma_1}}\right), \tag{19}$$

$$\overline{\Phi}(\xi,q) = \frac{1}{q}\exp\left(-\xi\sqrt{\frac{a_2 q}{q+\gamma_1}}\right), \tag{20}$$

$$\begin{aligned}\overline{v}(\xi,q) = {} & \frac{1}{2}\overline{\chi}\left(\xi\sqrt{M_4},q,-i\omega,M_3,d_1\right)+\frac{1}{2}\overline{\chi}\left(\xi\sqrt{M_4},q,i\omega,M_3,d_1\right)\\ & +Gr_3\overline{\chi}(\xi\sqrt{M_4},q,0,M_3,d_1)+R_0\overline{\chi}(\xi\sqrt{M_4},q,d_2,M_3,d_1)\\ & +R_1\overline{\chi}(\xi\sqrt{M_4},q,d_3,M_3,d_1)+Gm_3\overline{\chi}(\xi\sqrt{M_4},q,0,M_3,d_1)\\ & +R_2\overline{\chi}(\xi\sqrt{M_4},q,d_4,M_3,d_1)+R_3\overline{\chi}(\xi\sqrt{M_4},q,d_5,M_3,d_1)\\ & -Gr_3\overline{\chi}(\xi\sqrt{b_1},q,0,0,\gamma_1)-R_0\overline{\chi}(\xi\sqrt{b_1},q,d_2,0,\gamma_1)\\ & -R_1\overline{\chi}(\xi\sqrt{b_1},q,d_3,0,\gamma_1)-Gm_3\overline{\chi}(\xi\sqrt{a_2},q,0,0,d_1)\\ & -R_2\overline{\chi}(\xi\sqrt{a_2},q,d_4,0,d_1)-R_3\overline{\chi}(\xi\sqrt{a_2},q,d_5,0,d_1),\end{aligned} \tag{21}$$

$$\overline{\chi}(\xi,q,a,b,c) = \frac{1}{q+a}\exp\left(-\xi\sqrt{\frac{q+b}{q+c}}\right). \tag{22}$$

Eqs. (19) and (20) are written in simplified form

$$\overline{\theta}(\xi,q) = \overline{\chi}\left(\xi\sqrt{b_1},q,0,0,\gamma_1\right), \tag{23}$$

$$\overline{\Phi}(\xi,q) = \overline{\chi}(\xi\sqrt{a_2},q,0,0,\gamma_1), \tag{24}$$

where

$$Gr_3 = \frac{Gr_2{\gamma_1}^2}{d_2 d_3},\ Gm_3 = \frac{Gm_2{\gamma_1}^2}{d_4 d_5},\ Gr_2 = \frac{Gr_1}{\delta_1},\ Gm_2 = \frac{Gm_1}{\delta_6},\ R_0 = \frac{-{d_2}^2 Gr_2 + 2d_2 Gr_2\gamma_1 - Gr_2{\gamma_1}^2}{d_2-d_3},$$

$$R_1 = \frac{-{d_3}^2 Gr_2 - 3d_3 Gr_2\gamma_1 + Gr_2{\gamma_1}^2}{d_2-d_3},\ R_2 = \frac{-{d_4}^2 Gm_2 + 2d_4 Gm_2\gamma_1 - Gm_2{\gamma_1}^2}{d_4-d_5},$$

$$R_3 = \frac{{d_5}^2 Gm_2 - 3d_5 Gm_2\gamma_1 + Gm_2{\gamma_1}^2}{d_4-d_5},\ d_2 = \frac{\delta_4}{2}+\sqrt{\frac{{\delta_4}^2+4\delta_5}{2}},\ d_3 = \frac{\delta_4}{2}-\sqrt{\frac{{\delta_4}^2+4\delta_5}{2}},$$

$$d_4 = \frac{\delta_8}{2}+\sqrt{\frac{{\delta_8}^2+4\delta_9}{2}},\ d_5 = \frac{\delta_8}{2}-\sqrt{\frac{{\delta_8}^2+4\delta_9}{2}},\ Gr_1 = \frac{Gr\phi_2}{d_0},\ Gm_1 = \frac{Gm\phi_3}{d_0},\ \delta_1 = b_1 M_4,$$

$$\delta_2 = b_1 d_1 - M_4\gamma_1 - M_4 M_3,\ \ \delta_3 = M_4 M_3 \gamma_1,\ \ \delta_4 = \frac{\delta_2}{\delta_1},\ \ \delta_5 = \frac{\delta_3}{\delta_1},\ \ \delta_6 = a_2 - M_4,$$

$$\delta_7 = a_2 d_1 - M_4\gamma_1 - M_4 M_3,\ \ \delta_8 = \frac{\delta_7}{\delta_6},\ \ \delta_9 = \frac{\delta_3}{\delta_6}.$$

Now, inverse Laplace transform of Eqs. (21), (23) and (24) is:

$$\begin{aligned} v(\xi,\tau) =& \frac{1}{2}\chi\left(\xi\sqrt{M_4},\ \tau, -i\omega,\ M_3,\ d_1\right) + \frac{1}{2}\chi\left(\xi\sqrt{M_4}, \tau, i\omega, M_3, d_1\right) \\ &+ Gr_3\chi\left(\xi\sqrt{M_4}, \tau, 0\ , M_3, d_1\right) + R_0\chi\left(\xi\sqrt{M_4}, \tau,\ d_2, M_3, d_1\right) \\ &+ R_1\chi\left(\xi\sqrt{M_4}, \tau, d_3, M_3, d_1\right) + Gm_3\chi\left(\xi\sqrt{M_4}, \tau, 0\ , M_3, d_1\right) \\ &+ R_2\chi\left(\xi\sqrt{M_4}, \tau, d_4\ , M_3, d_1\right) + R_3\chi\left(\xi\sqrt{M_4}, \tau, d_5, M_3, d_1\right) - Gr_3\chi\left(\xi\sqrt{b_1}, \tau, 0, 0, \alpha\gamma_0\right) \\ &- R_0\chi\left(\xi\sqrt{b_1}, \tau, d_2, 0, \alpha\gamma_0\right) - R_1\chi\left(\xi\sqrt{b_1}, \tau, d_3, 0, \alpha\gamma_0\right) - Gm_3\chi(\xi\sqrt{a_2}, \tau, 0, 0, d_1) \\ &- R_2\chi(\xi\sqrt{a_2}, \tau, d, 0, d_1) - R_3\chi(\xi\sqrt{a_2}, \tau, d_5, 0, d_1), \end{aligned} \tag{25}$$

$$\theta(\xi,\tau) = \chi\left(\xi\sqrt{b_1}, \tau, 0, 0, \gamma_1\right), \tag{26}$$

$$\Phi(\xi,\tau) = \chi(\xi\sqrt{a_2}, \tau, 0, 0, \gamma_1), \tag{27}$$

where

$$\overline{\chi}(\xi, q; a, b, c) = \frac{1}{q+a}\exp\left(-\xi\sqrt{\frac{q+b}{q+c}}\right), \tag{28}$$

$$\chi(\xi,\tau,a,b,c) = e^{-a\tau-\xi} - \frac{\xi\sqrt{b-c}}{2\sqrt{\pi}}\int_0^\infty\int_0^\tau \frac{e^{a\tau}}{\sqrt{\tau}}\exp\left(a\tau - c\tau - \frac{\xi^2}{4u} - u\right)I_1\left(2\sqrt{(b-c)ut}\right)dt\,du. \tag{29}$$

Skin friction, Nusselt number and Sherwood number.

Skin friction, Nusselt number and Sherwood number are defined as:

$$C_f = \frac{1}{(1-\phi)^{2.5}}\frac{\partial v}{\partial \xi} + \beta\frac{\partial^2 v}{\partial\xi\partial\tau}\bigg|_{\xi=0}, \tag{30}$$

$$Nu = \frac{\partial\theta}{\partial\xi}\bigg|_{\xi=0}, \tag{31}$$

http://dx.doi.org/10.5772/intechopen.72863

$$Sh = \left.\frac{\partial \Phi}{\partial \xi}\right|_{\xi=0}. \tag{32}$$

Velocity field for regular grade-two fluid without mass transfer.

For $Gm = \phi = 0$ in Eq. (22) reduce to the following form:

$$\begin{aligned} v(\xi,\tau) =& \frac{1}{2}\chi\left(\xi\sqrt{M_4},\ \tau, -i\omega,\ M_3,\ d_1\right) + \frac{1}{2}\chi\left(\xi\sqrt{M_4}, \tau, i\omega, M_3, d_1\right) + Gr_3\chi\left(\xi\sqrt{M_4}, \tau, 0\ , M_3, d_1\right) \\ & + R_0\chi\left(\xi\sqrt{M_4}, \tau,\ d_2, M_3, d_1\right) + R_1\chi\left(\xi\sqrt{M_4}, \tau, d_3, M_3, d_1\right) - Gr_3\chi\left(\xi\sqrt{b_1}, \tau, 0, 0, \alpha\gamma_0\right) \\ & - R_0\chi\left(\xi\sqrt{b_1}, \tau, d_2, 0, \alpha\gamma_0\right) - R_1\chi\left(\xi\sqrt{b_1}, \tau, d_3, 0, \alpha\gamma_0\right), \end{aligned} \tag{33}$$

where $b_1 = \Pr\gamma_0,\ M_4 = \frac{M+\gamma_0}{d_0}, M_3 = \frac{M\gamma_1}{M+\gamma_0}, d_1 = \frac{\gamma_1}{d_0}, d_0 = 1 + \beta\gamma_0,\ Gr_1 = \frac{Gr}{d_0},$

which is quite identical to the solution of Sheikh et al. [19] for $\frac{1}{k} = R = 0$.

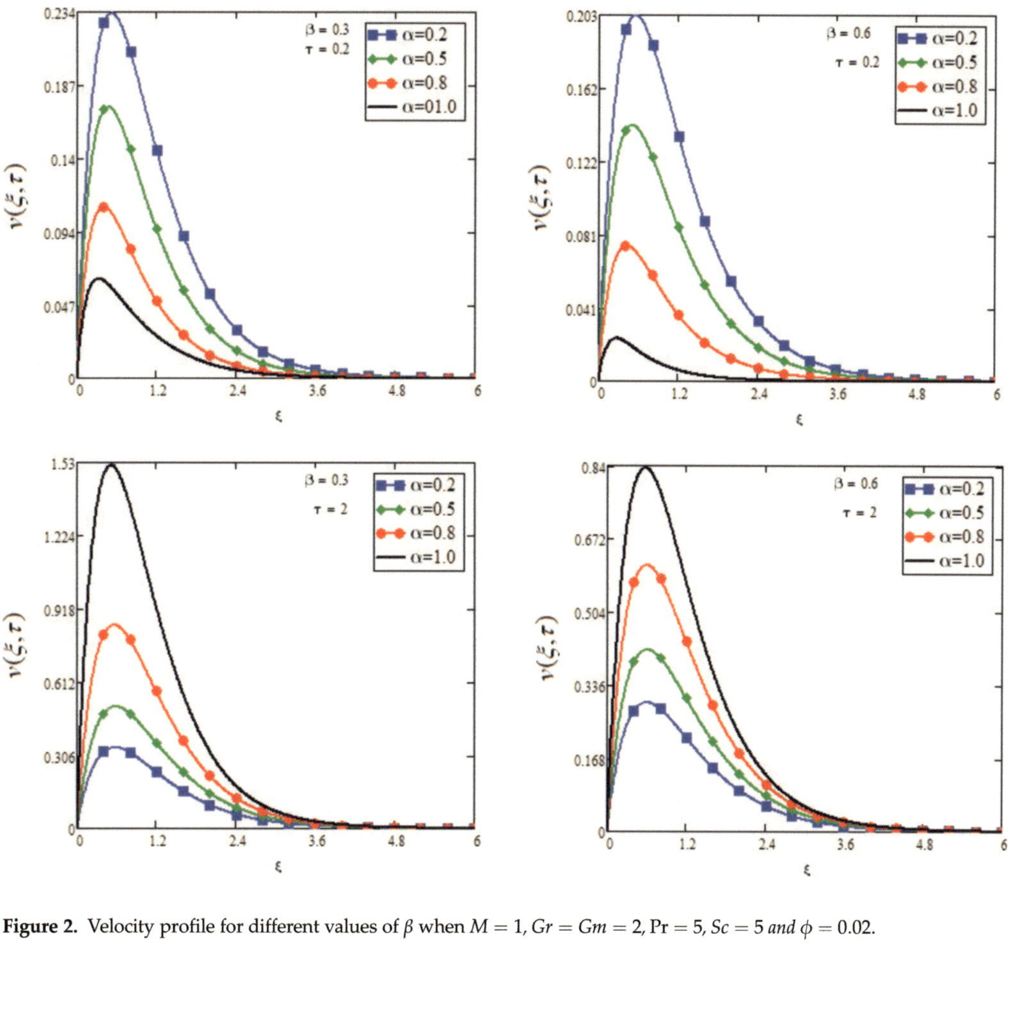

Figure 2. Velocity profile for different values of β when $M = 1, Gr = Gm = 2, \Pr = 5, Sc = 5$ *and* $\phi = 0.02$.

4. Graphical discussion

A fractional model for the outflow of the second-grade fluid with nanoparticles over an isothermal vertical plate is studied. The coupled partial differential equations with Caputo-Fabrizio time-fractional derivatives are solved analytically via Laplace transform method. Furthermore, the influence of different embedded parameters such as α, ϕ, β, M, t Gr, Gm and Sc is shown graphically.

Figures 2–7 depict the effect of α on $v(\xi, \tau)$ for two different values of time. It is clear from the figures that for smaller value of τ, when (τ = 0.2) fractional velocity is larger than classical velocity and for larger value of τ, when (τ =2) fractional velocity is less than classical velocity. Clearly, increasing values of α decrease $v(\xi, \tau)$.

Figure 2 represents the influence of β on both the velocity and microrotation profiles. A decreasing behaviour is observed for increasing values of β in both cases. In this figure, the comparison of second-grade fluid velocity with Newtonian fluid velocity is plotted. It is

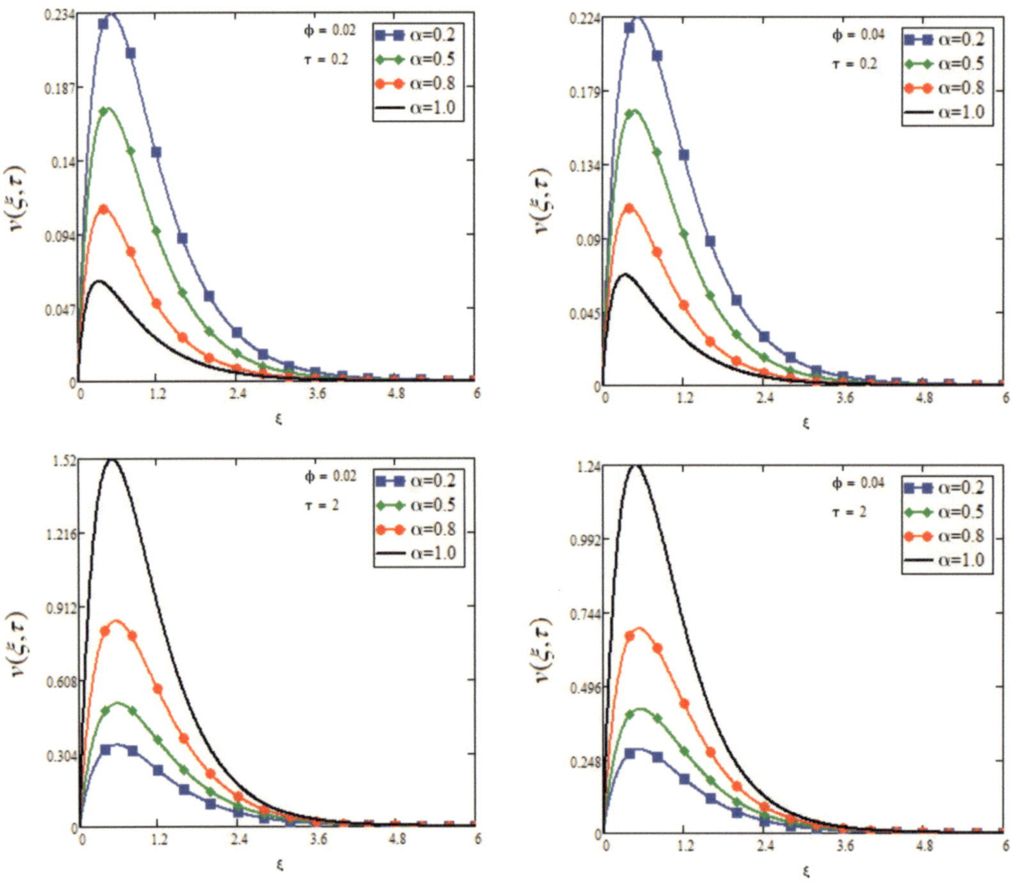

Figure 3. Velocity profile for different values of ϕ when $M = 1, Gr = Gm = 2, \Pr = 5, Sc = 5$ *and* $\beta = 0.3$.

http://dx.doi.org/10.5772/intechopen.72863

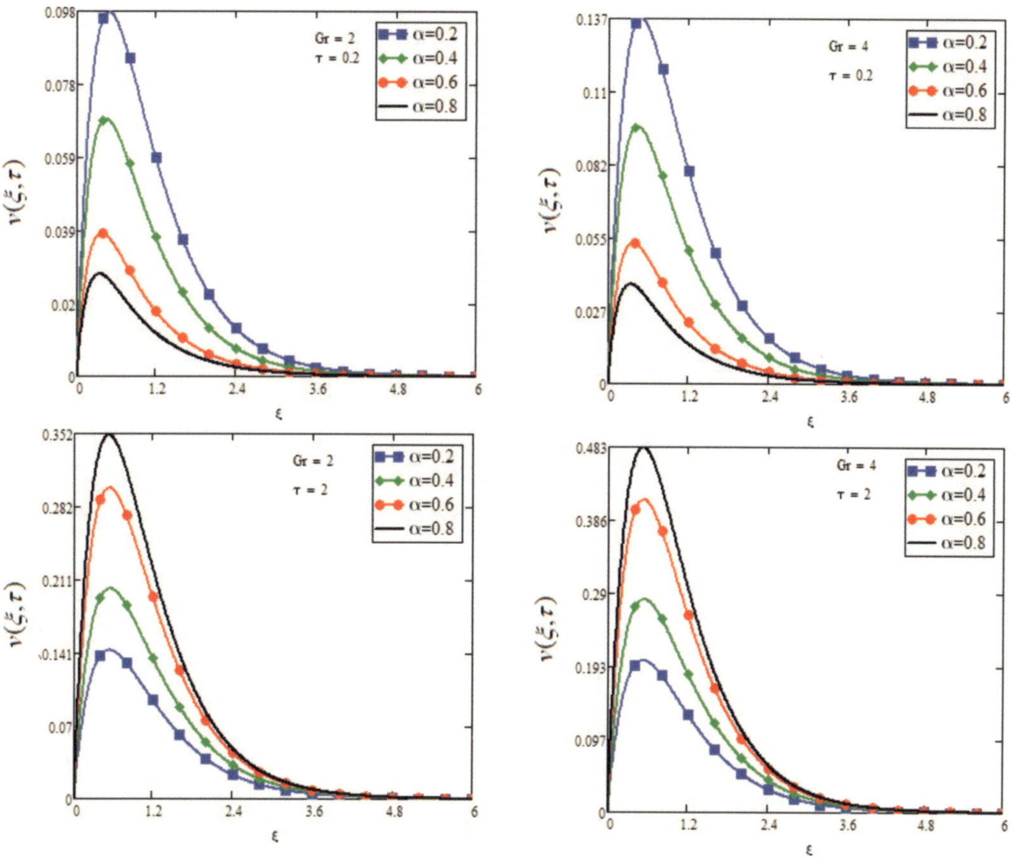

Figure 4. Velocity profile for different values of Gr when $M = 1, \phi = 0.02, Gm = 2, \Pr = 5, Sc = 5 \; and \; \beta = 0.3$.

obvious that the boundary layer thickness of second-grade fluid velocity is greater as compared to boundary layer thickness of Newtonian fluid. More clearly, the velocity of second-grade fluid is smaller than Newtonian fluid.

Figure 3 shows the influence of ϕ on the flow. It was found that the velocity of fluid decreases with the increase in ϕ due to the increase in viscosity. Because by increasing volume fraction, the fluid becomes more viscous, which leads to a decrease in the fluid velocity.

Figures 4 and **5** show the influence of thermal Grashof number Gr and mass Grashof number Gm on velocity and microrotation. Increasing values of both of these parameters are responsible for the rise in buoyancy forces and reducing viscous forces, which result in an increase in fluid velocity and magnitude of microrotation.

Figure 6 depicts the MHD effect on velocity. In this type of flows, magnetic force results in achieving steady state much faster than the non-MHD flows. Moreover, increasing values of M enhances the Lorentz forces, as a result decelerates the fluid velocity. **Figure 7** illustrates variations in velocity for different values of Schmidt number, Sc. It shows that velocity decreases

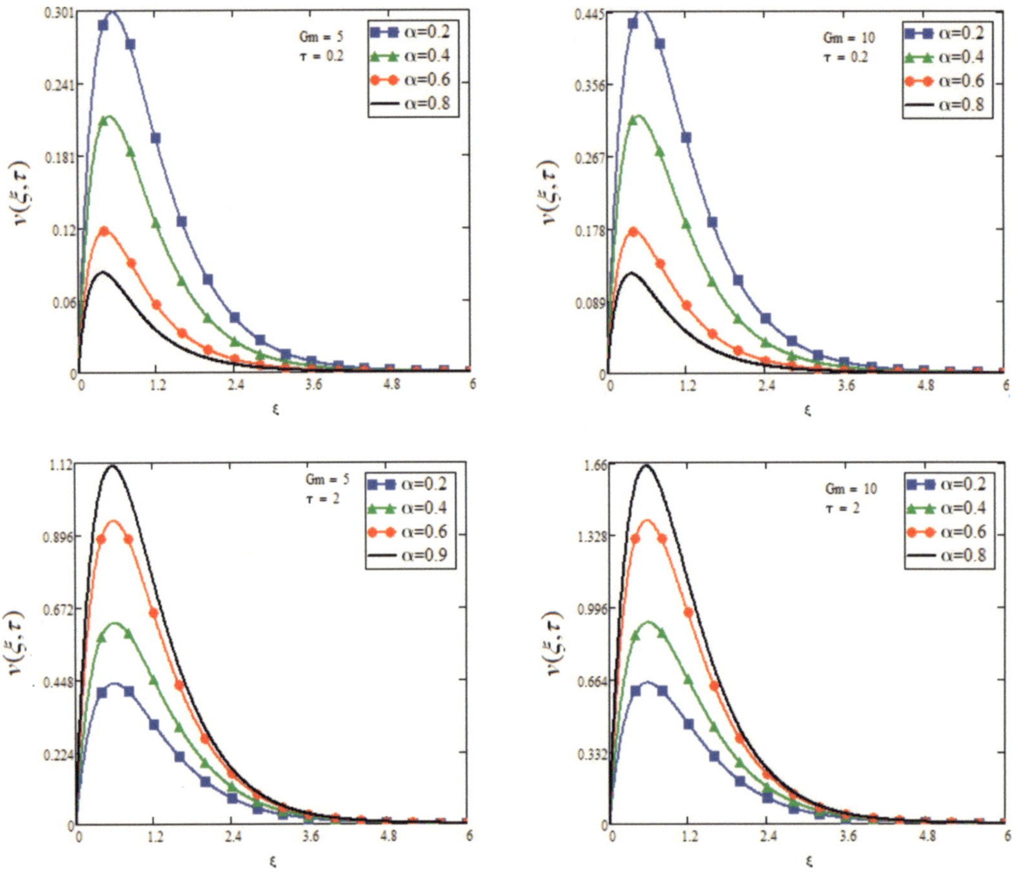

Figure 5. Velocity profile for different values of Gm when $M = 1, Gr = 2, \phi = 0.02, \mathrm{Pr} = 5, Sc = 5$ and $\beta = 0.3$.

when Sc value increases. The effect of Schmidt number on velocity is identical to that of the magnetic parameter. The influence of phase angle $\omega\tau$ on the velocity profile is shown in **Figure 8**. The velocity is showing fluctuating behaviour.

In order to show the effect of α, τ and ϕ on the temperature profile in **Figure 9**, it is found that temperature increasing with increasing value of ϕ. **Figure 10** shows the effect of α and τ on temperature profile. This figure shows the effect of α on the temperature profile for two different values of τ. For smaller value of τ $(\tau = 0.2)$, classical temperature is less than fractional temperature, and for larger value, when $\tau = 2$, then the graph shows opposite behaviour. **Figure 11** shows the comparison of present solution with published result of Sheikh et al. [19]. It is noted that in the absence of porosity and radiation, the present result is similar to those obtained in [19. See **Figure 9**], which shows the validity of our obtained results.

Variations in skin friction, Nusselt number and Sherwood number are shown in **Tables 2–4**. The effect of $\beta, \alpha, Gr, Gm, M, Sc, \phi, \omega\tau$ and τ on the skin friction is studied in **Table 2**. It is found

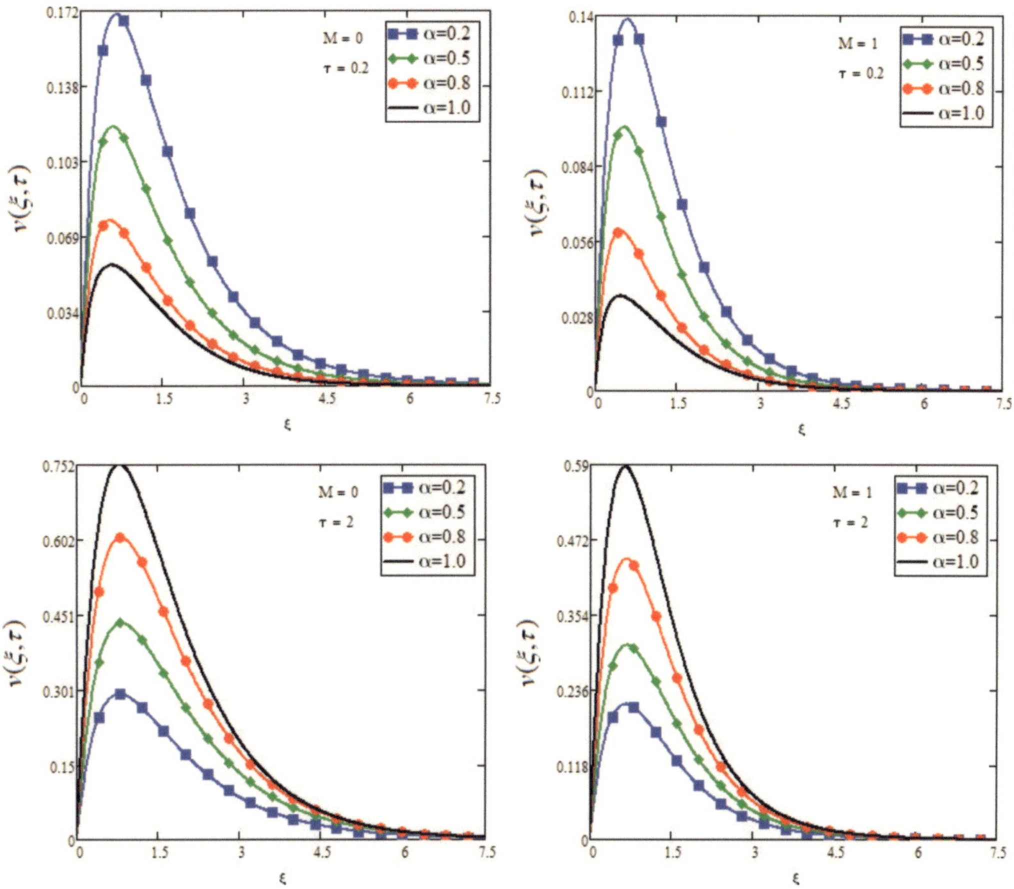

Figure 6. Velocity profile for different values of M when $\phi = 0.02$, $Gr = Gm = 2$, $\Pr = 5$, $Sc = 5$ and $\beta = 0.3$.

that skin friction increases when there is an increase in $\alpha, Gr, Gm, M, Sc, \omega\tau$ and τ.but it is noticed that for increasing value of β and ϕ, skin friction decreases. It is due to the fact that when ϕ increases, it gives rise to lubricancy of the oil. **Table 3** represents the effect of α, ϕ and τ on Nusselt number. As values of α, ϕ and τ increase, Nusselt number decreases. From **Table 4**, it is clear that when Sc increases, Sherwood number increases, and an increase in τ decreases Sherwood number.

5. Conclusion remarks

Unsteady MHD flow of generalized second-grade fluid along with nanoparticles has been analyzed. The exact solution has been obtained for velocity, temperature and concentration profile via the Laplace transform technique. The effects of various physical parameters are studied in various plots and tables with the following conclusions:

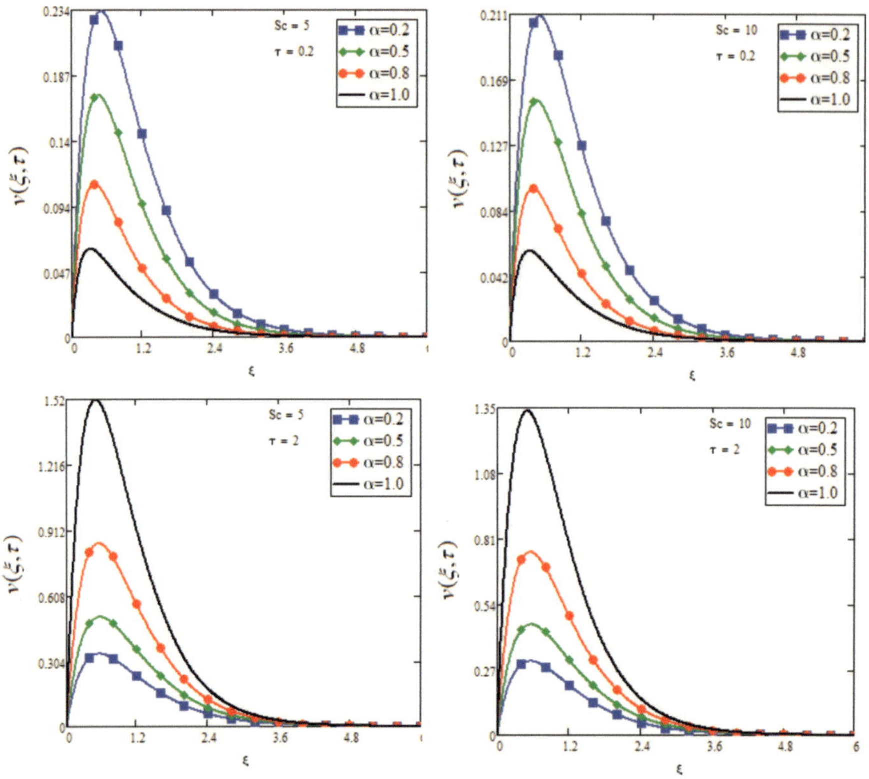

Figure 7. Velocity profile for different values of *Sc* when $M = 1, Gr = Gm = 2, \Pr = 5, \phi = 0.02$ *and* $\beta = 0.3$.

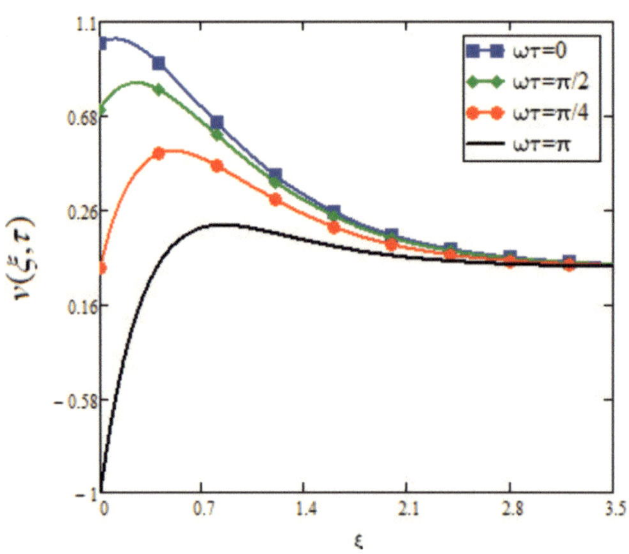

Figure 8. Velocity profile for different values of ω.

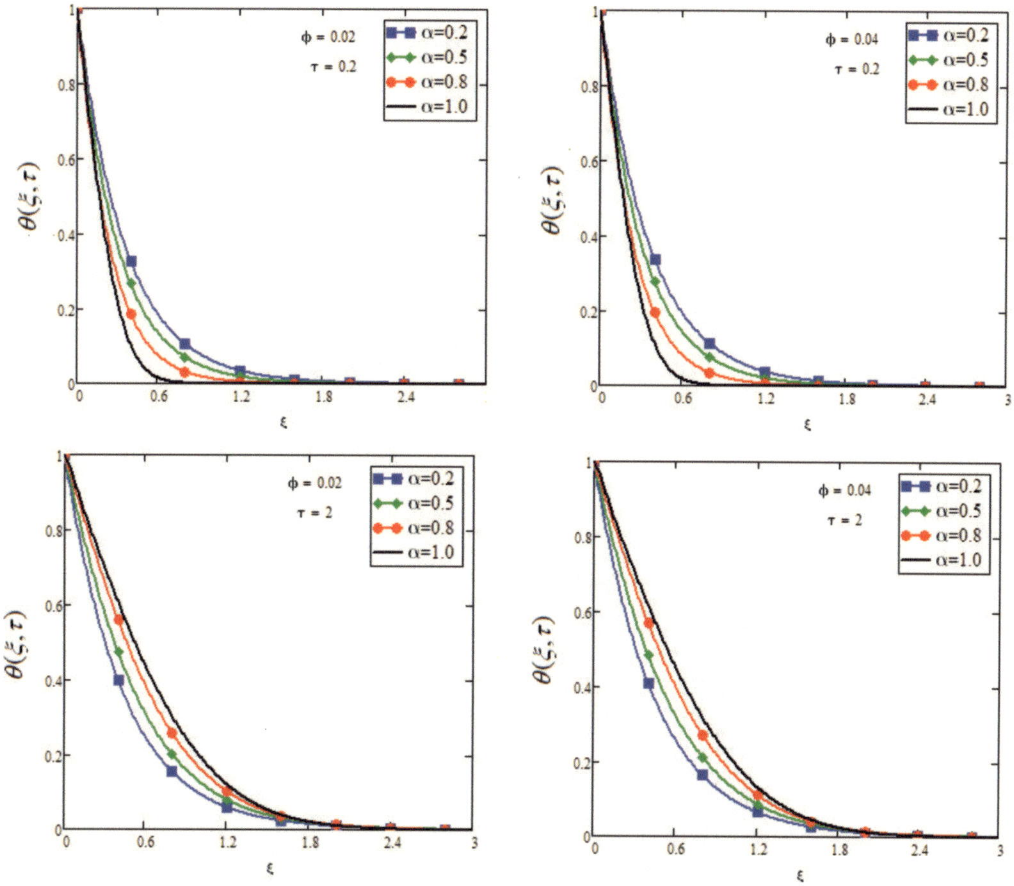

Figure 9. Temperature profile for different values of ϕ when $\Pr = 5$.

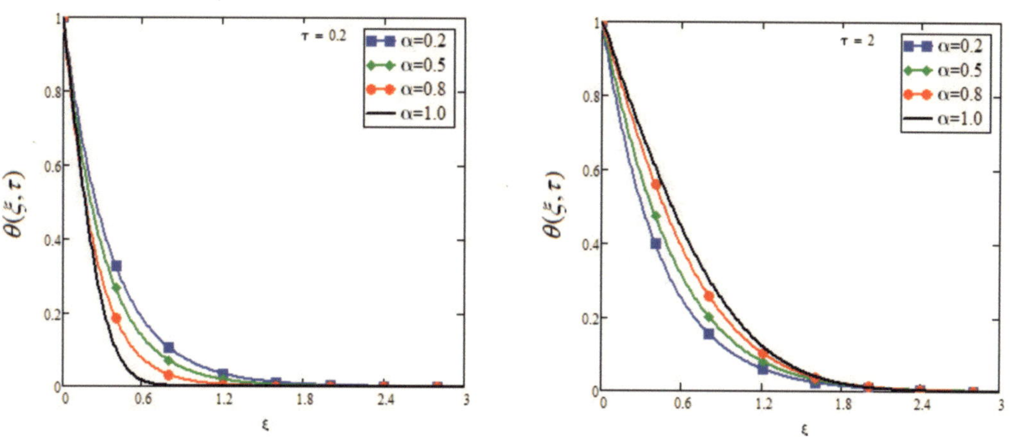

Figure 10. Temperature profile for different values of time parameter.

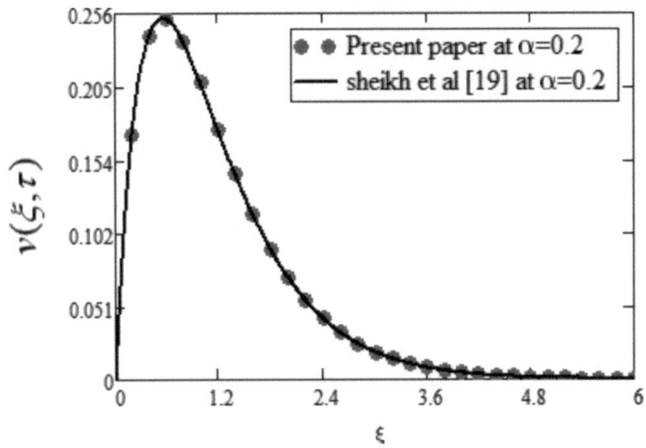

Figure 11. Comparison of this study with Sheikh et al. [19], when $\frac{1}{k} = R = 0$.

β	α	Gr	Gm	M	Sc	ϕ	$\omega\tau$	τ	C_f
0.3	0.2	5	5	1	5	0.02	$\frac{\pi}{2}$	2	2.867
$\overline{0.6}$	0.2	5	5	1	5	0.02	$\frac{\pi}{2}$	2	2.857
0.3	$\overline{0.4}$	5	5	1	5	0.02	$\frac{\pi}{2}$	2	3.66
0.3	0.2	$\overline{7}$	5	1	5	0.02	$\frac{\pi}{2}$	2	4.121
0.3	0.2	5	$\overline{\overline{7}}$	1	5	0.02	$\frac{\pi}{2}$	2	4.086
0.3	0.2	5	5	$\overline{2}$	5	0.02	$\frac{\pi}{2}$	2	2.869
0.3	0.2	5	5	1	$\overline{7}$	0.02	$\frac{\pi}{2}$	2	3.135
0.3	0.2	5	5	1	5	$\overline{0.04}$	$\frac{\pi}{2}$	2	0.599
0.3	0.2	5	5	1	5	0.02	$\overline{\pi}$	2	4.371
0.3	0.2	5	5	1	5	0.02	$\frac{\pi}{2}$	$\overline{4}$	3.719

Table 2. Effect of various parameters on the skin friction.

α	ϕ	τ	Nu
0.2	0.02	2	−1.923
$\overline{0.4}$	0.02	2	−1.606
0.2	$\overline{0.04}$	2	−1.872
0.2	0.02	$\overline{4}$	−1.526

Table 3. Effect of various parameters on the Nusselt number.

Sc	τ	Sh
5	2	–2.018
$\overline{7}$	2	–2.388
5	$\overline{4}$	–1.646

Table 4. Effect of various parameters on the Sherwood number.

- With increase in volume friction ϕ of nanofluid, lubricancy of the fluid increases.
- An increase in second-grade parameter β leads to a decrease in fluid velocity.
- The velocity profile shows different behaviour for fractional parameter for different values of time.
- In limited cases, the obtained solutions reduced to the solution of Sheikh et al. [19].

Author details

Farhad Ali[1,2,3]*, Madeha Gohar[3], Ilyas Khan[4], Nadeem Ahmad Sheikh[3], Syed Aftab Alam Jan[3] and Muhammad Saqib[3]

*Address all correspondence to: farhad.ali@tdt.edu.vn

1 Computational Analysis Research Group, Ton Duc Thang University, Ho Chi Minh City, Vietnam

2 Faculty of Mathematics and Statistics, Ton Duc Thang University, Ho Chi Minh City, Vietnam

3 Department of Mathematics, City University of Science and Information Technology, Peshawar, Khyber Pakhtunkhwa, Pakistan

4 Basic Engineering Sciences Department, College of Engineering Majmaah University, Majmaah Saudi Arabia

References

[1] Nonnenmacher TF, Metzler R. On the Riemann-Liouville fractional calculus and some recent applications. Fractals. 1995;**3**(03):557-566

[2] Sabatier J, Agrawal OP, Machado JT. Advances in fractional calculus (Vol. 4, no. 9). Dordrecht: Springer; 2007

[3] Laskin N. Fractional schrödinger equation. Physical Review E. 2002;**66**(5) 056108

[4] Naber M. Time fractional Schrödinger equation. Journal of Mathematical Physics. 2004; **45**(8):3339-3352

[5] Baleanu D, Golmankhaneh AK, Golmankhaneh AK. The dual action of fractional multi time Hamilton equations. International Journal of Theoretical Physics. 2009;**48**(9):2558-2569

[6] Mainardi F. Fractional Calculus and Waves in Linear Viscoelasticity: An Introduction to Mathematical Models. London, UK: World Scientific; 2010

[7] Hayat T, Nadeem S, Asghar S. Periodic unidirectional flows of a viscoelastic fluid with the fractional Maxwell model. Applied Mathematics and Computation. 2004;**151**(1): 153-161

[8] Qi H, Jin H. Unsteady rotating flows of a viscoelastic fluid with the fractional Maxwell model between coaxial cylinders. Acta Mechanica Sinica. 2006;**22**(4):301-305

[9] Meral FC, Royston TJ, Magin R. Fractional calculus in viscoelasticity: An experimental study. Communications in Nonlinear Science and Numerical Simulation. 2010;**15**(4): 939-945

[10] Qi H, Xu M. Unsteady flow of viscoelastic fluid with fractional Maxwell model in a channel. Mechanics Research Communications. 2007;**34**(2):210-212

[11] Carpinteri A, Mainardi F. Fractals and fractional calculus in continuum mechanics (Vol. 378). Verlag GmbH Wien: Spring; 2014

[12] Hilfer R. Threefold introduction to fractional derivatives. Anomalous transport: Foundations and applications. 2008:17-73

[13] Gorenflo R, Mainardi F, Moretti D, Paradisi P. Time fractional diffusion: A discrete random walk approach. Nonlinear Dynamics. 2002;**29**(1):129-143

[14] Ali F, Jan SAA, Khan I, Gohar M, Sheikh NA. Solutions with special functions for time fractional free convection flow of brinkman-type fluid. The European Physical Journal Plus. 2016;**131**(9):310

[15] Shahid N. A study of heat and mass transfer in a fractional MHD flow over an infinite oscillating plate. SpringerPlus. 2015;**4**(1):640

[16] Caputo M, Fabrizio M. A new definition of fractional derivative without singular kernel. Progress in Fractional Differentiation and Applications. 2015;**1**(2):1-13

[17] Shah NA, Khan I. Heat transfer analysis in a second grade fluid over and oscillating vertical plate using fractional Caputo–Fabrizio derivatives. The European Physical Journal C. 2016;**76**(7):1-11

[18] Ali F, Saqib M, Khan I, Sheikh NA. Application of Caputo-Fabrizio derivatives to MHD free convection flow of generalized Walters'-B fluid model. The European Physical Journal Plus. 2016;**131**(10):377

[19] Sheikh NA, Ali F, Khan I, Saqib M. A modern approach of Caputo–Fabrizio time-fractional derivative to MHD free convection flow of generalized second-grade fluid in a porous medium. Neural Computing and Applications. 1-11

[20] Choi SU. Enhancement thermal conductivity of fluids with nanoparticles. International Mechanical Engineering Congress and exposition. 1995

[21] Wu YY, Kao MJ. Using TiO_2 nanofluid additive for engine lubrication oil. Industrial Lubrication and Tribology. 2011;**63**(6):440-445

[22] Wang X, Xu X, Choi S, U S. Thermal conductivity of nanoparticle-fluid mixture. Journal of Thermophysics and Heat Transfer. 1999;**13**(4):474-480

[23] Garg J, Poudel B, Chiesa M, Gordon JB, Ma JJ, Wang JB, et al. Enhanced thermal conductivity and viscosity of copper nanoparticles in ethylene glycol nanofluid. Journal of Applied Physics. 2008;**103**(7):074301

[24] Lee JH, Hwang KS, Jang SP, Lee BH, Kim JH, Choi SU, Choi CJ. Effective viscosities and thermal conductivities of aqueous nanofluids containing low volume concentrations of Al_2O_3 nanoparticles. International Journal of Heat and Mass Transfer. 2008;**51**(11):2651-2656

[25] Abdullah MIHC, Abdollah MFB, Amiruddin H, Nuri M, Rashid N. Effect of hBN/Al_2O_3 nanoparticle additives on the tribological performance of engine oil. Jurnal Teknologi. 2014;**66**(3):1-6

[26] Winer WO. Molybdenum disulfide as a lubricant: A review of the fundamental knowledge. Wear. 1967;**10**(6):422-452

[27] Shafie S, Gul A, Khan I. Molybdenum disulfide nanoparticles suspended in water-based nanofluids with mixed convection and flow inside a channel filled with saturated porous medium. AIP Conference Proceedings. 2016, October;**1775**:030042

Chapter 11

Performance Evaluation Criterion of Nanofluid

Sudarmadji Sudarmadji, Bambang SAP and Santoso

Additional information is available at the end of the chapter

http://dx.doi.org/10.5772/intechopen.74610

Abstract

In this chapter, we will discuss the increase of heat transfer as well as the increase in pressure drop to determine whether nanofluid is feasible for use in practical applications. Addition of nanoparticles will change the thermal properties of the cooling fluid, by calculation with performance evaluation criterion (PEC). If PEC < 1, then the heat transfer performance is less than the pumping power, so the system is not feasible for use in increasing heat transfer. If PEC = 1, then the heat transfer performance is smaller equal to the pumping power so that the system does not have an impact on increasing heat transfer. If PEC > 1, then the heat transfer performance is higher than the energy used to drive the fluid or pumping power, and then it can be accepted as a solution to the problem of increasing heat transfer, so that the system is feasible for use in practical applications.

Keywords: nanofluid, performance evaluation criterion, heat transfer enhancement, pressure drop

1. Introduction

Today, energy becomes very important in human life; it is used to help human beings activity every day. Without energy, humans will be paralyzed and cannot do anything. Energy savings are a challenging topic to be investigated by scientists. The heat exchanger is a device widely used in the chemical, automotive, electronic, and food industries and involves heat transfer processes that directly influence the economy of these industries. Enhancement of heat transfer is one of the ways used for energy saving. Various methods have been done for the use of energy savings; both are the passive and active methods [1–3]. Passive enhancement methods include a surface coating, rough and finned surfaces, insertion devices, curved geometry, and nanofluid. Active method enhancement needed external energy to maintain the enhancement of the mechanism. Active methods include surface vibration, fluid fibration, electrohydrodynamics (EHD), and the use of the magnetic field. Over the past few decades, the effect of high electric fields on the rate of heat transfer is widely known as electrohydrodynamics (EHD), [4] and the effect of the magnetic field on magnetic iron oxide particle (Fe_3O_4) numerically has been investigated using control volume finite element method (CVFEM) [5]. However, when the available space is limited by the process, it is interesting to use the device with the same size or smaller with better performance. It can be achieved by modifying the cooling with higher thermal conductivity to enhance the heat transfer coefficient when compared with that of conventional fluids for the same geometry.

The objective of this chapter is to how to decrease the size of the thermal system or to increase their transferred thermal power using nanofluid. Nanofluids are colloidal suspensions of nanoparticles which are engineered to have the thermal conductivity higher than that of the base fluid and which can be used for this purpose [6, 7]. However, together with thermal conductivity enhancement, the viscosity is increased, and the gain in transferred heat is paid regarding pumping power. There is a competition between heat transfer rate and pumper power.

2. Nanofluids

The cooling process is the process of heat transfer from a heat source and removed to the environment (heat sink) at the lower temperature. In forced convection heat transfer occurs at high-temperature to low-temperature fluids, which are separated by pipe walls. Increased heat dissipation is done by increasing the convection surface area, i.e., using fins. This method has been abandoned because of the higher the dimensions, and the more massive the equipment consequently the price becomes expensive. The second way is to increase the fluid flow rate, but it will have an impact on the greater use of energy, which causes the system to be inefficient.

The new method to increase heat transfer is to improve thermal properties of fluids especially heat conductivity. This way is by adding solid particles with high thermal conductivity into the cooling fluid with low thermal conductivity as shown in **Figure 1**.

Started by Maxwell as the pioneer [8], adding solid particles of micro size into the fluid is expected to improve thermal properties of fluids, especially heat conductivity. Because the particle size is large enough, the particles quickly agglomerate and cause clogging the channel. Later, Choi [6] introduced the term nanofluid defined as colloids made of a base fluid and nanoparticle size (1–100 nm). The properties of the fluid increase especially the heat conductivity (k), viscosity (μ), and density (ρ) increase in proportion to the increased particle volume concentration, whereas the specific heat (Cp) decreases proportionally to the increase in volume concentration. The effect of nanoparticle concentration on the physical properties of nanofluid has been studied by describing the variation in the ratio of physical properties of nanofluid to pure water as a function of nanoparticle volume concentration [9]. **Figure 2** shows the effect of nanoparticles Al_2O_3 of nanoparticle concentration on the physical properties of nanofluid, i.e., viscosity, density, and thermal conductivity, which have increased, while the specific heat is slightly decreased compared to the base fluid.

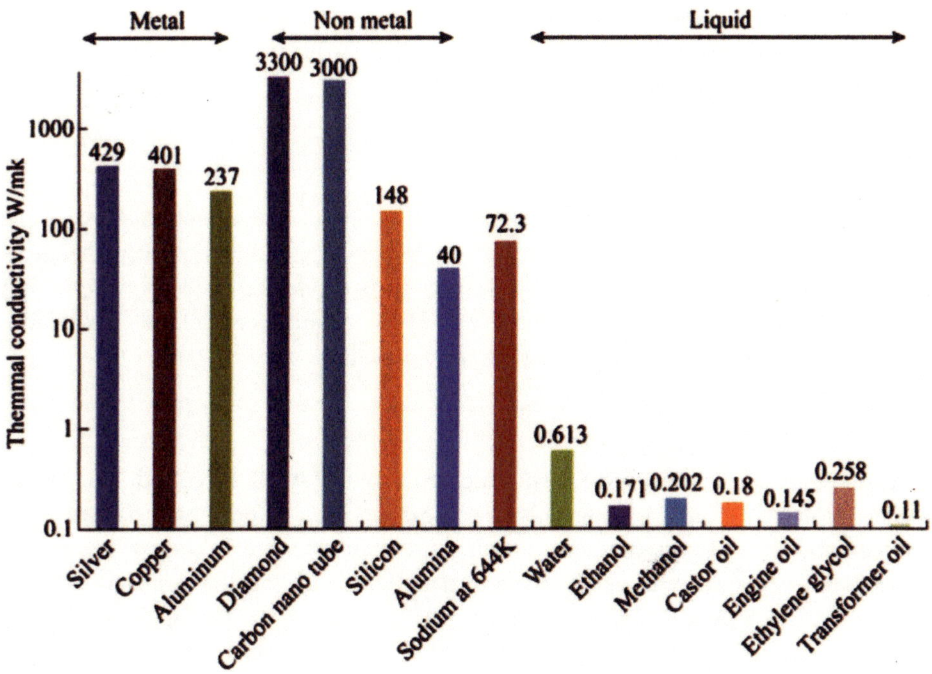

Figure 1. Comparison of thermal conductivity for different materials.

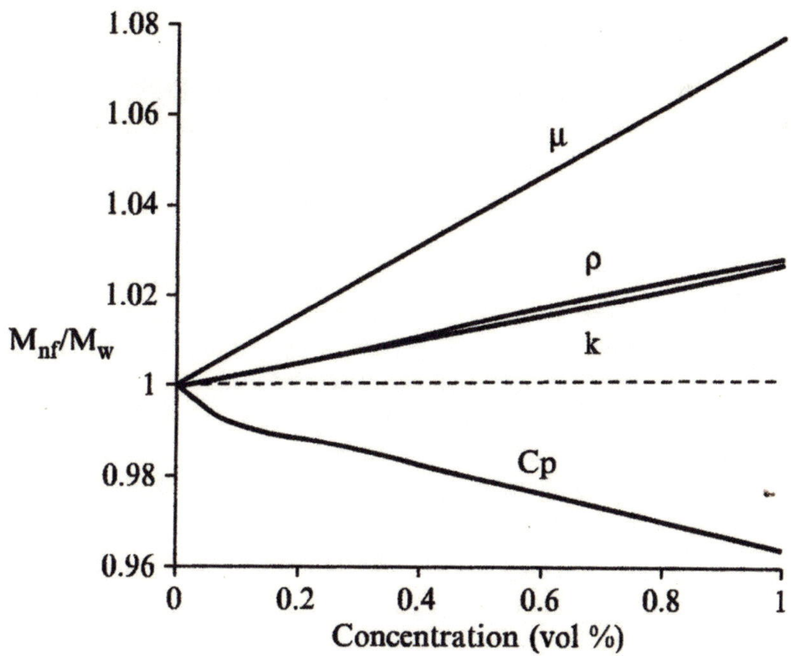

Figure 2. Dimensionless physical properties of nanofluids in comparison to those of pure water.

At first, the researchers only investigated the effect of particle volume concentration on thermal conductivity enhancement. Three possible approaches have been pursuing the study of nanofluid: experimental, empirical, and numerical.

2.1. Preparation methods of nanofluid

For producing high quality of nanofluids, some special conditions are prerequisites, e.g., stable suspension, permanent suspension, no particle agglomeration, and no chemical change of the fluids. Nanofluid preparation is a critical task with the use of nanoparticles for improving the thermal conductivity of base fluid. Fundamentally, there are two methods for producing nanofluids, i.e., (1) two-step method and (2) one-step method.

2.1.1. Two-step method

The two-step method is widely used in producing nanofluid synthesis considering the limitations of nanoparticle supplies commercially by some companies. In this means, nanoparticles are first created and then dispersed in the base fluids. Ultrasonic vibration is used to reduce the agglomeration of particles intensively. Making nanofluids utilizing the two-step processes is challenging because individual particles tend to quickly agglomerate. This agglomeration is due to attractive van der Waals forces between nanoparticles, and the agglomeration is a critical issue in all nanopowder technology, including nanofluid technology, and a crucial step to success in achieving high-performance heat transfer nanofluids [9]. The methods to prevent the agglomeration of particles, usually using a surfactant that regulates acidity (pH) of the base fluids.

2.1.2. One-step method

This method simultaneously generates and disperses the nanoparticles into the fluid base, while the first method deploys previously manufactured nanoparticles into the base fluid. Both methods involve reduction reactions or ion exchange.

Ions and other reactions products are then dispersed in the base fluid together with the nanoparticles since they are almost impossible to separate from their surroundings. For nanofluids containing high thermal conductivity, a one-step method is preferred to prevent particle oxidation. The advantage of a one-step technique is that nanoparticle agglomeration is the minimum, while the disadvantage is that very little nanofluid is produced. The one-step method has produced nanofluids in small quantities for research purposes only, and it is challenging to produce nanofluids commercially by this method [9]. They will be difficult to do for two reasons: firstly, a process that requires a vacuum significantly and slows the production of nanofluid, thereby limiting the production rate, and, secondly, producing nanofluids by this methods is expensive [10].

While most nanofluid productions to date have used one of the above techniques, other techniques are available depending on the particular combination of nanoparticle material and fluids [11]. The early studies on nanofluids focused on the measurement of the thermal conductivity. Later, more experiment regarding the convective heat transfer of nanofluids has been developed continuously.

3. Thermal properties of nanofluid

Thermophysical property, especially the thermal conductivity, is a vital issue in nanofluid heat transfer phenomena. Prediction of thermal conductivity has been a severe challenge until now because many parameters affect the thermal conductivity values. Temperature, type of the base fluid, nanoparticle material, shape, size, volumetric fraction, production, and mixing methods may significantly change the thermal conductivity. The literature research on thermal conductivity of nanofluids is a guide to understand how different parameters affect the value and what kind of thermal conductivity model is selected for the calculation heat transfer enhancement.

Secondly, the viscosity is also crucial in nanofluid heat transfer performance, as the usage of nanofluid viscosity also increases. Prediction of the viscosity of nanofluids is also a challenging topic because of its increase in the pumping power. The similar parameters that affect thermal conductivity affect viscosity value.

3.1. Thermal conductivity of nanofluid

Researchers have widely studied nanofluid thermal conductivity. Investigation the increase of analytical thermal conductivity a solid-liquid mixture by adding solid particles of micro-size balls into a liquid known as the Maxwell model.

$$k_{nf} = \frac{k_p + 2k_b + 2\varphi(k_p - k_b)}{k_p + 2k_b - \varphi(k_p - k_b)} k_b \tag{1}$$

where φ is the volume fraction of the nanofluid, k_b is the thermal conductivity of the base fluid, and k_p is the thermal conductivity of the nanoparticles.

Hamilton and Crosser [13] proposed a model for nonspherical particles by introducing a shape factor n given by $n = 3/\varphi$. The thermal conductivity is expressed as follows:

$$k_{nf} = \frac{k_p + (n-1)k_b - (n-1)\varphi(k_b - k_p)}{k_p + (n-1)k_b - \varphi(k_b - k_p)} k_b \tag{2}$$

A modified Maxwell's model was proposed by Xuan and Li [14] by considering the Brownian motion of the particles in the base fluid for the thermal conductivity enhancement given as

$$k_{nf} = \frac{k_p + 2k_b + 2(k_b - k_p)\varphi}{k_p + 2k_b - (k_b - k_p)\varphi} k_b + \frac{\rho\varphi C_p}{2}\sqrt{\frac{k_B T}{3\pi r_c \mu}} \tag{3}$$

where k_B is the Boltzmann constant, rc is the apparent radius of the cluster, and μ is a dynamic viscosity.

Figure 3 shows the variation of k_{nf}/k_b as a function of alumina volume fraction with and without CTAB along with its best fits with and without interfacial resistance. The k_{nf}/k_b value

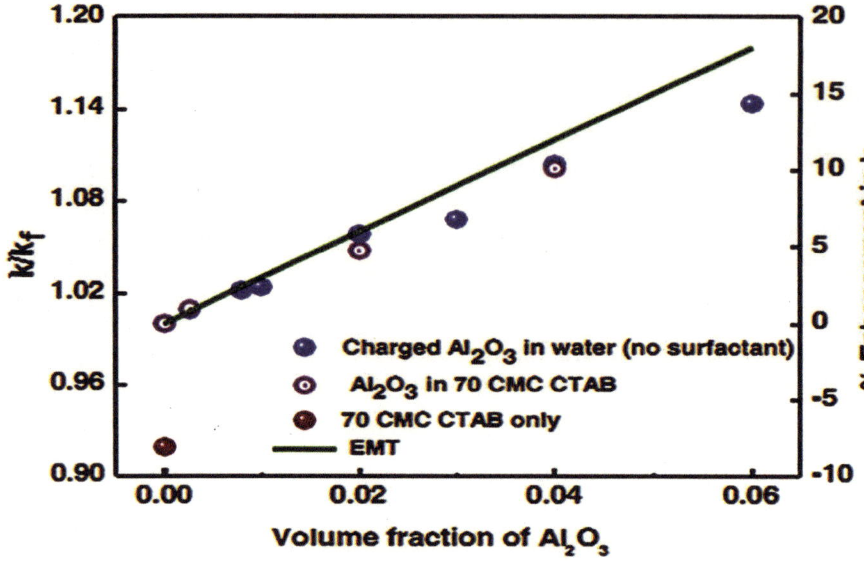

Figure 3. Variation of k_{nf}/k_b as a function of alumina nanoparticle volume fraction with and without CTAB and its best fit with an interfacial resistance.

of 70 CMC surfactant is also shown. The effective medium theory (EMT) fits on the experimental data indicated by the solid line, which shows a perfect agreement with empirical data. It can be seen that the value of knf/kb with pure surfactant was negative, while it was positive at all other concentrations of nanoparticles [15]. In general the higher the particle volume fraction, the higher the nanofluid thermal conductivity of nanofluids.

3.2. Viscosity of nanofluid

Besides the thermal conductivity of the nanofluid, another important thermophysical property is viscosity. Viscosity describes the internal resistance of a fluid to flow, and it is an essential property for all thermal applications involving fluids [10]. In laminar flow, the pressure drop is directly proportional to viscosity. Furthermore, convective heat transfer coefficient is influenced by viscosity. Hence, viscosity is as essential as thermal conductivity in engineering systems involving fluid flow. There has been a lot of research done about nanofluids but mostly related to heat transfer [16]. The increase in viscosity doubled, and energy is required to move the fluid to fourfold so that fluid viscosity plays a significant role in the use of energy in the cooling system.

Most of the viscosity enhancement studies obtained with the dispersion of nanoparticles in the base fluid correlated with the effect of volume fraction, size, and temperature were available in the literature. The rheological behavior of nanofluid categorized into four groups [17, 18], nanofluids with volume concentration less than 0.1 vol.% whose viscosity fits with the Einstein equation, semi-dilute nanofluids with 0.1–5 vol.% with aggregation of nanoparticles, semi-concentrated nanofluids with 5–10 vol.% with aggregation of nanoparticles, and concentrated nanofluid with 10 vol.% concentration, is out of the usual nanofluids [19–21].

Theoretical investigations.

There are some existing theoretical formulas to estimate the viscosity of nanofluid. Among them, equation suggested by Einstein [22] is a pioneer in determining the viscosity equation. The assumptions based on the linear viscous fluid containing spherical particles and low particle volume fractions ($\varphi < 0.02$). The suggested formula is as follows:

$$\mu_{nf} = \mu_b(1 + 2.5\varphi) \tag{4}$$

where μ_{nf} is the viscosity of nanofluid, μ_b is the viscosity of the base fluid, and φ is the volume fraction. It is a linear increase of the viscosity with increasing volume concentration. This formula has a limitation that is very small particle concentration. Later on, many researchers contributed to correct this formula [16].

In 1952, Brinkman [23] extended Einstein's formula to be used with moderate particle concentrations, and this correlation has more acceptance among the researchers. For particle concentrations less than 4%, the expression is as follows:

$$\mu_{nf} = \mu_b(1 - \varphi)^{2.5} \tag{5}$$

Considering the effect due to the Brownian motion of particles on the bulk stress of an approximately isotropic suspension of rigid and spherical particles, Batchelor [24] proposed the following formula in 1977:

$$\mu_{nf} = \mu_b\left(1 + 2.5\varphi + 6.5\varphi^2\right) \tag{6}$$

It is clear from the above two relations that, if the second or higher order of φ is ignored, then these formulas will be the same as Einstein's equation has been validated for a particle volume fraction up to $\varphi < 0.1$ [16].

Nguyen et al. [17] showed that both the Brinkman [23] and Batchelor [24] equations severely underestimate nanofluid viscosities, except at very low particle volume fractions (lower than 1%). They have proposed two correlations for nanofluids consisting 47 and 36 nm of Al_2O_3 nanoparticles with water, respectively, as follows:

$$\mu_{nf} = \mu_b \times 0.904e^{0,1842\varphi} \tag{7}$$

$$\mu_{nf} = \mu_b\left(1 + 0.025\varphi + 0.015\varphi^2\right) \tag{8}$$

Both of these models determine the viscosity by only considering base fluid viscosity and the particle volume fraction. Furthermore, they proposed a correlation for computing CuO water viscosity as shown in Eq. (9) [16]:

$$\mu_{nf} = \mu_b\left(1.475 - 0.319\varphi + 0.051\varphi^2 + 0.009\varphi^3\right) \tag{9}$$

Most of the equations have been developed to express viscosity as a function of volume fraction of nanoparticles. However, the temperature is an important factor in nanofluid viscosity, and, consequently, several equations have been created to investigate the temperature effect on viscosity. Some literature is available about the temperature effect over nanofluid viscosity [16]. Yang et al. [25] experimentally measured temperature effect of viscosity with four temperatures (35, 43, 50, and 70°C) for four nanofluid solutions taking graphite as nanoparticles. They experimentally showed that kinematic nanofluid viscosity decreases with the increase of temperature. Anoop et al. [26] studied the viscosity of CuO-ethylene glycol, Al_2O_3-ethylene glycol, and Al_2O_3-water for the temperature range of 20–50°C with a volume concentration of 0.5, 1, 2, 4, and 6 vol.%. They found that viscosity reduces with an increase in temperature. Investigation studies by Duangthongsuk and Wongwises [27] with TiO_2 and water for a temperature range of 15–50°C have found that viscosity of nanofluids decreases with the rise in temperatures.

A correlation between viscosity and temperature for pure fluids was presented by Reid et al. [28]:

$$\mu_{nf} = A\exp\left(\frac{B}{T}\right) \tag{10}$$

where A and B are the functions of concentrations and T is temperature that is written by Yaws [29] as

$$\log\left(\overset{\mu}{nf}\right) = A + BT^{-1} + CT + DT^2 \tag{11}$$

where A, B, C, and D are the fitting parameters.

Some correlations have also been suggested taking into account both temperature and volume fraction effects on viscosity [16]. In 2006, Kulkarni et al. [30] proposed correlations that relate viscosity of copper oxide nanoparticles suspended in water with a temperature range of 5–50°C:

$$\ln\mu_{nf} = A\left(\frac{1}{T}\right) - B \tag{12}$$

here, A and B are functions of volume fraction φ. This correlation is mainly for aqueous solution and is not applicable to nanofluids in the subzero temperature range.

The application for nanofluid with low viscosity (high temperature) and high thermal conductivity (small volume fraction) is promising for the future.

4. Heat transfer enhancement

Enhancement of heat transfer is a favorite and an important topic that is highly relevant to current and future energy systems and renewable energy systems as well as for energy conservation and environmental protection. The increase in heat transfer statement is usually expressed by increasing the heat transfer coefficient of a system. The purposes of improving the heat transfer rate are to reduce the size and simultaneously increase the capacity of the thermal system.

The thermal system is expressed by thermal performance, which is the increase of heat transfer coefficient (h) and hydraulic performance, that is, the amount of energy required to circulate fluid within the system. The thermohydraulic performance is used as the performance indicator of heat exchange tool. The heat exchanger performance test means comparing the characteristics of the heat transfer coefficient and pressure loss (pressure drop) on a device of the same dimension.

Heat transfer enhancement is the process of increasing the effectiveness of heat exchanger. It can be achieved when the heat transfer power of a given device increased or when the pressure losses generated by the device are reduced.

5. Performance evaluation criteria (PEC)

The main advantage of the nanofluid is it has a high thermal conductivity, which is used for improving the efficiency of the thermal system. Adding small particles to the base fluid liquid increases the viscosity of nanofluid [31], which also increases the pressure drop on the systems. Due to the increased pressure drop, the operational costs of a system will be high due to the increase in pumping power. Hence, the viscosity of nanofluid is a significant parameter for determining the feasibility of nanofluid for heat transfer applications, depending on the significant increase in both thermophysical-properties of thermal conductivity and increased viscosity [19].

As heat transfer and pressure drop are the most critical factors, they can be compared to several approaches. It is defined as the ratio of heat transferred to the required pumping power in the test section. To evaluate the benefits provided by the enhanced properties of the nanofluids studied, an energetic performance evaluation criterion (PEC) is defined as heat transfer and hydrodynamics are the most critical factors. They can be compared to a global energy approach using the PEC defined as the ratio of heat flow rate transferred to the required pumping power in the system [32]:

$$PEC = \frac{mC_p(T_i - T_o)}{v\Delta P} \tag{13}$$

5.1. Pressure drop

Additionally, the heat transfer rate of nanofluids increased due to increased thermal conductivity, and the pressure drop also increased due to the increase in the nanofluid viscosity.

The pressure drop ($\Delta p = p_1 - p_2$) is directly related to the pumping power to maintain flow; for laminar flow the pressure drop is shown in Eq. (14):

$$\Delta p = \frac{32\mu L V_{avg}}{D^2} \tag{14}$$

The pressure drop due to viscous effects represents an irreversible pressure loss, and it is called pressure loss Δp_L in circular tube that can be determined from the pressure drop using the Darcy-Weisbach equation below:

$$\Delta p_L = f \frac{L}{D} \frac{\rho U^2}{2} \tag{15}$$

But mass flow rate is $\acute{m} = \rho A U$, and the pressure drop from Eq. (15) is

$$\Delta p = f \frac{L}{D} \frac{\rho \left(\frac{\acute{m}}{AU}\right)^2}{2} = f \frac{L}{D} \frac{4\acute{m}\rho}{\pi D^2} = f \frac{L}{D^3} \frac{4\acute{m}\rho}{\pi} \tag{16}$$

Performance evaluation criteria by Lee [33] obtained a boundary line for laminar flow using thermal conductivity and viscosity of nanofluids to compared than base fluid. The method has been used to analyze the performance of nanofluid in a microchannel heat sink. A microchannel heat sink as a passive method can be a cooling device by dissipating heat into the surrounding air. The microchannel heat sink consists of N number of circular channels, each with diameter d, as shown in **Figure 4**. The total channel width W is constant ($W = N \times d$).

Assume a constant heat flux boundary condition for all channel, and the flow is hydrodynamically and thermally fully developed. The heat transfer flow rate for convection heat transfer as Newton's law of cooling is

$$Q = hA\Delta T \tag{17}$$

The difference in temperature between the surface wall temperature and the local bulk fluid temperature is

$$Q = T_w - T_m = \frac{Q}{N\pi dLh} = \frac{Q}{\left(\frac{w}{d}\right)dL} \frac{1}{kNu} \tag{18}$$

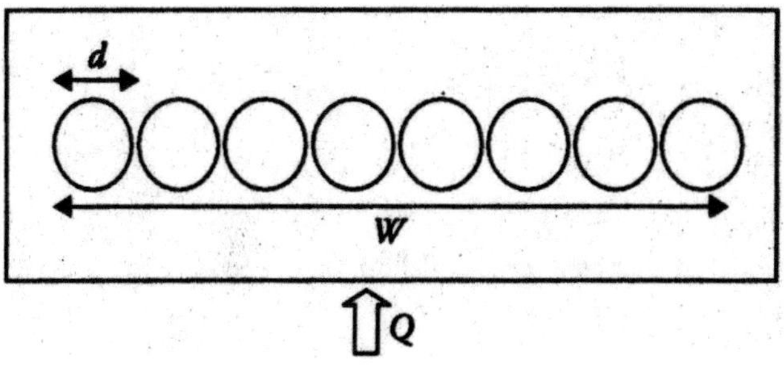

Figure 4. A schematic of a microchannel heat sink.

The pressure drop for a channel with length L as follows:

$$\Delta P = f\frac{L\rho U^2}{2d} \tag{19}$$

For laminar flow, $f = 64/\mathfrak{R}$

$$\Delta P = \frac{128\mu\left(\frac{V}{N}\right)L}{\pi d^4} = \frac{128\mu VL}{\pi W d^3} \tag{20}$$

The above equation can be written to d as follows:

$$d = \left(\frac{128\mu VL}{\pi W \Delta P}\right)^{1/3} \tag{21}$$

Therefore, at a fixed volume flow rate, fixed pumping power, and a given channel length, the following correlation can be obtained for laminar flow with constant a Nusselt number:

$$\Delta T = \frac{Q}{W\pi LNu}\left(\frac{128\mu VL}{W\pi\Delta P}\right)^{1/3} \infty \frac{\mu^{\frac{1}{3}}}{k} \tag{22}$$

If the nanofluid is more efficient than their base fluids, the difference between the wall and bulk fluid temperature should be smaller than the temperature using base fluids:

$$\Delta T_{bf} > \Delta T_{nf} \tag{23}$$

and

$$\frac{k_{nf}}{k_{bf}} > \left(\frac{\mu_{nf}}{\mu_{bf}}\right)^{1/3} \tag{24}$$

Eq. (24) shows that nanofluids are effective as long as the thermal conductivity enhancement is more than the one-third power of the viscosity enhancement for laminar flow regime. The boundary line for the performance of nanofluids in laminar flow regime is shown in **Figure 5a**. The thermal conductivity and viscosity ratios of Al_2O_3 and multiwalled carbon nanotube (MWCNT) nanofluids are measured in the work of Wu et al. [34] plotted in **Figure 5**. Alumina nanofluids can provide better performance than base fluids under laminar flow regime, whereas the tested MWCNT nanofluids are very unfavorable.

If a ratio nanofluid viscosity to the base fluid $\frac{\mu_{nf}}{\mu_{bf}}$ is equal to two, the ratio thermal conductivity of nanofluid to thermal conductivity base fluids must be more than 1259 as shown in **Figure 4a**.

Figure 6 shows the optimization of the preparation of nanofluid, starting with the measurement of the stability of nanofluid, if it fast agglomerates necessary additional treatment, for high stability by adding surfactant or surface modifier.

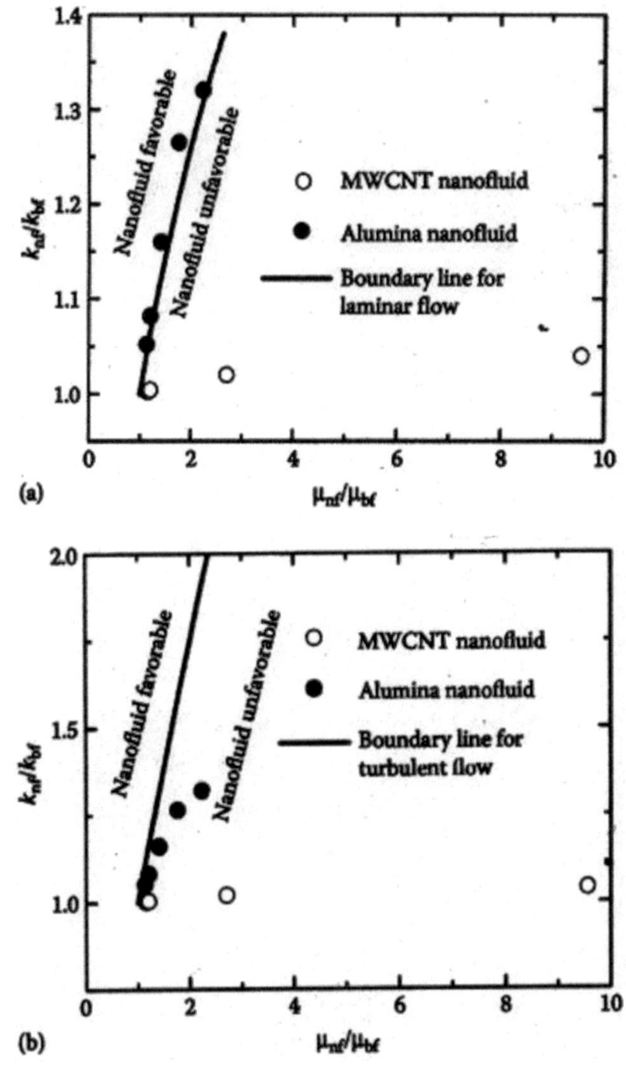

Figure 5. The boundary line for the performance of nanofluid at fixed volume flow rate and fixed pumping power: (a) laminar flow regime and (b) turbulent flow regime. The alumina and MWCNT nanofluids measured in Wu et al. correlation [32].

Viscosity measurements were performed to determine the viscosity value of nanofluid, for high viscosity nanofluid required special treatment to decrease viscosity, i.e., nanofluid used at the temperature above room temperature. Following the above discussion with an increasing temperature, the viscosity of nanofluid decreases.

When the viscosity of the nanofluid is low, and the thermal conductivity of the nanofluid is high, nanofluids can be used in practical applications.

http://dx.doi.org/10.5772/intechopen.74610

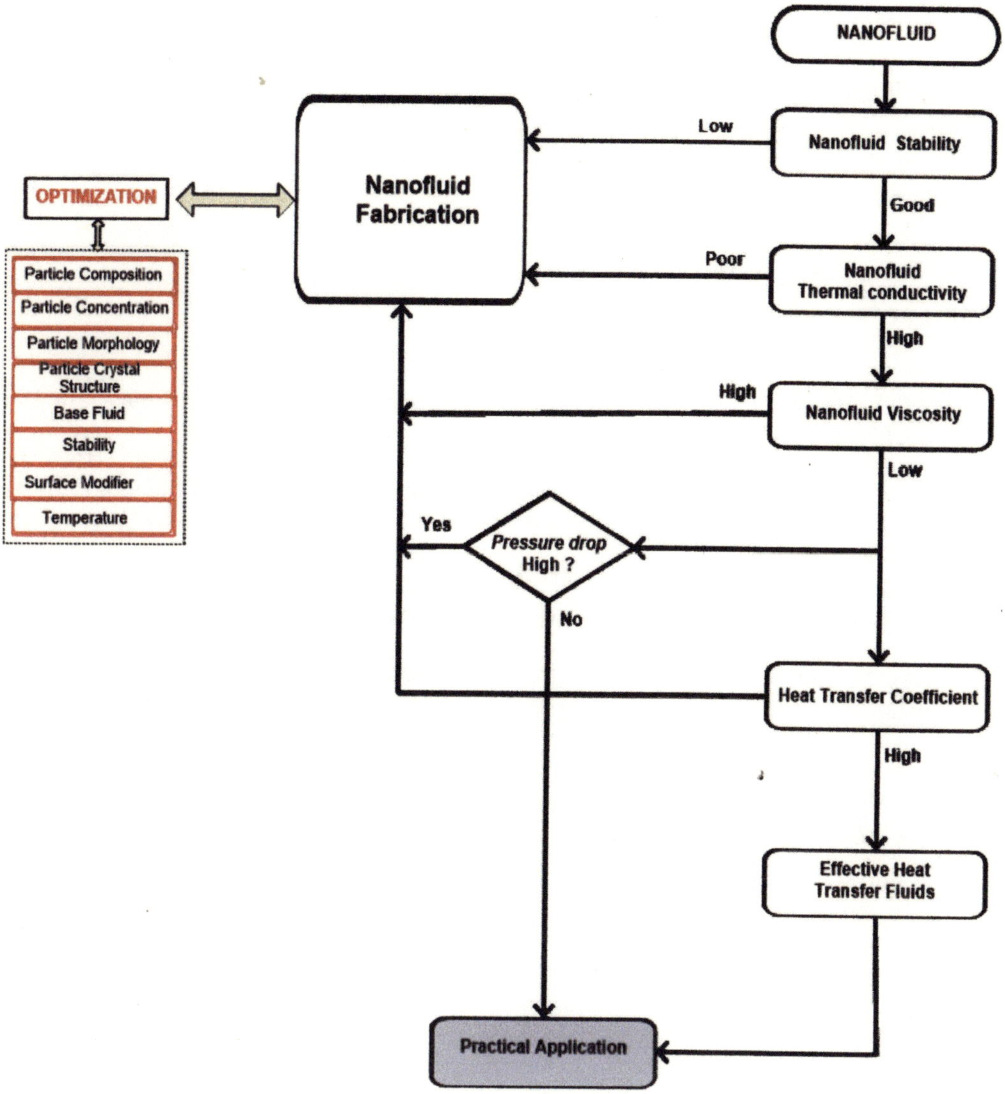

Figure 6. Optimization fabrication of nanofluid.

6. Conclusion

The above discussion can be concluded that the application for nanofluid with low viscosity (high temperature) and high thermal conductivity (small volume fraction) is promising for the future.

Nanofluid can be used as a fluid that has a high-performance increase of heat transfer suitable only for the cooling process, as hot fluids with low particle volume fraction.

Author details

Sudarmadji Sudarmadji*, Bambang SAP and Santoso

*Address all correspondence to: sudmaji@yahoo.co.id

State Polytechnic of Malang, Malang, Indonesia

References

[1] Bergles AE. Recent developments in enhanced heat transfer. Heat and Mass Transfer. 2011;**47**:1001-1008

[2] Bergles AE. Techniques to enhance heat transfer. In: Rohsenow WM, Hartnett JP, Cho YI, editors. Handbook of Heat Transfer. 3rd ed. New York: Mc-Graw Hill; 1998. p. 785-860

[3] Bergles AE, Morton HL. Survey and evaluation of techniques to augment convective heat and mass transfer. Cambridge: Massachusetts Institute of Technology; 1965

[4] Sheikholeslami M, Bhati MM. Active methods for nanofluid heat transfer enhancement by means of EHD. International Journal of Heat and Mass Transfer. 2017;**109**:115-122

[5] Sheikhleslami M, Hayat T, Alsaedi A. Numerical study for external magnetic source influence on water-based nanofluid convective heat transfer. International Journal of Heat and Mass Transfer. 2017;**106**:745-755

[6] Choi SUS. Enhancing thermal conductivity of fluids with nanoparticles, developments and application of non-Newtonian flows. Fedgazette. 1995;**231**(MD66):99-105

[7] Choi SUS. Nanofluids: A new field of scientific research and innovative applications. Heat Transfer Engineering. 2008;**29**(5):429-431

[8] Maxwell JC. A Treatise on Electricity and Magnetism. Oxford, UK: Clarendon Press; 1873

[9] Saha SK, Tiwari M, Sunden B, Wu Z. Nanofluids, Next Generation Heat Transfer. In: Advanced Heat Transfer. Switzerland: Springer; 2016. 128 p. DOI: 10.1007/978-3-319-29480-3

[10] Hua YW, David F, Jules R, Stephen C. Review and comparison of nanofluid thermal conductivity. Heat Transfer Engineering. 2008;**29**(5):432-460

[11] Wang XQ, Mujumdar A. A review on nanofluid part II: Experimental and applications. Brazilian Journal of Chemical Engineering. 2008;**25**(4):631-648

[12] Bontemps A, Reberio JP, Ferroulliat S, Gruss JA, Sariano O, Wang B. Experimental Study of Convective Heat Transfer and Pressure Drop of SiO_2/Water Nanofluid Part 1: Nanofluid Characterization-Imposed Wall Temperature; 2008

[13] Hamilton RL, Crosser OK. Thermal conductivity of heterogeneous two-component systems. Industrial and Engineering Chemistry Fundamentals. 1962;**1**(3):187-191

[14] Xuan Y, Li Q. Investigation convective heat tranfer and flow features of nanofluid. Journal Heat Transfer. 2003;**125**(1):151

[15] Angayarkanni SA, Philips J. Review on thermal properties of nanofluid: Recent development. Advances in Colloid and Interface Science. 2015;**225**:146-176

[16] Mahbulbul IM, Saidur R, Amalina MA. Latest development on the viscosity of nanofluids. International Journal of Heat and Mass Transfer. 2012;**55**:875-885

[17] Nguyen HAC, Desgranges F, Roy G, Galanis N, Mare T, Boucher S. Temperature and particle-size dependent viscosity data for water-based nanofluids – Hysteresis phenomenon. International Journal of Heat and Fluid Flow. 2007;**28**(6):1492-1506

[18] Ghadini A, Saidur R, Metselaar HSC. A review of nanofluid stability properties and characterization in stationary conditions. International Journal of Heat and Mass Transfer. 2011;**54**(17–18):4051-4068

[19] Rebay M, Kakac S, Cotta R. Thermophysical properties of nanofluid. In: Microscale and Nanoscale Heat Transfer. New York: Taylor and Francis; 2016

[20] Chandrasekar M, Suresh S, Bose AC. Experimental investigation and theoretical determination of thermal conductivity and viscosity of Al_2O_3/water nanofluid. Experimental Thermal and Fluid Science. 2010;**34**(2):210-216

[21] Chen H, Ding Y, Lapkin A. Rheological behavior of nanofluids containing tube/rod-like nanoparticles. Power Technology. 2009;**194**:132-141

[22] Einstein A. A new determination of the molecular dimensions, from Annalen der Physik. 1906;**19**(4):289-306

[23] Brinkman H. The viscosity of concentrated suspensions and solutions. Journal of Chemical Physics. 1952;**20**:571

[24] Batchelor G. The effect of Brownian motion on the bulk stress in a suspension of spherical particles. Journal of Fluid Mechanics. 1977;**83**(1):97-117

[25] Yang GWY, Zhang ZG, Grulke EA, Anderson WB. Heat transfer properties of nanoparticle-in-fluid dispersions (nanofluids) in laminar flow. International Journal of Heat and Mass Transfer. 2005;**48**(6):1107-1116

[26] Anoop KB, Kabelac S, Sundararajan T, Das SK. Rheological and flow characteristics of nanofluids: Influence of electroviscous effects and particle agglomeration. Journal of Applied Physics. 2009;**106**(3):34909

[27] Duangthongsuk W, Wongwises S. Measurement of temperature-dependent thermal conductivity and viscosity of TiO_2–water nanofluids. Experimental Thermal and Fluid Science. 2009;**33**(4):706-714

[28] Reid R, Prausnitz J, Sherwood T. The Properties of Gases and Liquids. 4th ed. New York: McGraw-Hill; 1987

[29] Yaws CL. Physical Properties: A Guide to the Physical, thermodynamic and transport property data of industrially important chemical compounds. New York: Mc Graw-Hill, 1977

[30] Kulkarni DP, Das DK, Chuchwu GA. Temperature dependent rheological property of copper oxide nanoparticles suspension (nanofluid). Journal of Nanoscience and Nanotechnology. 2006;**6**(4):1150-1154

[31] Mewis J, Wagner NJ. Colloidal Suspension Rheology. Cambridge, U.K: Cambridge University Press; 2012

[32] Ferrouilliat S,Ferro Bontemps A, Gruss JA. Hydraulic and heat transfer study of SiO_2/water nanofluids in horizontal tubes with imposed wall temperature boundary International Journal of Heat and Fluid Flow. 2011;**32**(April):424-439

[33] Lee JH. Convective performance of nanofluids for electronic cooling. ProQuest: Standford University; 2009

[34] Wu Z, Feng Z, Bengt S, Warso L. A comparative study on thermal conductivity and rheology properties of alumina and multiwalled carbon nanotube nanofluids. Frontiers in Heat and Mass Transfer. 2014;**5**(18):1-10

Chapter 12

Microfluidic Membrane Filtration Systems to Study Biofouling

Ishita Biswas, Aloke Kumar and Mohtada Sadrzadeh

Additional information is available at the end of the chapter

http://dx.doi.org/10.5772/intechopen.75006

Abstract

In wastewater treatment, the membrane functions as a semipermeable barrier that restricts transport of undesired particulates. A major problem related to membrane filtration processes is fouling of membranes by colloidal particles, organic matter, and biomaterials. Among the various types of fouling, biofouling is one of the most severe, as it is a dynamic process. Even a few surviving cells that adhere to the membrane surface multiply exponentially at the expense of biodegradable substances in the feed solution. To analyze the mechanism of biofouling, membrane cell is typically considered as a black-box, where only the input and the output can be measured and put into use for analysis. Microfluidic devices are being used to study and understand the nature, properties, and evolution of biofouling. A primary advantage of a microfluidic membrane is the ability to conduct real-time observations of biofilm. This chapter presents an overview of the biofouling in membrane processes and different fabrication technique of microfluidic membrane systems.

Keywords: biofilm, biofouling, microfabrication, microfiltration membrane, microfluidics

1. Introduction

1.1. Membrane biofouling

Pressure-driven membrane processes can be used to filter a wide range of small materials, ranging from monovalent ions and dissolved organic matter to biological substances. They have become very popular for treating sea and waste water. However, they face the problem of fouling on a continuous basis. Fouling is the unwanted accumulation of substances on the membrane surface. There are five types of fouling including scaling (by divalent ions), heavy metal fouling, organic fouling, colloidal fouling, and biofouling [1, 2]. Among these fouling types, biofouling is the most severe since it is a dynamic process and is also the most confronted one, and can contribute as much as 45% of the total fouling [1].

Biofouling due to biofilms (matrix-encapsulated bacterial colonies) and colloidal materials act as the main components of membrane fouling [2]. Moreover, biofilms have a significant impact on the membranes used for different types of water filtration such as brackish and seawater. Once a cell is attached to the membrane surface, it decreases membrane permeability by forming a gel layer [3]. Biological substances always remain in the membrane. Even if 99.9% of these materials are removed by pre-treatment, the remaining 0.1% can grow exponentially by using biodegradable substances in the feed (waste) water [4].

1.2. Biofouling due to bacterial colonization

Biofouling occurs due to the adsorption of the biological cells on a membrane surface [5]. Biological organisms are usually identified by their length scale. Microorganisms, which lie within very small length scales (1–200 μm), include bacteria, fungi, and algae. Furthermore, length scale >200 μm is referred to macro-organisms such as larvae, barnacles, hydroids, tubeworms, mussels, and bivalves [1]. Bacteria are a common biofouling agent and are found extensively in nature. Bacterial colonization of a surface is an extremely complex process, where several phenomena can take place at multiple length and time scales [6–8]. Colonization on the surface starts with adhesion of bacteria to a solid-liquid interface. The interaction of bacteria with the surface leads to the formation of extracellular polymeric substance (EPS), where bacterial cells are embedded in a matrix. These matrix-encapsulated, surface-associated bacterial communities are referred to as a biofilm [9, 10]. EPS, the binding material of biofilms, is composed of long-chain biomolecules such as polysaccharides, nucleic acids, protein, DNA and lipids [11–14]. Biofilms can play an important role in chronic infections [1]. Moreover, they are prevalent in industrial and shipping environment, causing significant problems related to environmental impacts and health risks [15].

2. Background

2.1. Biofouling due to biofilm on membrane

Biofilm is one of the most challenging issues in membrane technology [16–18]. The adsorption of bacteria cell on the membrane surface depends on membrane properties such as membrane materials, hydrophobicity, and roughness [19]. The adhesive nature of EPS is considered as the most severe problem in membrane biofouling [20, 21]. Biofilm on the membrane surface reduces the permeate flux and salt rejection [22–24]. In membrane technology, the flux and salt rejection are the two primary criteria for characterizing of membrane performance. The more the flux and salt rejection, better the membrane performance is. The volume flux (J) of porous membrane is usually calculated by Hagen-Poseuille equation where the pores are assumed to have the same radius,

$$J = \frac{\varepsilon r^2}{8\eta\tau}\frac{\Delta P}{\Delta x} \tag{1}$$

Where, Δx is the membrane thickness, ΔP is pressure difference across the membrane, η is the viscosity, τ is tortuosity, r is the radius of the pore, and ε is the porosity of the membrane. Porosity can be calculated by,

$$\varepsilon = \frac{n_p \pi r^2}{A_m} \tag{2}$$

A_m is the membrane surface area and n_p is the number of pores. Tortuosity is defined by:

$$\tau = \frac{(2 - \varepsilon)^2}{\varepsilon} \tag{3}$$

Matin et al. Provided a list of typical bacteria species that can cause biofilm formation on the membrane surface as well as a reduction in flux decline and salt rejection due to the formation of biofilm on the membrane surface [25]. They observed that, without bacterial adhesion, the membrane was able to reject (R) 98.2% salt. The rejection was decreased by 4.6% because of the biofouling on the membrane.

Biofilm is a complex structure due to the viscoelastic nature of EPS that can lead to the formation of memory effect in a material [10, 26]. Rheological measurement of the biofouling layer on the membrane surface is required to understand the EPS nature. Patsios et al. [27] performed some rheological measurements of the biofouling layer on the membrane. They obtained nonlinear behavior of shear stress and strain of the EPS. They claimed that EPS shows more elastic nature than viscous on the membrane surface. The storage modulus G´, the elastic part, was higher than the loss modulus G´´ that is the viscous component. [25].

2.2. Microfluidic approach in biofouling study

In wastewater treatment, microfiltration membranes with the pore sizes lying between 0.1 and 10 μm are used to remove bacteria. Membranes are usually part of an opaque setup, where only the input and the output can be measured. Advancements in micro-/nano-technologies, for example, microfluidic devices can be employed to study membrane processes at the pore-scale. An example of this is the use of microfluidic-based membrane mimics, which are being used to explore a wide variety of membrane related issues, including biofouling. An essential advantage of microfluidic membrane mimics in studying biofouling is that they make real-time microscopy of biofouling possible. **Figure 1** shows a basic schematic difference between membrane filtration mode and microfluidic approach. The pillars are shown in **Figure 1b** are solid in structures and usually made of polydimethylsiloxane (PDMS). The gap between the pillars is considered as the pore. The coverslip is used to seal the device.

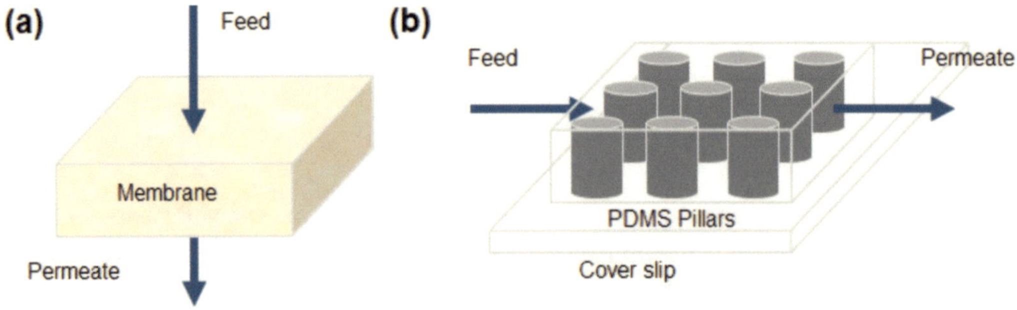

Figure 1. Schematics of (a) membrane filtration where feed is wastewater and permeate is the clean water and (b) microfluidic filtration mode.

Design and fabrication are the initial steps to work with microfluidic devices. Different types of fabrication techniques include photolithography, electron lithography, hot embossing and injection molding, etc. Photolithography is a common technique when feature sizes larger than 1 μm are desired. The nanoscale feature can also be fabricated by e-beam lithography where the minimum resolution could go down to 10 nm [28, 29].

2.3. Bacterial streamer due to biofouling

The impact of hydrodynamic flow on biofilms is the large time-dependent deformations that can result in nonlinear phenomena. An example of such phenomena is the bacterial streamer. Streamers form in flowing water and attach to the surface by the upstream "head" while the downstream "tail" can oscillate [6, 10, 30–32]. Streamers in a microfluidic system are typically tethered at one end to the pillar walls while the rest of the body is suspended in the downstream direction. Their filamentous structure can extend significantly with the flow [6, 33, 34]. Drescher et al. [35] revealed that streamers can cause a sudden and rapid clog in the fluid flow system in comparison with the biofilm attached to the surface. Surface hugging biofilms have a very modest effect on the flow rate whereas; streamers can drastically decrease the flow rate in a very short period [31].

Rusconi et al. [36] reported streamer formation in the microfluidic channel under laminar flow conditions. They observed formation of a single streamer in the middle of the channel connecting the inner corners of the channel. They also claimed that secondary flows in the curved edge of the channel were responsible for the location of the streamer, which was located at the mid-plane. They further investigated the streamer formation behavior by changing the radius of the curvature of a zigzag microchannel and discovered that streamer formation depends on the geometric angle of microchannel [37].

Valiei et al. [6] observed streamers through the height of the channel with 50 × 8 array of micro-pillars and mentioned it as a 'web' of the streamers. They claimed that flow rate has a significant impact on the number of streamer formation. A higher number of streamer formations was reported in the middle of channel height. **Figure 2** shows the formation of bacterial streamers in a microfluidic device with an array of micropillars. The white arrow indicates the

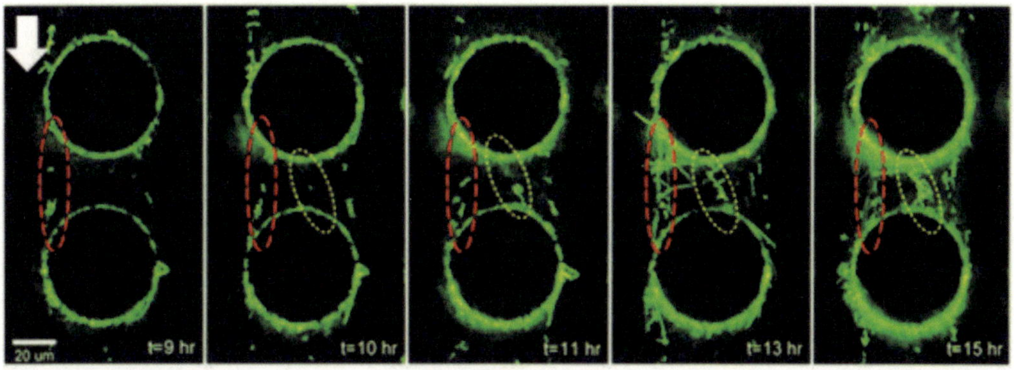

Figure 2. Streamer formation in a microfluidic channel. Figures are reproduced with permission from Ref [6].

flow direction, and the red and yellow ellipses show the formed streamers attached between two pillars. As can be seen, the thickness of the streamers increased with the increase of time of the streamer. Fluorescence microscopy was used to capture the image where only bacteria cells are visible (green). The fluid media and the EPS appear dark.

Marty et al. [33, 34] studied the effect of different pore sizes and filtration modes on the lengths of streamers that formed in a microfluidic membrane mimic system. They fabricated a microfluidic device with 25 straight, interconnected and staggered PDMS pillars to observe the nature of biofouling in a membrane mimic. The width and height of pillars were 10 and 50 μm respectively, and the mimic membrane pore size was 10 μm. They found that flow configuration and presence of tortuosity in a microchannel has a significant impact on streamer formation.

3. Basic overview of fabrication techniques

3.1. Membrane fabrication

Membrane process is an emerging separation technology. The membrane itself is the heart of a membrane process. It can be classified as polymeric and inorganic, porous and dense, isotropic and anisotropic, hydrophilic and hydrophobic, etc. **Figure 3** gives an overview of types and preparation process of the polymeric membranes. Phase inversion (phase separation) and track etching are the most widely used techniques for the preparation of porous membranes [38].

In phase inversion process method, the polymer is transformed in a controlled manner from liquid to solid state by changing the thermodynamic state of the polymer, solvent and the solution [38, 39]. Symmetric porous phase inversion membranes are made by using water vapor as the coagulant. For making asymmetric membranes by phase inversion temperature increase and a liquid nonsolvent is used to precipitate the polymer (**Figure 3**). In track etching method, a high energy particle radiation is applied to the polymeric film, to damage the polymeric matrix and create tracks. By etching the polymeric material along the track uniform cylindrical

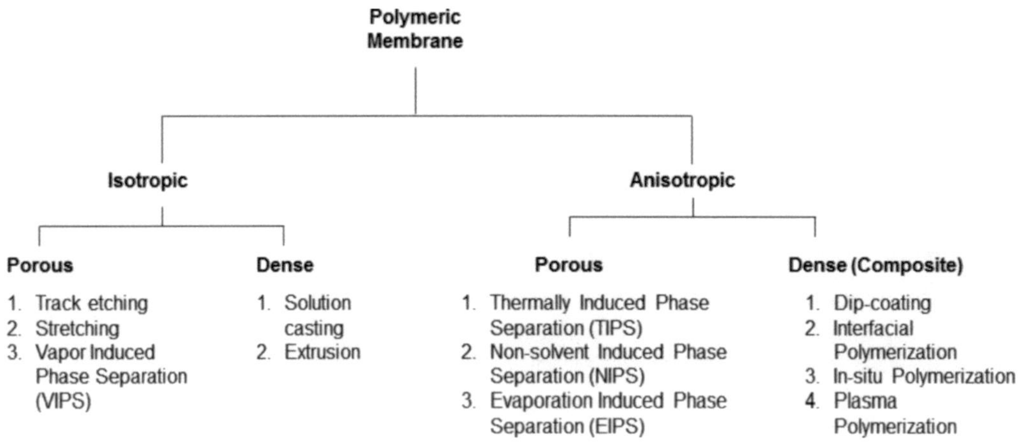

Figure 3. Preparation methods of polymeric membrane.

pores can be obtained. Dense membranes (symmetric and asymmetric) are mainly synthesized by solution casting and interfacial polymerization of two monomers on a substrate. A detailed explanation of membrane preparation techniques is available in the literature [38].

Membrane process	Polymer used in the fabrication process	Fabrication technique	Pore size	Pressure range (bar)	Flux range ($l.m^{-2}.h^{-1}.bar^{-1}$)	Application
Microfiltration (MF)	Polyvinylidene fluoride (PVDF), poly (tetraflurethylene) (PTFE), polypropylene (PP), Polyethylene (PE), polyethersulfone (PES)	Phase inversion, stretching track etching	Porous 10^{-1}–10 μm	0.1–2.0	>50	Separation of macromolecular to cellular size particles (Bacteria/ fat and some proteins)
Ultrafiltration (UF)	Polyacrylonitrile (PAN), polysulfone (PS), poly (phthazine ether sulfone ketone) (PPESK), poly (vinyl butyral) PVDF PES	Phase inversion, solution wet-spinning	Porous 10^{-2}–10^{-1}μm	1.0–5.0	10–50	Separation of molecular to macromolecular size particles (all proteins)
Nanofiltration (NF)	Polyamides, polysulfones, polyols, polyphenols	Interfacial polymerization, layer-by-layer deposition Phase inversion	Porous 10^{-3}–10^{-2}μm	5.0–20	1.4–12	Separation of Ionic molecular size particles (Lactose)
Reverse osmosis (RO)	Cellulose acetate/ triacetate aromatic polyamide, polypiperzine, polybenziimidazoline	Phase inversion Solution casting	Dense/ Porous 10^{-4}–10^{-3}μm	10–100	0.05–1.4	Separation of ions (all minerals)

Table 1. Summary of different types of pressure-driven membrane processes [38–41].

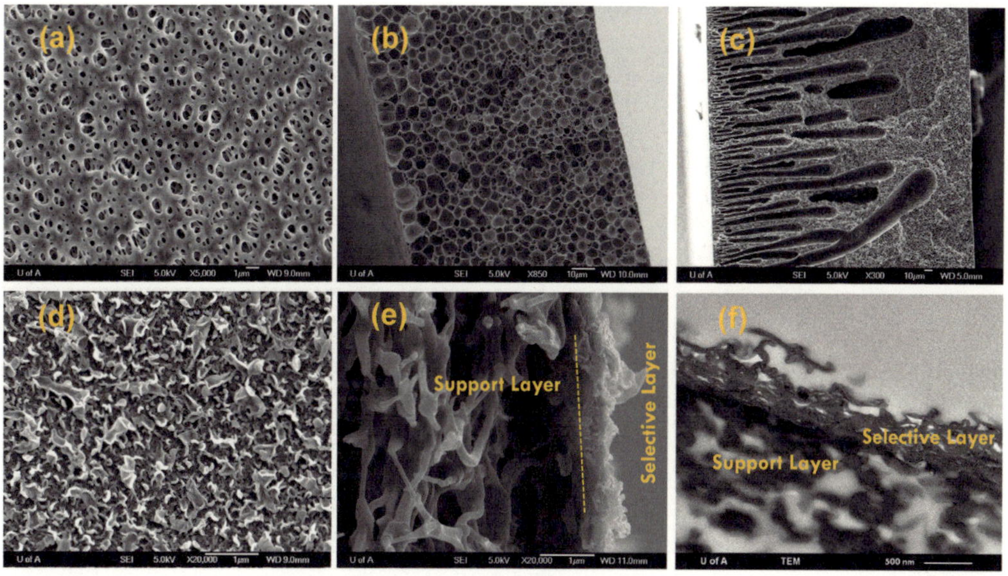

Figure 4. (a) Surface SEM image of (a) a phase inversion porous membrane, (b) cross-sectional SEM image of an isotropic phase inversion membrane, (c) cross-sectional SEM image of an anisotropic phase inversion membrane, (d) surface SEM image of thin film composite (TFC) dense membranes, (e) cross-sectional SEM image of a TFC membrane, and (f) cross-sectional SEM image of a TFC membrane.

A summary of different types of pressure-driven membrane processes with their fabrication technique, separation principle, pore morphology, pressure and flux ranges are given in the **Table 1**. Scanning electron microscopy (SEM) images of different types of membranes are presented in **Figure 4**.

3.2. Microfluidic device fabrication

There are many types of fabrication techniques available for making micro/nano devices such as photolithography, etching, soft lithography, hot embossing, injection molding, E-beam lithography, and micro-stereolithography. Photolithography and etching are two popular fabrication techniques. Soft lithography is a well-known method for microfabrication. McDonald et al. [42] fabricated microfluidic system with PDMS by a soft lithography technique to make 20–100 μm microfluidic structure. This technique has also worked well on hydrogel polymers (calcium alginate) to fabricate microfluidic network of 100 μm wide and 200 μm deep and 25 × 25 μm cross-section [43]. A complex structure with feature sizes larger than 20 μm can be achieved by using rapid prototyping [44]. The fabrication of 500–2000 μm diameters and 200–1000 μm height cylindrical columns [45] is possible by hot embossing technique. A schematic diagram of a microfluidic device is shown in **Figure 5**. This device is used to observe the biofilm behavior and the change of hydrodynamics of the fluid flow through the channel [6]. The chip has one inlet and one outlet and is made by traditional photolithography using polydimethylsiloxane (PDMS).

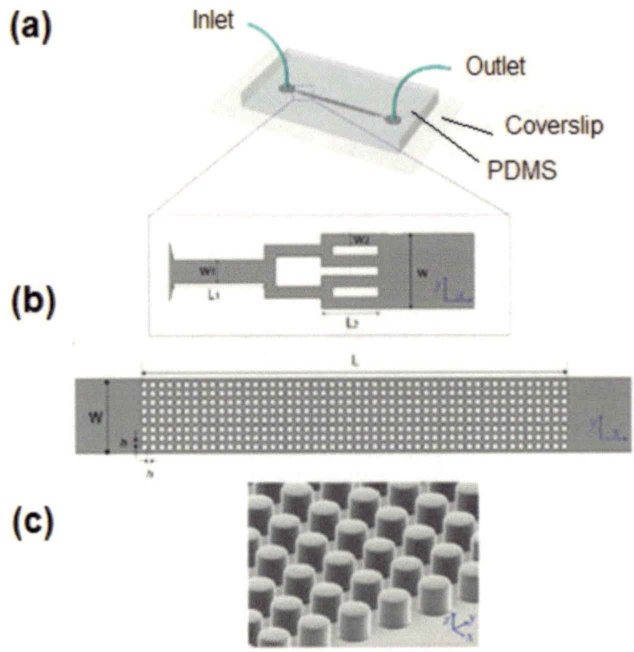

Figure 5. (a) Schematic of an microfluidic chip, (b) top view of the micropillar array, and (c) SEM image of micropillars of the equal diameter of 50 μm. Figures are taken with permission from Ref. [6].

4. Membranes in microfluidic devices

4.1. Direct incorporation of membranes into microfluidic devices

The commercial membrane can be incorporated into the microfluidic devices directly. The membrane can be fabricated as per the requirement by following the traditional membrane fabrication techniques described above and then bonded to the microfluidic chip. Russo et al. [46] directly incorporated polymeric membrane into silicon-based lab-on-chip device. Silicon substrate coated with a thin nitride film was used to serve as a support structure for the track-etched membrane. Patterning was conducted by UV exposure through chrome glass mask and CF_4 reactive ion etching to transfer the pattern to the nitride layer. The process was repeated on the other side of the wafer by using the second mask with pores on it. The membrane was finally incorporated into the PDMS device.

The membrane can be placed between two microfluidic chips and make a sandwiched structure. This is also another way of using a membrane directly in the microfluidic devices. By using this technique, a three-dimensional microfluidic network was designed by Ismagilov et al. [47] to investigate the interactions of chemical and biochemical reagents. They used a polycarbonate membrane between two PDMS microfluidic devices to make the sandwiched structure.

Membrane integrated with microfluidic device plays an essential role in the medical application [48–51]. To study the complex phenomena inside the vascular system different types of membrane with the microfluidic devices are used. A microfluidic device was fabricated by sandwiching polyester membrane between microfluidic chips and used to study the interaction of cancer cells with a vascular endothelium and to prevent the metastatic disease [49]. A membrane with microfluidic device was also used to demonstrate the lungs injury [50] by toxic substances [51]. Huh et al. made a microfluidic airway system with an approximate diameter of respiratory bronchioles (narrowest airways of the lungs) to explore the cellular-level lung injury [50]. To make a sandwich structure of a membrane in a microfluidic device, bonding of the membrane and the device is a critical issue to deal with the leakage. PDMS mortar film, which is made by mixing PDMS and toluene, can be used to effectively make the bond [52, 53]. Young et al. [52] fabricated such kind of devices to measure the biomolecule permeability across the porous membrane. PDMS prepolymer was cured, and 3 mm diameter holes were punched through the cured PDMS. PDMS mortar layer was then generated on a glass support, and the PDMS substrate was placed on the support so that the holes were not in contact with the mortar layer. On the other side, the membrane was pressed down into the mortar layer. Finally, the membrane was placed between two substrates and bonded with PDMS mortar layer. Using an additional PDMS separator with the membrane can be another way to prevent the leakage [54].

4.2. Membrane fabrication as a part of the microfluidic device fabrication

The membrane can be fabricated as a part of a microfabrication process instead of using the traditional membrane fabrication technique. Karnik et al. fabricated a composite membrane of copper, aluminum, spin-on-glass (SOG), and palladium for the water gas shift reaction experiment [55]. Silicon nitride was deposited on both sides of silicon wafers by chemical vapor deposition process. A thin layer of aluminum acted as an adhesive layer of the palladium. Photolithography and wet etching were used to pattern holes on the copper-aluminum layer and to obtain a microchannel. Ookawara et al. [56] fabricated a microchannel as a microseparator for oil and water separation. They made 10 mm curved radius and 112 μm width slits on 80 μm thick SUB308 plates by photolithography. A stack was made by putting the plates with and without slits in turn and diffusion bonded to make microchannel feature. Heyderman et al. [28] fabricated nanopore membrane chip by combining the techniques of hot embossing and photolithography. Silicon (Si) master mold with nanopore arrays was fabricated by using electron beam lithography, and the pores were replicated on PMMA by a hot embossing technique. Various etching processes were used to transfer the pores on Si_3N_4 to fabricate the final nanopore membrane. Though they used PMMA resist with chromium, Si, and silicon nitride (Si_3N_4), the final membrane they obtained was made of Si_3N_4. The pores diameter varied from 100 nm to 450 nm of Si_3N_4 nanopore membrane. To analyze and separate the biological cell Dong et al. fabricated micromachined separator with soft magnetic micropillar arrays that could act as a membrane to observe the performance of the cell separation [57]. A membrane with embedded channel was used to study the hydrodynamic behavior and the fouling formation on the membrane during filtration of synthetic wastewater made of polystyrene particle [58]. They used square-shaped silica capillaries to template the membrane. Polyvinylpyrrolidone (PVP) and N-methylpyrrolidone (NMP) were used as polymer and solvent for membrane preparation, respectively. The silica capillaries were glued to a glass plate and the polymer solution was cast on the glass plate at room temperature. The structured membrane was then kept in a vapor bath and tap water bath for coagulation and phase separation. After making the final structured membrane the silica capillary is placed in the channel of the membrane. The membrane was then placed between two lamination sheets to seal the chip.

4.3. Microfluidic membrane mimic

The microfluidic membrane mimic can be defined as a part of the device with pillars or curvature or any designed structured and the tiny gap between the structures that serve as a porous membrane. The design of the microseparator can be changed to mimic the different pore sizes and shapes for the membrane study in a microfluidic device. For instance, Hassanpourfard et al. [59] designed and developed a detailed fabrication protocol for making microfluidic device that mimics the porous media to study the biofilm formation. Bacchin et al. [60] used different shapes of PDMS microseparator to ensure the uniform flow of the suspension over the width of the filtering part and to study the fouling. Derekx et al. [61] investigated the fouling behavior in a PDMS microfluidic mimic membrane by

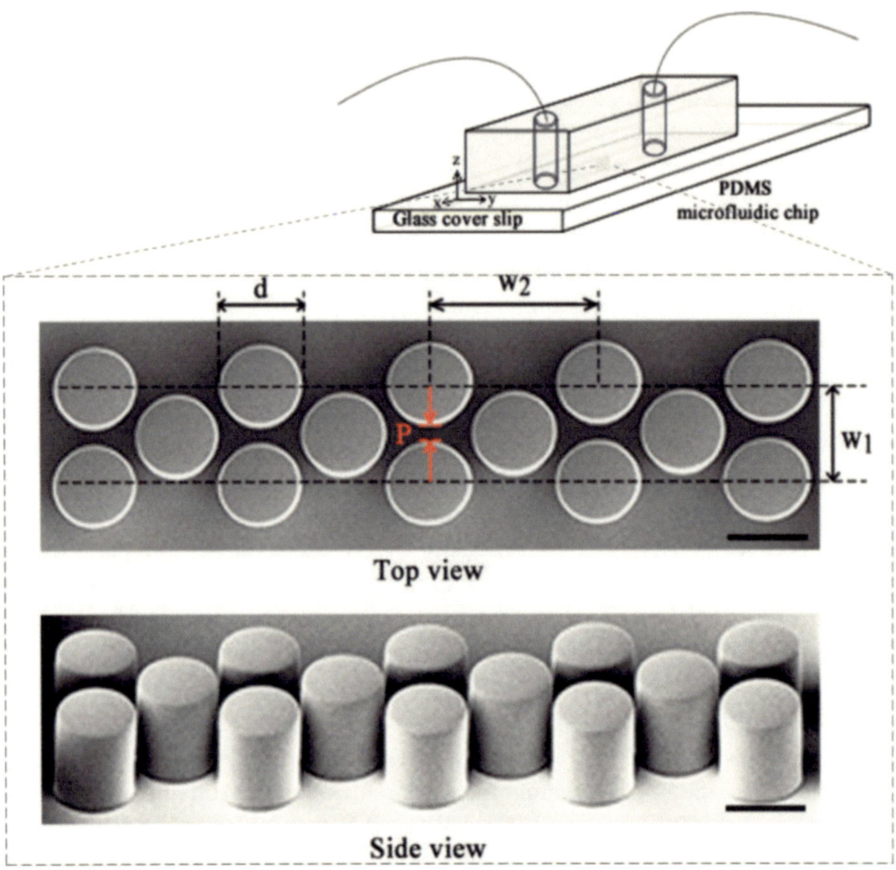

Figure 6. Schematic of microfluidic device with the mimic membrane structure [30]. The dimensions are d = 50 μm, w_1 = 60 μm, w_2 = 104 μm and P = 10 μm. The scale bar is 50 μm.

the experiment and computer simulation. The research on microfluidic membrane mimic has been mainly focused on fouling phenomena in porous media. For instance, Marty et al. [34] fabricated microfluidic devices with straight, interconnected and staggered channels to observe the biofouling nature in the microfluidic device due to biofilm. They studied the

Different materials and polymers used in the fabrication process	Membrane pore size	Fabrication technique of membrane	Incorporate membrane in microfluidic device	Different types of applications
Cellulose acetate [46]	MWCO: 350 Da [46]	Casting	Direct casting and sandwich the membrane in between the microfluidic devices	1. Biological analysis 2. Investigate chemical or biochemical interaction 3. Medical Application 4. Fouling characterization
Polyetherimide (PEI), Polyvinylpyrrolidone (PVP), and N-methylpyrrolidone (NMP) [58]	3–8 μm [58]			
Polycarbonate [47, 53]	0.1–1 μm vertical pore 10 μm thick [47]	Commercial membrane		
Polyester [49, 50, 53]	400 nm [49, 50] 10 μm thick and 3 μm and 20 μm pore [53]			
Polyamide [54]	RO: MWCO: 200DA [54]			
Polyethylene terephthalate (PET) [48]	8 μm [48]	Track etching		
Cyclopore polycarbonate regular and thin clear, nuclepore polycarbonate [52]	1 μm [52]			
PDMS [51]	10 μm thick and 10 μm effective diameter [51]	Soft lithography		
Copper, aluminum and palladium [55]	60, 200 and 500 nm [55]	Composite membrane and MEMS fabrication	Membrane fabrication as a part of microfluidics device fabrication	1. Oil-water separator 2. Magnetic micro separator 3. Fouling analysis 4. Biofouling study
PMMA, Si3N4, Si, Si3N4 and Cr [28]*	Micro-slit, 112 μm [28]	Hot embossing & photolithography		
SUS304 Plate[56]	500, 330, 140 nm [56]			
PDMS [33, 34, 60]	width: 10 μm, Length: 200 μm or 170 μm and Depth: 50 μm [33, 34] Constriction: 20 μm smallest width: 50 μm [60]	Soft lithography		
Cellulose ester [61]	5 μm [61]	Commercial membrane		

Table 2. Summary of different types of microfluidic membrane device fabrication.

effect of different pore sizes and dead-end and pseudo cross-flow filtration modes on the biofouling during filtration. In subsequent work, they also reported that pore tortuosity and secondary flows have a significant impact on biofouling formation in the mimic system [33]. In the pseudo filtration mode, they did not work on the effect of pressure difference on the biofouling formation during filtration.

Biswas et al. [30] designed a microfluidic membrane mimic by using photolithography technique to investigate the biofouling under different flow condition. The minimum pore size considered was 10 μm and the micropillars were distributed in a staggered pattern. **Figure 6** shows the schematic of their microfluidic device with the mimic membrane structure [30]. Transparent PDMS microsystem is used to mimic the membrane to study the bacteria transfer in the porous interface. The diameter and depth (in z-direction) of the micropillars are 50 μm. Their primary focus was to study the deformation mechanism of bacterial streamer that occur at the downstream location of the membrane during filtration process. They did not focus on the effect of pressure on the biofouling formation. **Table 2** shows a summary of different microfluidic membrane fabrication techniques with pore information and their application.

5. Conclusion

Membrane processes have been widely used in various industries for water and gas treatment. Pressure-driven membrane processes for water treatment are typically categorized by their rejection ability into MF, UF, NF, and RO. Biofouling on the membrane surface is the most severe fouling among all fouling phenomena including colloidal fouling, scaling, and organic material fouling. The dynamic behavior and viscoelastic nature of biofouling make it more complicated. Hence, it is very important to observe the real-time phenomenon that is occurring during biofouling. Microfluidic devices have therefore become essential tools to study the biological growth in a flow regime. Integrating membranes with microfluidic devices has become very popular over the past decade. There are several ways to incorporate membrane into the microfluidic device. The commercial membrane can be bonded to the device directly, or the membrane can be fabricated as a part of a fabrication process. Microfluidic devices equipped with membranes have been widely used in the medical application to study the complex permeability of macromolecular, drug or other protein. Such devices have recently used to study the fouling phenomenon in porous media. In this chapter, a thorough literature review was also provided about the microfluidic membrane filtration for biofouling study.

Acknowledgements

The authors gratefully acknowledge the financial support provided by the Natural Sciences and Engineering Research Council of Canada (NSERC) and Canada's Oil Sands Innovation Alliance (COSIA). AK acknowledges support from the Saroj Poddar Young Investigator Grant.

http://dx.doi.org/10.5772/intechopen.75006

Author details

Ishita Biswas[1], Aloke Kumar[2] and Mohtada Sadrzadeh[1]*

*Address all correspondence to: sadrzade@ualberta.ca

1 Department of Mechanical Engineering, 10-367 Donadeo Innovation Center for Engineering, Advanced Water Research Lab (AWRL), University of Alberta, Edmonton, AB, Canada

2 Department of Mechanical Engineering, Indian Institute of Science, Bangalore, India

References

[1] Komlenic R. Rethinking the causes of membrane biofouling. Filtration & Separation. 2010;**47**(5):26-28

[2] Peña N et al. Evaluating impact of fouling on reverse osmosis membranes performance. Desalination and Water Treatment. 2013;**51**(4-6):958-968

[3] Nguyen T, Roddick FA, Fan L. Biofouling of water treatment membranes: A review of the underlying causes, monitoring techniques and control measures. Membranes (Basel). 2012;**2**(4):804-840

[4] Flemming H-C et al. Biofouling—The Achilles heel of membrane processes. Desalination. 1997;**113**(2):215-255

[5] Flemming HC, et al. Springer Series on Biofilms 2009. 4

[6] Valiei A et al. A web of streamers: Biofilm formation in a porous microfluidic device. Lab on a Chip. 2012;**12**(24):5133

[7] Kumar A et al. Microscale confinement features can affect biofilm formation. Microfluidics and Nanofluidics;**14**(5):895-902

[8] Toole GO, Kaplan HB, Kolter R. Biofilm formation as microbial development. Annual Review of Microbiology. 2000;**54**:49-79

[9] Costerton JW et al. Microbial biofilms. Annual Review of Microbiology. 1995;**49**:711-745

[10] Karimi A et al. Interplay of physical mechanisms and biofilm processes: Review of microfluidic methods. Lab on a Chip. 2015;**15**(1):23-42

[11] Bazaka K et al. Bacterial extracellular polysaccharides. Adv. Exp. Med. Biol. 2011;**715**:213-226

[12] More TT et al. Extracellular polymeric substances of bacteria and their potential environmental applications. Journal of Environmental Management. 2014;**144**:1-25

[13] Friedman BA et al. Structure of exocellular polymers and their relationship to bacterial flocculation. Journal of Bacteriology. 1969;**98**(3):1328-1334

[14] Flemming HC, Neu TR, Wozniak DJ. The EPS matrix: The "house of biofilm cells". Journal of Bacteriology. 2007;**189**(22):7945-7947

[15] Bixler GD, Bhushan B. Biofouling: Lessons from nature. Philosophical Transactions. Series A, Mathematical, Physical, and Engineering Sciences. 2012;**370**(1967):2381-2417

[16] Bereschenko LA et al. Molecular characterization of the bacterial communities in the different compartments of a full-scale reverse-osmosis water purification plant. Applied and Environmental Microbiology. 2008;**74**(17):5297-5304

[17] Mayer C et al. The role of intermolecular interactions: Studies on model systems for bacterial biofilm. International Journal of Biological Macromolecules. 1999;**26**(1):3-16

[18] Flemming HC. Biofouling in water systems--cases, causes and countermeasures. Applied Microbiology and Biotechnology. 2002;**59**(6):629-640

[19] Park N et al. Biofouling potential of various NF membranes with respect to bacteria and their soluble microbial products (SMP): Characterizations, flux decline, and transport parameters. Journal of Membrane Science. 2005;**258**(1-2):43-54

[20] Herzberg M, Elimelech M. Biofouling of reverse osmosis membranes: Role of biofilm-enhanced osmotic pressure. Journal of Membrane Science. 2007;**295**(1-2):11-20

[21] Herzberg M, Kang S, Elimelech M. Role of extracellular polymeric substances (EPS) in biofouling of reverse osmosis membranes. Environmental Science & Technology. 2009; **43**(12):4393-4398

[22] Miura Y, Watanabe Y, Okabe S. Membrane biofouling in pilot-scale membrane bioreactors (MBRs) treating municipal wastewater: Impact of biofilm formation. Environmental Science & Technology. 2007;**41**(2):632-638

[23] Huang LN, Wever HD, Diels L. Diverse and distinct bacterial communities induced biofilm fouling in membrane bioreactors operated under different conditions. Environmental Science & Technology. 2008;**42**(22):8360-8366

[24] Barnes RJ et al. Nitric oxide treatment for the control of reverse osmosis membrane biofouling. Applied and Environmental Microbiology. 2015;**81**(7):2515-2524

[25] Matin A et al. Biofouling in reverse osmosis membranes for seawater desalination: Phenomena and prevention. Desalination. 2011;**281**:1-16

[26] Barai P, Kumar A, Mukherjee PP. Modeling of Mesoscale variability in biofilm shear behavior. PLoS One. 2016;**11**(11):e0165593

[27] Patsios SI et al. A novel method for rheological characterization of biofouling layers developing in membrane bioreactors (MBR). Journal of Membrane Science. 2015;**482**:13-24

[28] Heyderman LJ et al. High volume fabrication of customised nanopore membrane chips. Microelectronic Engineering. 2003;**67-68**:208-213

[29] Srijanto BR et al. Nanostructured silicon membranes for control of molecular transport. Journal of Vacuum Science and Technology. B, Nanotechnology & Microelectronics. 2010; **28**(6):C6PC6P48-C6PC6P52

[30] Biswas I et al. Nonlinear deformation and localized failure of bacterial streamers in creeping flows. Scientific Reports. 2016;**6**:32204

[31] Hassanpourfard M et al. Bacterial floc mediated rapid streamer formation in creeping flows. Scientific Reports. 2015;**5**:13070

[32] Das S, Kumar A. Formation and post-formation dynamics of bacterial biofilm streamers as highly viscous liquid jets. Scientific Reports. 2014;**4**:7126

[33] Marty A et al. Impact of tortuous flow on bacteria streamer development in microfluidic system during filtration. Biomicrofluidics. 2014;**8**(1):014105

[34] Marty A et al. Formation of bacterial streamers during filtration in microfluidic systems. Biofouling. 2012;**28**(6):551-562

[35] Drescher K et al. Biofilm streamers cause catastrophic disruption of flow with consequences for environmental and medical systems. Proceedings of the National Academy of Sciences of the United States of America. 2013;**110**(11):4345-4350

[36] Rusconi R et al. Laminar flow around corners triggers the formation of biofilm streamers. Journal of the Royal Society Interface. 2010;**7**(50):1293-1299

[37] Rusconi R et al. Secondary flow as a mechanism for the formation of biofilm streamers. Biophysical Journal. 2011;**100**(6):1392-1399

[38] Mulder M. Basic Principles of Membrane Technology. 2nd ed. Barendrecht, Boston, London: Kluwer Academic Publishers, Springer; 2007

[39] Vogelaar L, Barsema JN, van Rijn CJM, Nijdam W, Wessling M. Phase separation Micromolding—PSμM**. Advanced Materials. 2003;**15**(16):1385-1389

[40] Lalia BS et al. A review on membrane fabrication: Structure, properties and performance relationship. Desalination. 2013;**326**:77-95

[41] Sagle A, Freeman B. Fundamentals of Membranes for Water Treatment. The Future of Desalination in Texas. 2004;**2**:1-17

[42] Mcdonald JC, Duffy DC, Anderson JR, Chiu DT, Wu H, Schueller OJA, Whitesides GM. Review general fabrication of microfluidic systems in poly (dimethylsiloxane). Electrophoresis. 2000;**21**:27-40

[43] Cabodi M, Choi NW, Gleghorn JP, Lee CSD, Bonassar LJ, Stroock AD. A microfluidic biomaterial. Journal of the American Chemical Society. 2005;**127**(40):13788-13789

[44] Qin D, Xia Y, Whitesides GM. Rapid prototyping of complex structures with feature sizes larger than 20 μm**. Advanced Material Communication. 1996;**8**(11):917-919

[45] Yu-Chuan S, Liwei L, Pisano AP. A water-powered osmotic microactuator. Journal of Microelectromechanical Systems. 2002;**11**(6):736-742

[46] Russo AP et al. Direct casting of polymer membranes into microfluidic devices. Separation Science and Technology. 2004;**39**(11):2515-2530

[47] Ismagilov RF et al. Microfluidic arrays of fluid-fluid diffusional contacts as detection elements and combinatorial tools. Analytical Chemistry. 2001;**73**:5207-5213

[48] Srigunapalan S et al. A microfluidic membrane device to mimic critical components of the vascular microenvironment. Biomicrofluidics. 2011;**5**(1):13409

[49] Song JW et al. Microfluidic endothelium for studying the intravascular adhesion of metastatic breast cancer cells. PLoS One. 2009;**4**(6):e5756

[50] Huh D et al. Acoustically detectable cellular-level lung injury induced by fluid mechanical stresses in microfluidic airway systems. Proceedings of the National Academy of Sciences of the United States of America. 2007;**104**(48):18886-18891

[51] Huh D et al. Reconstituting organ-level lung functions on a chip. Science. 2010;**328**(5986):1662-1668

[52] Young EWK, Watson MWL, Srigunapalan S, Wheeler AR, Simmons CA. Technique for real-time measurements of endothelial permeability in a microfluidic membrane Chip using laser-induced fluorescence detection. Analytical Chemistry. 2010;**82**(1):808-8016

[53] Chueh B-h et al. Leakage-free bonding of porous membranes into layered microfluidic Array systems. Analytical Chemistry. 2007;**79**:3504-3508

[54] Kaufman Y et al. Microfluidic NF/RO separation: Cell design, performance and application. Journal of Membrane Science. 2012;**396**:67-73

[55] Karnik SV, Hatalis MK, Kothare MV. Towards a palladium micro-membrane for the water gas shift reaction: Microfabrication approach and hydrogen purification results. Journal of Microelectromechanical Systems. 2003;**12**(1):93-100

[56] Ookawara S, Ishikawa T, Ogawa K. Applicability of a miniaturized micro-separator/classifier to oil-water separation. Chemical Engineering & Technology. 2007;**30**(3):316-321

[57] Dong T et al. A smart fully integrated micromachined separator with soft magnetic micropillar arrays for cell isolation. Journal of Micromechanics and Microengineering. 2010;**20**(11):115021

[58] Ngene IS et al. A microfluidic membrane chip for in situ fouling characterization. Journal of Membrane Science. 2010;**346**(1):202-207

[59] Hassanpourfard M et al. Protocol for biofilm streamer formation in a microfluidic device with micro-pillars. Journal of Visualized Experiments. 2014;(90)

[60] Bacchin P et al. Colloidal surface interactions and membrane fouling: Investigations at pore scale. Advances in Colloid and Interface Science. 2011;**164**(1-2):2-11

[61] Derekx Q et al. Numerical and experimental study of fouling in microfluidic channels and microfiltration membranes. Procedia Engineering. 2012;**44**:54-56